AF685863

Topologies of the Flesh

SERIES IN CONTINENTAL THOUGHT

Editorial Board

Steven Crowell, Chairman, Rice University
Elizabeth A. Behnke
David Carr, Emory University
John J. Drummond, Fordham University
Lester Embree, Florida Atlantic University
Burt C. Hopkins, Seattle University
José Huertas-Jourda, Wilfrid Laurier University
Joseph J. Kockelmans, Pennsylvania State University
William R. McKenna, Miami University
Algis Mickunas, Ohio University
J. N. Mohanty, Temple University
Thomas Nenon, University of Memphis
Thomas M. Seebohm, Johannes Gutenberg Universität, Mainz
Gail Soffer, New School for Social Research
Elizabeth Ströker, Universität Köln †
Richard M. Zaner, Vanderbilt University

International Advisory Board

Suzanne Bachelard, Université de Paris
Rudolf Boehm, Rijksuniversiteit Gent
Albert Borgmann, University of Montana
Amedeo Giorgi, Saybrook Institute
Richard Grathoff, Universität Bielefeld
Samuel Ijsseling, Husserl-Archief te Leuven
Alphonso Lingis, Pennsylvania State University
Werner Marx, Albert-Ludwigs Universität, Freiburg †
David Rasmussen, Boston College
John Sallis, Boston College
John Scanlon, Duquesne University
Hugh J. Silverman, State University of New York, Stony Brook
Carlo Sini, Università di Milano
Jacques Taminiaux, Louvain-la-Neuve
D. Lawrence Wieder
Dallas Willard, University of Southern California

Topologies of the Flesh

A Multidimensional Exploration of the Lifeworld

STEVEN M. ROSEN

OHIO UNIVERSITY PRESS ATHENS

Ohio University Press, Athens, Ohio 45701
www.ohio.edu/oupress
© 2006 by Steven M. Rosen

Printed in the United States of America
All rights reserved

Ohio University Press books are printed on acid-free paper ∞ ™

13 12 11 10 09 08 07 06 5 4 3 2 1

Library of Congress Cataloging-in-Publication Data

Rosen, Steven M.
Topologies of the flesh : a multidimensional exploration of the lifeworld / Steven M. Rosen.— 1st ed.
p. cm. — (Series in Continental thought ; 33)
Includes bibliographical references and index.
ISBN 0-8214-1676-6 (alk. paper)
1. Phenomenology. 2. Topology. I. Title. II. Series.
B829.5.R67 2006
142'.7—dc22

2005037873

To Lesley, David and Jana,

Jonas and Hannah

CONTENTS

ILLUSTRATIONS

FIGURES

TABLES

PREFACE

1. THE WAY INTO THE LIFEWORLD

When I peel myself away from this computer screen long enough to turn my head and consider what appears below my window, at once I notice the commingling of vividly colored flowers arrayed in beds amidst the background foliage of the front lawn. And beyond the roofs of houses across the road, I can see the white chop of the windswept wavelets in Vancouver harbor, and the layered mountains that enfold the inlet in shades of blue and gray. But the pull of cyberspace, and of modern technology in general, does seem irresistible. The high-powered abstractions of this realm relentlessly draw my attention. In imposing themselves on my awareness, the world of concrete life is relegated to the background and overshadowed. I am hardly alone in my tendency to succumb to the lure of technology and other heady possibilities on the contemporary scene, and so to become oblivious to the earth in which I dwell. Participating in modern culture renders the lifeworld peripheral. But it is precisely this world that I intend to explore in the present book.

To be sure, that is easier said than done. For one thing, the eclipse of the lifeworld actually long predates the advent of modern technology. The Renaissance was a critical juncture. It was then that there arose a more "individualistic, and rational understanding of nature" (Gebser 1985, 15), one involving a greater sense of detachment from the world and concomitant inclination to objectify that world, accompanied by a more abstract experience of the space and time in which objects were situated (Heidegger 1962/1977). Yet the repression of the lifeworld was well in progress even before the Renaissance. Phenomenological ecologist David Abram (1996) observes that the concealment of the sensuous realm had already begun with the coming to prominence of alphabetic language in ancient Hebrew and Greek cultures. Could I really expect, then, to look out of my window at the flowers, ocean, and mountains and directly experience the innocence and purity of the primal lifeworld?

Has human perception not been veiled by millennia of cultural conditioning that has had the effect of distancing us from nature? So reentering the lifeworld is certainly not simply a matter of walking away from my computer to "smell the roses." Instead it seems I must find a way of going back to a long forgotten mode of knowing and being.

But should we really want to "go back"? Was the separation from the lifeworld simply a regrettable mistake? I do not think so. It is certainly true that, in the primordial lifeworld, self and other, subject and object, were not dualistically split off from each other as they later came to be. But neither were they consciously fused. Instead subject and object tended to be *con*fused; there was a limited ability consciously to differentiate them. Therefore, pre-Renaissance awareness is not something to be idealized. According to philosopher Owen Barfield, this "kind of knowledge . . . was at once more universal and less clear" (1977, 17). The cultural philosopher Jean Gebser (1985) and communications theorist Walter Ong (1977) make it plain that pre-Renaissance experience was less lucidly focused than the mode of awareness that succeeded it. The decisive separation of subject and object served the interest of creating sharper understanding, a greater capacity for reflection and intellectual achievement; in that way it helped to fulfill humankind's potential. So, far from being merely a pathological departure from an ideal state of affairs, the transition to well-differentiated consciousness was both necessary and beneficial. It does seem, then, that we should not wish simply to go *back* to the primal lifeworld.

However, is there any denying that, in *today's* world, the splitting from nature has progressed to the point where it not only has reduced the quality of our lives but threatens the very life of our planet? The more detached we have become from nature, the more insensitive to it we have grown. And the more insensitive, the more we have tended to regard it as nothing but dead matter, there at our disposal, held in reserve for our indiscriminate use. The conviction that nature's processes can be manipulated by us through our technologies, controlled arbitrarily for our own ends—such a view of nature seems largely responsible for the all too well-known state of affairs prevailing today: noxious wastes of every kind seeping into the earth, polluting the oceans and atmosphere, endangering countless animal species; natural resources becoming exhausted with impending shortages of food and energy; ecological balances being disrupted; the syndrome of drought/famine/disease steadily worsening. And because we never really cease to be a part of the natural world from which we distance ourselves, our estrangement from nature brings an estrangement from ourselves and from each other. As a consequence, "fragmentation is now very widespread, not only throughout society, but also in each individual" (Bohm 1980, 1). Psychopathology is rampant and the social fabric unravels. Family and church disintegrate. Ethnic conflicts rage around the world. International banditry and terrorism grow to alarming proportions. Nuclear weapons proliferate out of control.

Where, then, do we presently stand vis-à-vis the lifeworld? We do not wish simply to go *back* to it, yet it seems we cannot survive much longer in the toxic environment that has resulted from cutting our ties to it. Is there any way out? I suggest that there is, though the path in question is difficult and oddly circuitous. I venture to say that we can (re)turn to the lifeworld not simply by departing from the world of abstraction, but by going so far into it that, in a manner of speaking, we "come out on the other side"!

In attempting to clarify this enigmatic proposition, let me first point out that we could not simply depart from abstraction even if we wanted to. The reason is that that is what abstraction is all about: simple departures. The word "abstract" is from the Latin *abstractus,* "dragged away, pp. of *abstrahere,* to draw from or separate."[1] Abstraction, then, is about separating, drawing boundaries to set things apart from each other in a categorical manner. Under the dualistic rule of abstraction, we strictly adhere to the logic of either/or: we are either here or there, inside or outside, different or the same, mental or physical—*abstract or concrete.* From this we can see that any attempt to leave abstraction behind, to cross its outer boundary and pass into the concrescence of the lifeworld, is certain to be frustrated by the fact that all such crossings are themselves acts of abstraction. So the true end of abstraction cannot *merely* be an end, since any "clean break" of this sort would only testify to the fact that abstraction was actually still taking place! Like the proverbial Chinese finger puzzle, all efforts to break free of abstraction leave us squarely within it, for that is what abstraction essentially entails: the effort to break free, to produce clean breaks.

Still, while abstraction evidently possesses no categorical limit, no exterior boundary whose crossing would simply bring it to an end, might it not possess an *interior* boundary? Instead of seeking to break out of abstraction, suppose we were to move in the other direction. If we went further with abstraction, went all the way *inside* it, following its own trajectory to its point of fulfillment, might we not then be able to "exit" on the "other side"?

The strange nonlinearity of such a movement is intimated in Heidegger's essay "The End of Philosophy" (1964/1977, 373–92). It would seem that the lofty abstractions of philosophy could not be further removed from the concreteness of the senses. Philosophical thinking indeed is a prime exemplar of the kind of high-flying intellectual reflectiveness that has obscured our bond with the earth. By the "*end* of philosophy," does Heidegger mean the termination of such ratiocination, coupled perhaps with a descent into the lifeworld? It is clear that he does not. Rather, "The end of philosophy proves to be the triumph of the manipulable arrangement of a scientific-technological world" (377). That is, philosophy ends with its transformation into the modern sciences—sciences that have now been brought to culmination, and whose objectifications and abstract analyses apparently have brought us as far away as we could possibly be from the world of lived experience. However, in philosophy's realization of its "most extreme possibilities" (375), Heidegger indicates that one possibility may have been overlooked:

> But is the end of philosophy in the sense of its evolving into the sciences also already the complete actualization of all the possibilities in which the thinking of philosophy was first posited? Or is there a *first* possibility for thinking apart from the *last* possibility which we characterized (the dissolution of philosophy in the technologized sciences), a possibility from which the thinking of philosophy would have to start, but which as philosophy it could nevertheless not experience and adopt? (377)

If there were such a "first possibility for philosophical thinking"—one that was unrealizable throughout the history of philosophy but can be broached now that philosophical abstraction has reached its climax in the technological sciences—then the essential task of thinking would be to think that possibility.

But *what is* that possibility? Heidegger alludes to it later in his essay when he asks why the notion of "openness" he has been discussing has always been misunderstood: "Is it because man's ecstatic sojourn in the openness of presencing is turned only toward what is present and the presenting of what is present? But what else does this mean than that presence as such . . . remains unheeded?" (390). In speaking of being "turned only toward what is present and the presenting of what is present," Heidegger apparently is referring to the exclusive preoccupation with *object* and *subject* (respectively). Though we have been engaged in an "ecstatic sojourn in the openness of presencing," this prereflective movement has been obscured in favor of a mode of reflection in which the subject presents to himself only what is present, the objects that are cast before him. Presencing per se, "presence as such," is the first possibility for thinking that has gone unheeded through the whole course of Western philosophy. Elsewhere Heidegger refers to such presencing as *Being*. Philosopher Carol Bigwood notes in her reading of Heidegger that "Being is not a being, not God, an absolute unconditional ground or a total presence, but is simply the living web within which all relations emerge" (1993, 3). In other words, Heideggerian Be-ing is none other than the dynamic world of life process, the lifeworld. And, evidently, it is only at the *end* of philosophy, where the abstract splitting of subject and object has reached its culmination and has created the greatest degree of estrangement from the lifeworld, that—having followed the natural trajectory of abstraction to its "last possibility"—we can now (re)turn to the "first possibility" for thinking: the thinking of the concrete lifeworld, which in fact is the source of the abstraction to begin with (that "from which the thinking of philosophy would have to start," as Heidegger puts it).

Let me emphasize that Heidegger is not suggesting that we merely renounce thinking in favor of unmediated experience. Yet, while he does urge that we *think* Being, the kind of thinking he has in mind is unusual to say the least. Heidegger wants us to think in the *original* meaning of that word. Today, "A thought usually means an idea, a view or opinion, a notion"; in contemporary science and philosophy, thinking signifies "logical-rational representations" (1954/1968, 138). Noting the etymological consanguinity of "thinking" with "thanking," Heidegger claims that the modern understanding of thinking is an "impoverished" version of what earlier involved not merely an intellectual act but also a heartfelt giving of thanks (139). Spiegelberg (1982) summarizes Heidegger's radical interpretation of thinking as "an intent and reverent meditation with the whole of our being . . . heart as well as . . . intellect" (402). Only through a thinking that is also a whole-bodied thanking can we truly think Being, think the lifeworld in a way that does not merely objectify it but gratefully embraces it as that to which we owe our very existence.

It is true, however, that Heidegger tended toward a certain nostalgia for the past that had the effect of seeming to valorize it. Granting that our modern way of thinking one-sidedly favors abstraction and thus estranges us from the lifeworld, is contemporary rationality really just an "impoverished" form of an earlier, more complete kind of thinking to which we must now return? Or did prescientific thought actually not constitute an undifferentiated form of cognition in which mind and heart were to some extent confused? To repeat, re-inhabiting the lifeworld should not entail a going back that would simply negate the forward progress we have made. Nor *could* it really do so. The movement into abstraction cannot simply be reversed, since any such attempt to *cut off* abstraction would in fact be nothing more than an act of abstraction itself. So it is clear that, in reentering the lifeworld, while abstraction per se must be surpassed, it cannot just be dropped.

I suggest there is but one sort of boundary that will permit us to pass effectively beyond abstraction: the "interior boundary" hinted at above. This is the boundary or limit of *limitative thinking itself*. A paradox is involved here. Abstraction's inner boundary is its natural point of termination, its true end. Yet we have seen that the true end of abstraction cannot *merely* be an end, a "clean break." In order for abstraction truly to end, there is no avoiding paradox—an end that also is *not* an end, a boundary that is not one. Thus, while we do "come out on the other side" in crossing the inner horizon of abstraction, this movement beyond abstraction is at once a movement within it. Such is the peculiar logic that governs the transition to the lifeworld. Only by remaining *within* abstraction can we radically surmount it. Like the movement from one side of a Moebius strip to the other that paradoxically keeps us on the *same* side, our passage from abstraction to concrescence at once maintains the former (the Moebius strip in fact will play a pivotal role in the topological work of this book). Of course, the *supremacy* of the abstract is not maintained. What we realize instead is an internal harmony of abstraction and concrescence in which the prior meaning of each term changes profoundly.

To be sure, such a paradox boggles the mind. Nevertheless, if our aim is to exceed the one-sided rule of abstraction so we can re-inhabit the lifeworld, it seems the abstract mind needs to be boggled. But while this is a necessary requirement, it is not sufficient. Merely setting these abstract words against themselves is not enough. Beyond the bare assertion of paradox in enigmatic words such as those I have used, the paradox needs to be articulated more fully by being fleshed out. Only then can the lifeworld really come to life. Accordingly, what I seek to realize in the pages that follow is the embodiment of paradox. To that end, I will make use of *topology,* a field of study that is "rooted in the body" (Sheets-Johnstone 1990, 42)—as we will see in subsequent chapters.

2. PREVIEW OF THE CHAPTERS

In chapter 1, the topological method of exploring the lifeworld is introduced by placing the mathematical discipline of topology in historical perspective and identifying the core assumptions common to its modernist and postmodern applications. The investigation culminates with the understanding that a different approach to topology is required for engaging with the lifeworld, a phenomenological rendering that does justice to the paradox of Being. The new topological initiative is carried forward in chapter 2 through the work of the French philosopher Maurice Merleau-Ponty. Merleau-Ponty's key ontological concept of *the flesh of the world* is topologically embodied via a phenomenological reading of the Klein bottle (the three-dimensional counterpart of the Moebius strip). But a further step is required in making the fleshly lifeworld a concrete reality. However suggestive the topological narrative may be, it is evidently not enough to write *about* the realm of "wild Being" (Merleau-Ponty 1968, 211) and so assume the customary posture of authorial detachment and anonymity. If Being's actual presence is to be secured in the ontological text, rather than merely predicating Being—signifying it in such a way that it is implicitly projected as exterior to the author's semiotic act—the author must signify Being topologically by signifying *himself.* The self-signification of the text is taken up in the final section of chapter 2.

The first two chapters comprise part I of the book. This part is devoted to the topological realization of the three-dimensional lifeworld.

In part II, we recognize the existence of lower-dimensional lifeworlds and explore their interrelationships in depth. Chapter 3 introduces the lower dimensions via a late lecture by Heidegger on the ontological nature of time ("Time and Being," 1962/1972). Here the Klein bottle, the Moebius surface, and two other paradoxical structures are shown to be members of a closely related topological family, each member of which embodies a dimension of the flesh in its own right. In chapters 4 and 5, the diachronic or developmental aspect of topological Being is examined and we see how the several dimensions of the flesh engage in dialectical processes of individuation in which they are organically transformed in relation to one another. To facilitate understanding of how this happens, a metaphor of nativity is invoked, with lower dimensions of Being seen as playing the role of "midwife" in the "birthing" of the higher, "motherly" dimensions.

Having introduced the lower topological dimensions in chapters 3 through 5, their concrete realization is carried forward in the next three chapters. The process is enacted in two stages. First, the relatively abstract treatment of lower dimensionality is fleshed out in chapters 6 and 7 by giving the dimensions more tangible content. Whereas three-dimensional Being is associated with the human *cogito* or thinking subject, the lower-dimensional orders of the flesh are related to noncognitive, nonhuman lifeworlds of ontological action. But, again, writing *about* wild Being does not suffice if Being is to make its presence felt in the text as a living reality. To realize lower-dimensional Being in this manner, the author must once more signify Being by signifying himself. Exploring the question of self-signification in chapter 8, we discover that the written text will need to be accompanied by texts of "greater density," i.e., texts mediated not by written words but by palpable images, sounds, and root intuitions.

3. ACKNOWLEDGMENTS

The present book advances work on topological phenomenology initiated in two previous volumes. The first of these, *Science, Paradox, and the Moebius Principle* (Rosen 1994), is a book of my essays in which an earlier version of topological phenomenology is applied to various problems in science and philosophy. In my recent volume, *Dimensions of*

Apeiron (Rosen 2004), the role of topological phenomenology is explored philosophically in the broad context of historical and cultural change.

I would like to acknowledge the encouragement and support I have received from a number of individuals in the course of preparing this book. I am gratefully indebted to Arnold Berleant, David Dichelle, John Dotson, Eugene T. Gendlin, Lloyd Gilden, Marketa Goetz-Stankiewicz, Brian D. Josephson, Koichiro Matsuno, Yair Neuman, Milan Pomichalek, David Roomy, Lesley Brooke Rosen, Raymond Russ, Marlene A. Schiwy, Ernest Sherman, W. J. Stankiewicz, Louise Sundararajan, Geo N. Turner, and John R. Wikse. Much appreciated is Steven Crowell's patient stewardship of this project, the always helpful attention of David Sanders in the production phase, and the meticulous editing of Ed Vesneske, Jr., and John Morris. For their assistance in preparing illustrations, I give my thanks to Shelley MacDonald, Beth Pratt, and Mark Lewental. And I want to thank Martin Gardner and Paul Ryan for their kind permission to use their topological drawings in chapter 2 of this book.

PART I

Topology and Dimensional Flesh

CHAPTER ONE

INTRODUCTION TO TOPOLOGICAL PHENOMENOLOGY

In attempting to bring the lifeworld to life, my primary tool will be *topology.* This is the study of *topos,* a Greek word for "place." The lifeworld is clearly no empty space but possesses something of the quality of a *place.* The distinction between these two terms is instructive. "Space" is defined as "distance extending without limit in all directions . . . a boundless, continuous expanse . . . within which all material things are contained."[1] In his etymological study, Partridge relates "space" (from the Latin, *spatium*) to "*patere,* to lie open . . . wide-open, large" (1958, 644). "Place," on the other hand, has a more concrete meaning. Among other definitions, it is a "particular area or locality"; a "residence" or "dwelling"; a "particular . . . part of the body"; "the part of space occupied by a person or thing."[2] In keeping with the concreteness of *topos,* Sheets-Johnstone is able to demonstrate that, whereas the Euclidean study of space involves practices that are largely disembodied, "topology . . . is rooted in the body" (1990, 42). Topology, then, will be the discipline I will employ in exploring the embodied lifeworld.

But this is not a book about topology per se. Accordingly, I will make no attempt to provide a comprehensive account of the various applications of topology in various fields at various times. Yet the topological work undertaken in this volume should be better appreciated if placed in the context of other broad approaches to topology, and related to the overall history of the field.

Despite my intention of using topology to probe the concrete lifeworld, this area of inquiry is conventionally regarded as a branch of the most abstract of abstract sciences: mathematics. Here topology is generally defined as "the study of those properties of geometric figures that remain unchanged even when under distortion, so long as no surfaces are torn."[3] Let us briefly consider the history of this discipline.

1. MODERN TOPOLOGY IN HISTORICAL PERSPECTIVE

For two thousand years, Euclidean assumptions about the nature of space had gone largely unchallenged. Then, in the early nineteenth century, doubts were raised about Euclid's parallel postulate. From this there arose non-Euclidean systems, such as the hyperbolic and elliptic geometries. In the course of the nineteenth century, the Euclidean approach was further surpassed through the development of projective geometry, and, beyond that, the highly general study of topology was inaugurated.

The generality of a geometry can be understood in terms of the range of transformations it permits. In Euclidean geometry, a figure such as a circle can be transformed by moving it from one location to another, but the transformation must be "rigid"; that is, for the postulates of Euclid to hold up, the metrical properties of the circle cannot be disrupted by stretching or distorting the figure. Projective geometry is somewhat less restrictive. Here metrical relations can be changed to a limited extent; the circle, for example, can be transformed into an ellipse. However, with projective geometry, we could not go so far as to transform the circle arbitrarily into any shape we wished. For that we require topology. Topological transformations can be performed without regard to the size or shape of the mathematical object being changed as long as the object retains its continuity. In popular accounts of topology, the great flexibility of this "rubber sheet" geometry is often demonstrated via the example of changing a doughnut into a coffee cup. Despite the concrete differences between real-world doughnuts and coffee cups, their topological counterparts—abstractly taken as but continuous surfaces with single holes—are regarded to be the "same" object. This illustrates the key point for our purposes: that, historically, topology has primarily served as a tool of mathematical abstraction.

Now, modern mathematics is a prime example of the twentieth-century cultural movement known as *modernism*. It is modernism that we find at the "end of philosophy" (see preface). Ambitious and totalizing, imbued with the spirit of science, this movement aspires to gain complete objective knowledge of nature by means of abstract analysis, and, on this basis, to bring nature under cognitive human control. Beyond mathematics and the natural sciences, the impact of modernism has been felt in the social sciences, the arts and humanities, and the popular culture at large. While mainstream topology is surely a modernist enterprise, my use of this discipline as a means of reanimating the lifeworld clearly cannot be.

If the influence of modernism was on the rise in the first half of the twentieth century, it leveled off in the second, and modernism began to be questioned.[4] Around this time, uncertainty was burgeoning in a number of scientific disciplines—a development led by modern physics, "our culture's paradigm for all knowledge" (de Quincey 2002, 54). Mathematics was no exception. Its completeness and consistency had been fundamentally challenged by Gödel's theorem, and its foundations were shaken by deep conflicts among its several schools (formalism, logicism, intuitionism, etc.). The mathematician Morris Kline provides a telling account of the situation, summed up succinctly in the title of his book: *Mathematics: The Loss of Certainty* (1980).

Today, at the outset of the twenty-first century, the challenge to modernism continues. In the humanities and human sciences, certain novel applications of topology reflect the postmodern tenor of the times. In this regard, the neo-Freudian psychoanalytic theorist Jacques Lacan was a transitional figure.

Undoubtedly, Freud himself was a modernist, seeking as he did to achieve an objective scientific grasp of the human psyche. Was Lacan also a modernist? The Marxist theorist Louis Althusser conveys this impression in his claim that the effect of Lacan's work is to "give Freud's discovery its measure in theoretical concepts by defining as rigorously as is possible today the unconscious and its 'laws,' its whole object" (1969, 204). Lacan indeed turned to the science of linguistics, seemingly in order to put psychoanalysis on a more rigorous footing. Noting Freud's emphasis on the importance of language in the functioning of the unconscious (as evidenced, for example, in the wordplay often prevalent in dreams), Lacan contended that the unconscious does not merely *make use* of language; rather, it is a language unto itself. Since "the unconscious is structured as a language" (1966/1970, 188), could it not be precisely formulated in terms of the relationships between signifiers (graphic marks) and the meanings they signify? Lacan apparently attempted just such a formulation in his article, "The Function of Language in Psychoanalysis" (1968).

Moreover, Lacan seemed not content to stop with a merely linguistic clarification of psychoanalysis. In an ostensive effort to achieve an even higher level of precision, he appealed to that modernist discipline par excellence, mathematics. The signifying process that constituted the unconscious discourse of the human subject was to be spelled out "definitively" via the use of topology. To this end, Lacan presented a diagram of a Moebius strip:

> The diagram can be considered the basis of a sort of essential inscription at the origin, in the knot which constitutes the subject. This goes much further than you might think at first, because you can search for the sort of surface able to receive such inscriptions. You can perhaps see that the sphere, that old symbol for totality, is unsuitable. A torus, a Klein bottle, a cross-cut surface, are able to receive such a cut. And this diversity is very important as it explains many things about the structure of mental disease. If one can symbolize the subject by this fundamental cut, in the same way one can show that a cut on a torus corresponds to the neurotic subject, and on a cross-cut surface to another sort of mental disease. (1966/1970, 192–93)

The foregoing passage is cited by the physicists Alan Sokal and Jean Bricmont (1999) in their scathing critique of Lacan's use of topology, and his handling of mathematics in general. Taking it for granted that Lacan indeed had aspired to "mathematize" psychoanalysis, the authors begin by declaring that "We shall not enter . . . into the debate concerning the purely psychoanalytic part of Lacan's work. Rather, we shall limit ourselves to an analysis of his frequent references to mathematics" (18). Sokal and Bricmont then proceed to demonstrate in detail Lacan's many "misuses" and "abuses" of mathematical ideas. It must be said that Lacan left himself open to such criticism. It does appear, at least on the surface, that he sought to employ mathematics for the purpose of clarifying the unconscious and defining it with greater exactitude. As far as I know, he never explicitly questioned mathematics per se. Apparently, then, Lacan was playing a modernist game and therefore was subject to its rules, which means that any absence of clarity should count against him. But let us consider what he actually said about the language of the unconscious, and what this indicates for the topological language he purportedly used to "clarify" it.

In a key lecture published in *The Languages of Criticism and the Sciences of Man* (1966/1970), Lacan says: "The unconscious has nothing to do with instinct or primitive knowledge or preparation of thought in some underground. It is a thinking with words, with thoughts that escape your vigilance, your state of watchfulness" (189). Language essentially involves repetition. "The unconscious subject" engaged in this linguistic process "is something that tends to repeat itself" (191). Only by repeating itself, by replicating its act of signification, can the subject hope to affirm its existence. But this repetition is never a repetition of what is the same: "in its essence repetition as repetition of the symbolical sameness is impossible" (192). Consequently, the decentering of the subject is unavoidable. In reproducing itself, the subject alienates itself and this "necessitates the 'fading,' the obliteration, of the first foundation of the subject, which is why the subject, by status, is always presented as a divided essence" (192).

For Lacan, language in general "is constituted by a set of signifiers. . . . The definition of this collection of signifiers is that they constitute what I call the Other" (193). The "sphere of language" is thus comprised of an "otherness":

> All that is language is lent from this otherness and this is why the subject is always a fading thing that runs under the chain of signifiers. For the definition of a signifier is that it represents a subject not for another subject but for another signifier. This is the only definition possible of the signifier as different from the sign. The sign is something that represents something for somebody, but the signifier is something that represents a subject for another signifier. The consequence is that the subject disappears. (1966/1970, 194)

In other words, with Lacan's post-structuralist approach to language, the sign—which had constituted for the structuralist a fixed relationship between a signifier and its signified meaning—now dissolves into an evanescent flux of differences wherein the subject loses its substance, becoming ghostlike and ephemeral. In its repetition, the subject surely *desires* to substantiate itself, but its signification diverges, leading only to further signification in a never-ending series of displacements and slippages. Here, in the open-ended play of language, identity gives way to difference and solidity evaporates. How, then, can we have clear-cut definitions, equations, proofs, or any of the other positivistic appurtenances of modernist mathematics?

It is in this decidedly *post*modern (post-structuralist) context that Lacan makes use of topology. Contrasting the Moebius strip with the sphere ("that old symbol for totality"), he employs the Moebius signifier not to establish mathematical identity but to illustrate the spontaneous emergence of *difference:* whereas movement upon a sphere keeps us on the same side of the surface, movement on the Moebius diverges, carrying us over to the *other* side (I will clarify the properties of this paradoxical structure in subsequent chapters). Lacan was indeed speaking the language of the unconscious, where—as Freud well knew—wit (*witz*) plays an essential role. I suggest that, at bottom, Lacan's use of topology involved something of a *joke,* since it demonstrated "precisely" the inescapable *imprecision* of language. Sokal and Bricmont evidently did not get the humor. Perhaps these exemplars of modernist culture can be likened to poorly trained linguistic anthropologists who fail to gather a sufficient corpus from the "alien culture" they are seeking to investigate. Rather than dealing with the whole body of Lacan's work, Sokal and Bricmont declare at the outset that they "shall limit [themselves] to an analysis of [Lacan's] frequent references to mathematics." In thus extracting Lacan's work from its postmodern context to suit their own purposes, they can see it only through modernist eyes and consequently miss its playful nature.

One wonders, however, whether Lacan himself fully appreciated the joke. If he was only telling a joke, he certainly told it with a serious face. In the final decade of his career, he became more and more obsessed with topological abstraction to the point where his attempts at "mathematizing" psychoanalysis were beginning to alienate many of his own followers. Yet he persisted in a manner that seemed hardly consistent with one who is simply jesting. Perhaps he was engaged in what Sartre termed "self-deception" (1943/1975, 299ff.). That is, at times when Lacan was focusing on the positivity of mathematics, he was aware of the negative, post-structuralist side, but only peripherally, in a way that allowed him to dismiss it. If the negative were either completely opaque or wholly transparent to him, there would be no self-deception; in the former case, he simply could not lie about the negation of mathematical certainty, whereas, in the latter, he could at least not lie to *himself.* The game of self-deception depends on the "translucency" (Sartre, 302) of the member of the dyad upon which one is not focusing. At the very moment that Lacan is proclaiming the mathematical rigor of his "topological psychoanalysis" (as Sokal and Bricmont refer to it), the ephemerality of this "rigor" is diffusely filtering through to him. But, in order to banish ambiguity on those occasions, he chooses to blind himself to it.

It is difficult to determine whether Lacan was involved in such self-deception, whether he was dead serious about mathematics, or was simply jesting. If he *was* joking in his quasi-mathematical application of topology, the joke appears to become more obvious in the post-mathematical applications of topology evident today. In the writings of Michel Serres, Gilles Deleuze and Félix Guattari, Brian Massumi, Stephen Perrella, and others, "mathematics" appears to be employed in such a manner that the old foundations of mathematics are swept away. But we are going to see that the joke in fact may have another turn. As a first step in assessing the challenge of postmodern topology, let us digress to examine what is perhaps the core tenet of its modernist predecessor, a principle whose philosophical roots can be traced back to Plato.

2. CORE ASSUMPTIONS OF MODERNIST TOPOLOGY

In the *Timaeus,* Plato states that "we must make a threefold distinction and think of that which becomes, that in which it becomes, and the model which it resembles" (1965, 69). The first term refers to any particular object that is discernible through the senses. The "model" for the transitory object is the "eternal object," i.e., the changeless form or archetype. This perfect form is *eidos,* a rational idea or ordering principle in the mind of the Demiurge. Using his archetypal thoughts as his blueprints, the Divine Creator fashions an orderly world of particular objects and events. As for "that *in* which [an object] becomes," Plato speaks of the "receptacle," describing it as "invisible and formless, all-embracing" (70). It is the vessel used by the Demiurge to *contain* the changing forms without itself changing (69). Plato goes on to characterize the receptacle as *space* (71–72).

Note, however, that while the receptacle was *supposed* to contain change without itself changing, it actually did not function with perfect efficiency, as Plato himself admitted. Being subject to "fleeting potencies and constantly changing tensions" (Graves 1971, 71), the receptacle was given to porosity; at any time, it could "spring a leak." In other words, the inhomogeneity of Platonic space made it susceptible to being ruptured, to losing its continuity. Not until centuries later, with the philosophy of Descartes, was the idea of spatial continuity brought to fruition.

Descartes related the continuum to the concept of *extension*. Consider, as an illustration, the simplified space represented by a line segment. In the Cartesian approach, it is intuitively self-evident that the line, however short, has extension. It must then be continuous: it can possess no holes or gaps in it, since, if the point-elements composing it were not densely packed, we would not have a line at all but only a collection of extensionless points. The quality of being extended implies the infinite density of the constituent point-elements.

Yet, at the same time, intuitive reflection discloses the paradox that the absence of gaps in the continuum not only holds this classical space together but also permits it to be *indefinitely divided*. Without a gap in the line to interrupt the process, there is no obstacle to the endless partitioning of it into smaller and smaller segments. As a consequence, though the points constituting this continuum indeed are densely packed, they are distinctly set apart from one another. However closely positioned any two points may be, a differentiating boundary permitting further division of the line always exists. As Milič Čapek put it in his critique of the classical notion of space, "no matter how minute a spatial interval may be, it must always be an *interval* separating two points, each of which is *external* to the other" (1961, 19).

The infinite divisibility of the extensive continuum also implies that its constituent elements themselves are unextended. Consequently, the point-elements of the line can have no internal properties, no structure of their own. An element can have no boundary that would separate an interior region of it from what would lie on the outside; *all* must be "on the outside," as it were. In other words, the Cartesian line consists, not of internally substantial, concretely bounded entities, but only of abstract boundedness as such (Rosen 1994, 92). Sheer externality alone holds sway—what Heidegger called the "'outside-of-one-another' of the multiplicity of points" (1927/1962, 481). Moreover, whereas the point-elements of classical space are utterly unextended, when space is taken as a whole, its extension is unlimited, infinite. Although I have used a finite line segment for illustrative purposes, the line, considered as a dimension unto itself, actually would not be bounded in this way. Rather than its extension being terminated after reaching some arbitrary point, in principle, the line would continue indefinitely. This means that the sheer boundedness of the line is evidenced not only locally in respect to the infinitude of boundaries present within its smallest segment; we see it also in the line as a whole inasmuch as its infinite boundedness would be infinitely extended. Of course, this understanding of space is not limited to the line. Classically conceived, a space of any dimension is an infinitely bounded, infinitely extended continuum.

Naturally, it would be a category mistake to interpret the infinitude of classical space as a characteristic of what is *object.* Space is not an object but is the "receptacle" of the objects, the changeless context within which objects are manifested. This distinction, initially made by Plato, is reflected in the thinking of Kant, who held that perceptions of particular objects and events are contingent, always given to variation, but that perceptual awareness is organized in terms of an immutable intuition of space. In the words of B. A. G. Fuller and Sterling McMurrin, Kant took the position that "no matter what our sense-experience was like, it would necessarily be smeared over *space* and drawn out in *time*"[5] (1957, part 2, 220). Implied here is the categorial separation of *what* we observe—the circumscribed objects—from the *medium* through which we make our observations. We observe objects *by means of* space; we do not observe space. It is within the infinite boundedness of space that particular boundaries are formed, boundaries that enclose what is concrete and substantial. The concreteness of what appears within boundaries is the particularity of the object. In short, an object most essentially is that which is bound*ed,* whereas space is the contextual bounded*ness* that enables the finite object to appear.

The spatial context is what mediates between object and *subject.* The latter (personified by the Demiurge in the *Timaeus*) is the third term of the classical account and corresponds to what is *un*bounded. That an object possesses boundaries speaks to Descartes' characterization of it as *res extensa,* "an extended thing": what has extension will be bounded. In contrast, the subject is *res cogitans,* a "thinking thing." Entirely without extension in space, the subject has no boundaries or parts. As a consequence, it is indivisible (etymologically, this is equivalent to stating that the subject is an *individual*).

The crux of classical cognition, then, the axiomatic base serving as its unquestioned point of departure, is the self-evident intuition of *object-in-space-before-subject.* The object is *what* is experienced, the subject is the transcendent perspective *from* which the experience is had, and space is the continuous medium *through* which the experience occurs. The relationship among these three terms is that of categorial separation.

The classical formula is built into modern mathematics at a fundamental level. It is true that *topological* mathematics has great flexibility compared with geometric disciplines such as the Euclidean and projective geometries. In "rubber sheet" geometry, we can turn doughnuts into coffee cups with impunity. Yet however we may turn, twist, or deform a topological object to vary its concrete appearance, from the perspective of the mathematical analyst, the object will "look the same." That is, the doughnut and coffee cup, when regarded abstractly as continuous surfaces with single holes, are entirely equivalent (as noted above). Of course, the subject's conferral of abstract unity upon the varying appearances of the object is mediated by the third term in the classical formula, viz. the analytical space or continuum that contains the topological transformations without itself being transformed.

3. POST-LACANIAN APPLICATIONS OF TOPOLOGY

It is the old philosophical formula grounding modernist mathematics that appears to be challenged in Lacan's postmodern use of topology. The Lacanian Moebius strip is no well-defined topological object but a signifier whose radical divergence from mathematical identity seems to disrupt the whole relation of object-in-space-before-subject ("the subject disappears," says Lacan; 1966/1970, 194). At least as subversive are the post-Lacanian applications of topology mentioned above at the end of section 1. In these works, the fluid deformations of objects that conventional mathematics had contained in its analytical space now overspill their borders and are brought to bear on space itself. Michel Serres, for example, concretizes space, brings it down into the material realm where it becomes "pliable, tearable, stretchable . . . topological" (Serres 1994, 45). Space is presently "a mobile confluence of fluxes" (Serres and Latour 1995, 122) rather than a static container and is thus made susceptible to the topological vagaries and vicissitudes of history. Serres reaches similar conclusions about *time*. Consider Steven Connor's assessment of Serres on this count:

> In the foldings and refoldings of the fabric of time, the idea of an invariant surface on which the folding might be taking place, or of the clock which would tick off the time which elapses while it takes place are mere fictions. The truth yielded by the topological apprehension of time is that there is no such invariant background. . . . Serres sees the river [of time] running chaotically through a landscape that itself forms as it moves. (2002)

Perhaps the most influential figures in post-mathematical topology are the cultural theorists Gilles Deleuze and Félix Guattari. In their magnum opus, *A Thousand Plateaus* (1987), Deleuze and Guattari begin by implicitly contrasting their own approach with the totalizing tendencies of classical thought: "All *we* talk about are multiplicities, strata and segmentarities, lines of flight and intensities, machinic assemblages and their various types" (4; italics added). These diversities and divergences are ceaselessly at play as flows on surfaces, on planes without depth, and are all-pervasive in nature and culture. Deleuze and Guattari warn us not to look for any fixed meaning beneath, behind, or within this restless profusion of activity; as they see it, we can only describe how it functions, how patterns form, deform, and capriciously dissolve. Giving their own book as an example of such machinations, they say, "A book exists only through the outside and on the outside. A book itself is a little machine" (4).

The authors caution us against the subtle persistence of the impulse toward stasis and totalization, particularly as manifested in modernism. James Joyce, for instance, shatters the old linear unity of the word only to "posit a cyclic unity of the sentence, text, or knowledge." Friedrich Nietzsche demolishes "the linear unity of knowledge, only to invoke the cyclic unity of the eternal return." In these examples, "unity is consistently thwarted and obstructed in the object, while a new type of unity triumphs in the subject . . . a higher unity . . . in an always supplementary dimension to that of its object" (1987, 6). "In truth," declare Deleuze and Guattari, "it is not enough to say, 'Long live the multiple,' difficult as it is to raise that cry. . . . The multiple *must be made,* not by always adding a higher dimension, but rather in the simplest of ways . . . with the number of dimensions one already has available—always $n - 1$ (the only way the one belongs to the multiple: always subtracted). Subtract the unique from the multiplicity to be constituted; write at $n - 1$ dimensions" (6). Toward the end of their book, Deleuze and Guattari call for a "topology of multiplicities" (483).

Brian Massumi, the translator of *A Thousand Plateaus,* closely aligns himself with the authors and carries their work forward in his online essay, "Strange Horizon" (1998). What is primary for Massumi is movement, flux—the twistings, turnings, and undulations by which we continually renew and transmute ourselves. "The space of experience is . . . a topological hyperspace of transformation," says Massumi, and living topologically entails "newness . . . the emergence of unforeseen experiential form and configuration, inflected by chance. . . . The [lived] body is . . . always provisional because always in becoming." Classically, space and time appear as independent variables that contain and constrain change. Massumi contrasts this with the topological view in which "space and time are dependent variables." Rather than being governed by "logics of presence or position that box things in three-dimensional space strung out along a time line . . . logics of transition are needed: qualitative topologics." Massumi concludes that "the life of the body, its lived experience . . . cannot be contained in Euclidean space and linear time. They must be topologically described." Here, "Formal topologies are not enough." We require "ontogenetic" topologies that express the "continuing becoming" of experience. Massumi emphasizes that there is no "one topological figure, or even a specific formal non-Euclidean geometry, that corresponds to the body's space-time of experience or some general 'shape' of existence. Topologies . . . are modeling tools." To Massumi, dynamic topology is no mere metaphor for lived experience; instead it is a "biogram" that is literally interwoven with that experience. "If there is a metaphor at play," says the author, it is "rather mathematical representation that is the metaphor" (1998).

In a similar post-mathematical vein, architectural theorist Stephen Perrella introduces the topological notion of the *hypersurface:*

> In [conventional] mathematics, a hypersurface is a surface in hyperspace, but in the [present] context . . . the mathematical term is existentialised. . . . This reprogramming is motivated by cultural forces that have the effect of superposing existential sensibilities onto mathematical and material conditions, [as especially seen in] the recent topological explorations of architectural form. The proper mathematical meaning of the term hypersurface is . . . challenged by an inherently subversive dynamic. (n.d.)

Perrella explains that, "Instead of meaning higher in an abstract sense, 'hyper' means altered. . . . In an existential context, hyper might be understood as arising from a lived-world conflict as it mutates the normative dimensions of three-space" (n.d.). The author notes that

> the dominance of the mathematical model is becoming contaminated because the abstract realm can no longer be maintained in isolation. The defection of the meaning of hypersurface, as it shifts to a more cultural/existential sense, entails a reworking of mathematics. . . . This defection is a deconstruction of a symbolic realm into a lived one. . . . If [mathematical] ideals, as they are held in a linguistic realm, can no longer support or sustain their purity and dissociation, then such terms and meanings begin, in effect, to "fall from the sky." . . . Topological "space" differs from Cartesian space in that it imbricates temporal events within form. Space then, is no longer a vacuum within which . . . objects are contained, space is instead transformed into an interconnected, dense web of particularities and singularities . . . a material/immaterial flux of actual discourse. (n.d.)

Generally speaking, the new initiatives in topology pose a significant challenge to the order of abstract thinking that has constrained us for centuries. In so doing, they help pave the way for revitalizing the lifeworld. But let me now consider a misgiving I have about this postmodern approach.

4. FROM POSTMODERN TOPOLOGY TO TOPOLOGICAL PHENOMENOLOGY

Post-mathematical topology rightly questions the peremptory divisions so prevalent in the classical and modernist viewpoints. Deleuze and Guattari's notion of flowing intensities, for example, seeks to transgress the rigid boundaries of substantival thought in an attempt to offer a dynamic account of life process. In this regard, feminist theorist Elizabeth Grosz applauds Deleuze and Guattari's reconception of "bodies outside the binary oppositions imposed on the body by the mind/body, nature/culture, subject/object and interior/exterior oppositions" (1994, 164). Nevertheless, while Deleuze and Guattari apparently engage in a "global rejection of binary oppositions"[6] at one level of their discourse, in fact they wind up tacitly maintaining opposition on another, less explicit level. Here the fundamental oppositions of identity and difference, being and becoming, unity and multiplicity, are upheld. For, in each case, Deleuze and Guattari negate the first member of the pair and come down decisively on the side of the second ("subtract the unique from the multiplicity," they urge). The unambiguous negation of one term and affirmation of the other in the content of their assertions reinforces the binary splitting of terms at the implicit level of linguistic form. Evidently, when it comes to such basic oppositions, ambiguity is not something that can consistently be tolerated.

The same kind of "meta-dualism" is indicated in the writings of Brian Massumi. Although he beautifully demonstrates the need to reconceive lifeworld experience via a "strange one-sided topology" that surpasses the old dichotomies by working at a "paradoxically creative edge" (1998), Massumi seems to lose *his* edge at the meta-level. Here movement is unambiguously favored over stasis, matter over mind, difference over identity, becoming over being, the many over the one. Elizabeth Grosz shows a similar inclination. After effectively illustrating how we can surpass mind-body dualism by using the Moebius strip to express "the inflection of mind into body and body into mind" (1994, xii), she winds up privileging "the fields of difference, the trajectories of becoming" (210) over identity and being (in the Platonic heritage, becoming is associated with the multiplicity of the body and being with the unity of the mind). Such one-sided reactions to the totalizing tendencies of modernism seem to be a defining characteristic of postmodern thought in general. The "one-sidedness" in question certainly is not of the Moebius kind. Rather than genuinely challenging the old "purity and dissociation" (Perrella n.d.) of modernist philosophy by consistently applying topological paradox to the most basic philosophical dichotomies, postmodernism tends to slip into a sort of "*reverse* purism." Pure being is replaced by a mode of becoming that is every bit as pure. The world is no abstract unity, we are told; instead it is "sheer multiplicity." Yet, in either case, the world *is;* it is predicated (positively or negatively) in unambiguous terms, set off by an exterior (non-paradoxical) boundary, rendered a well-defined object in the analyst's space of discourse, one from which the post-mathematical subject stands aloof. At this level of analysis, the old formula of object-in-space-before-subject prevails and nary a "Moebius edge" can be found. It is true that, in terms of its *content,* the object is no longer a fixed thing, but is an "ever-changing multiplicity." Yet, in the subtler sphere of linguistic form, we do have an *object,* something that is non-topologically segregated from its binary opposite, viz. "changeless unity." Like all objects, the objects in question are well contained within the linguistic continuum serving as the subject's means of analysis. We can see here the operation of a dialectic in which postmodernism's propensity simply to negate the earlier tradition in fact tacitly maintains it *because* the method of simple negation is being employed. And it is in covertly preserving the old ontological formula that true access to the lifeworld continues to be barred.

Perhaps you recognize the "Chinese finger puzzle" at play in the postmodern struggle to break free from classical and modernist restraints. As I noted in the preface, all efforts to free ourselves from the abstraction of modernity by simply opposing it leave us squarely within it, since the sine qua non of abstraction is simple opposition. But I also intimated that, while abstraction has no exit, no exterior boundary, it may indeed possess an *interior* horizon, a paradoxical threshold at its innermost depths that leads beyond it. We can now better appreciate that the boundary in question is Moebius-like. In fairness to thinkers like Grosz and Massumi, they do make effective use of such a "strange one-sided topology" (Massumi 1998). What I am suggesting is that the application needs to be implemented in a consistent and thoroughgoing manner, with every effort being made not to lose one's "topological edge." From my own experience, I know all too well how difficult this is to achieve, and it would not surprise me to learn that there are places in this very text in which I myself lose my edge. As a dweller in a glass house, I must be careful, then, about the stones I hurl. We are all challenged to avoid limiting our applications of topological paradox to the surface of our discourse while allowing our deepest assumptions and forms of expression to remain tacitly governed by the old trichotomous formula (object-in-space-before-subject). To prepare ourselves for re-inhabiting the lifeworld, we must keep our topological edge "all the way down." Differently put, the expression of topological paradox must become *ontological,* lest the *old* ontological outlook continue its domination from below. Let us explore what this means.

It is through paradox that one challenges the traditional formula. Rather than saying, "X is" or "X is not," one says, "X is not-X." This is no mere affirmation or denial of a predicated content, but predication's denial of itself. In asserting that "X is not-X," the customary subject/predicate format *is* being used ("X" is the subject, "is not-X" is the predicate), but in a manner whereby the content that this sentence expresses calls the form into question. The paradoxical statement amounts to a declaration that the syntactical boundary condition that would delimit X cannot effectively do so. Simple predicative boundary assignment is thwarted, so that even though X implicitly is being posited as distinct from that which is external to it, at the same time it is inseparable from it. The statement "X is not-X" boggles our minds because the human mind is a reflective organ whose principal function is to draw clear-cut boundaries. Nevertheless, if we are to fully appreciate what is required for reentering the lifeworld, we must distinguish two orders of paradox.

Consider a commonly cited example of a paradoxical statement: "Everything I say is false." Evidently, this assertion is true if it is false, and false if it is true! Applying the general formula for paradox, X = not-X, to the particular case, the term X stands for the truthfulness of the assertion (it is both true and false). Following Heidegger (1927/1962, 31), we may call this order of paradox *ontical:* its key characteristic is that its opposing terms ("truth" and "falsity") are particular objects of thought, entities already projected before the subjectivity of the thinker. While the well-known "liar's paradox" certainly subverts the boundary between these objects, it "comes too late" to directly affect what Heidegger would call the *ontological* boundary, the division prereflectively established between the object(s) that are reflected upon and the existential subject that does the reflecting. It is this latter division that lies at the root of predication. Therefore, to confound predicative boundary-drawing in the most radical way, paradox must be taken beyond the merely ontical level and expressed more primordially; it must be brought to bear on the bounding of subject and object that "precedes" any mere division among particular objects. Thus, in saying "X is not-X," one must mean, "*I am not-I,*" with "I" taken as ontological: not just a *particular* (i.e., objectified) subject, a specifiable individual with a personal history and personal characteristics, a given ontical being. Rather than being some object of reflection, the "I" in the formula for paradox must be the prereflectively established subject that reflects.

Of course, the instant we install the prereflectively chosen "I" in the formula, it passes over into the domain of the *reflected upon,* itself becoming but an object now cast before a newly established subject not included in the formula. In thus formulating ontological paradox, the paradox becomes ontical. From this it should be clear that the rule of predication will not be radically challenged by the mere *formulation* of paradox. To write and think "I am not-I" in the usual manner of writing and thinking is to continue to predicate. Not that reflective predication would simply be suspended in realizing ontological paradox. The "I" or thinking subject would indeed still be reflecting upon itself, and, by virtue of the fact that it *was* reflecting, it would be making itself into an other, a "not-I." And yet, it would be doing this without *just* abstracting itself, without turning itself into *merely* what is other, thus cutting itself off from its prereflective roots. In realizing ontological paradox, the "I" would continue in the reflective posture, standing outside itself; but, at the same time, it also would be standing *within.* The philosopher Eugene Gendlin intimated the possibility of such a curious stance. In his essay "Words Can Say How They Work" (1993), Gendlin (expanding upon Heidegger's [1927/1962] notion of *Befindlichkeit* or "moody understanding") suggested that our words originate from a prereflective bodily source that continues to operate in the midst of our speaking, so that, at least in principle, we can both speak reflectively *about* this source and directly engage it. I propose that it is by bringing together the reflective and prereflective that the way is paved for re-inhabiting the lifeworld.

Nevertheless, old habits are slow to die. It is the reflective mode of consciousness that has long been dominant. Therefore, although the abstract words written on this page may be rooted in a prereflective source that continues to operate even as we read, it is abstraction that prevails. We may *reflect* on our concrete prereflectivity and its paradoxical relation to the reflective, yet we find it difficult vividly to feel its living presence at work in our midst.

This is where topology enters the picture. We are seeing that ontological paradox must be *embodied* (Rosen 1997), that the "I am not-I" must be fleshed out, made into a concrete reality. Topology is an indispensable tool in this regard. As I noted previously, "topology . . . is rooted in the body" (Sheets-Johnstone 1990, 42); "no matter how abstract it may become," asserts Steven Connor, "topology remains fundamentally bodily" (2002). That is why it is so helpful when Grosz expresses the relationship between subject and object in topological terms, and when Massumi speaks of "experience [being] doubled back on itself like a Möbius strip" (1998). What I am emphasizing for my part is that topological paradox must retain its edge all the way down into the ontological roots of our discourse. It would therefore not suffice for me to topologize the subject-object relation in the content of my writing while maintaining a non-topological posture in the underlying manner in which I express this content. In so limiting myself merely to *predicating* the topological linkage of subject and object, I fail to make ontological paradox a concrete reality. Such predication of the subject-object coupling in fact turns the copula into an *object,* an ontical content, from which I—this predicating subject—am decoupled. The concrete realization of ontological paradox thus requires that she who predicates makes her own presence felt topologically, rather than receding into an anonymity that serves to reinforce the old ontological posture. As Massumi puts it, "How can the literate become literal and the literal literate, in two-way, creative interference? Most of all, how can this [topological] looping be accomplished openly and without . . . the imposition of an authorial 'voice' or 'vision' aimed at grounding a sea-tossed world?" (1998). I take this as another way of asking, How can we employ topology to reanimate the lifeworld in practice, not just in theory? In subsequent chapters, no question is more critical than this, nor is any more difficult to address.

The application of topology to ontological paradox is what I mean by *topological phenomenology.* But doesn't phenomenology constitute a "closed loop of 'intentionality'"? Isn't "every phenomenological event . . . like returning home . . . without the portent of the new"? With these words, Massumi (1998) rejects the phenomenological approach. For Massumi, phenomenology is a "domesticating, self-satisfied subjectivism" that shuns the newness of becoming and takes refuge in the closed circle of being. I entirely agree that if we wish to participate in the vitality of the lifeworld, we cannot privilege being over becoming. Yet we have found that privileging becoming has the same effect: it implicitly bolsters the traditional ontological posture wherein the lifeworld is eclipsed. Therefore, rather than falling into a one-sided espousal of either being or becoming, we must stay in the dialectical saddle, keeping our topological edge.

But *what about* phenomenology? *Does* it tend to privilege being over becoming? The question in fact cannot be answered without further consideration of what "being" means. In a certain sense of this term, Edmund Husserl, founder of phenomenological philosophy, did appear to favor being. According to Quentin Lauer, Husserl's translator, his "philosophy is primarily concerned with the essence of what is (*das Seiende*) rather than with the essence of being (*das Sein*). . . . His problem, then, is not to discover the essence of either being or truth [*Wahrheit*], but rather to guarantee that knowledge is of 'what is' and that it is true [*wahr*]" (1965, 46). In other words, because of his modernist preoccupation with obtaining "absolute scientific certainty," Husserl was unconcerned with being *qua* being; his abiding interest was in being *qua* apodictic knowing. It is this epistemic, or subjectivist, essentially Cartesian, being—this "transcendental *cogito*"—that Husserl privileges over becoming.

In contrast to Husserl, his student, Martin Heidegger, took the question of the "Being of beings" as philosophy's foremost concern. In the preface we saw that, for Heidegger, "Being is not a being, not God, an absolute unconditional ground or a total presence, but is simply the living web within which all relations emerge" (Bigwood 1993, 3). That is to say, *Be-ing* constitutes the dimension of dynamic life process, the lifeworld. Here neither being nor becoming can be favored one-sidedly, since these modalities have not yet been torn asunder. (Recall, though, from the preface, that Heidegger himself did occasionally lapse into a nostalgic valorization of "original being.") It is the pre-trichotomous, inherently ambiguous, ontologically hybrid lifeworld that classical, modernist, and postmodern philosophies have all skipped over in their drive for one kind of "purity" or another. In the whole of Western philosophy, only ontological phenomenology appears to have had the stomach for pursuing the paradox of being and becoming (subject and object, identity and difference . . .) all the way down into its lifeworld roots.[7] Rather than seeking simply to eliminate the ambiguity, onto-phenomenology would consciously embody it, transform it into diaphanous flesh.

In the topological elaboration of ontological phenomenology that is to follow, we will explore the primordial ambiguity native to the *topos*. Here, among other things, we will find (1) that Being grows and develops like a living organism, (2) that there is more than one dimension of Being, (3) that these ontological dimensions develop in relation to each other, and (4) that new dimensions of Being arise from the dialectical interplay of the old. Most importantly, we shall discover that topological Being is *us*.

CHAPTER TWO

THE TOPOLOGY OF THE FLESH

I. PHENOMENOLOGY AND THE FLESH OF THE WORLD

The phenomenological movement is rooted in the nineteenth-century existentialist writings of thinkers like Søren Kierkegaard, Friedrich Nietzsche, and Fyodor Dostoevsky. It takes its contemporary form through the work of its principal figures: Edmund Husserl, Martin Heidegger, and Maurice Merleau-Ponty. In the present context, the latter two thinkers will be of special interest since we are going to focus on *ontological* phenomenology. It is onto-phenomenology that questions most deeply the old trichotomous formula, *object-in-space-before-subject.*

Phenomenology generally regards the activities of the detached Cartesian subject as idealizing objectifications of the world that conceal the concrete reality of the *lifeworld.* Obscured by the lofty abstractions of post-Renaissance Western culture, this earthly realm of lived experience is inhabited by subjects that are not anonymous, that do not fly above the world, exerting their influence from afar. In the lifeworld, the subject is a fully situated, full-fledged participant engaging in transactions so intimately entangling that it cannot rightly be taken as separated either from its objects or from the worldly context itself. As Heidegger puts it, the down-to-earth, living subject is a *being-in-the-world* (1927/1962), a being involved in

> a much richer relation than merely the spatial one of being located in the world. . . . This wider kind of personal or existential "inhood" implies the whole relation of "dwelling" in a place. We are not simply located there, but are bound to it by all the

> ties of work, interest, affection, and so on. (Macquarrie 1968, 14–15)

It is clear that all three terms of the classico-modernist formulation are affected by the phenomenological move. We found in chapter 1 that, traditionally, space is the sheer *boundedness* serving as the medium for the *unbounded* subject's operations upon *bounded* objects. The basic principle here is that of external relationship. That is, objects that are simply external to each other appear within a spatial context of *sheer* externality (the "'outside-of-one-another' of the multiplicity of points"; Heidegger 1927/1962, 481), and are operated upon by agents acting from outside of space. In the contrasting phenomenological approach, all relations are *internal.* Notwithstanding the Platonic/Cartesian idealization of the world, in the underlying lifeworld there is no object with boundaries so sharply defined that it is closed off completely from other objects. The lifeworld is characterized instead by the *transpermeation* of objects, by their mutual interpenetration, by the "reciprocal insertion and intertwining of one in the other," as Merleau-Ponty put it (1968, 138). With objects thus related by way of mutual containment, no *separate* container is required to mediate their relations, as would have to be the case with externally related objects. Phenomenological understanding supersedes the classical relationship of container and contained (of sheer boundedness and the merely bounded). Objects no longer are to be thought of as contained in space like things in a box, for, in intimately containing each other, they contain themselves.

At the same time, it must also be understood that, in the lifeworld, there can be no categorial division of object and *subject.* The lifeworld subject, far from being the disengaged, high-flying deus ex machina of Descartes, finds itself down among the objects, is "one of the visibles" (Merleau-Ponty 1968, 135), is itself always an object to some *other* subject, so that the simple distinction between subject and object is confounded and "we no longer know which sees and which is seen" (139). This placement of the subject among the objects is of course no materialistic reduction of it to a *mere* object (an inert lump of matter). Rather, the phenomenological grounding of the subject is indicative of the ambiguous interplay of subject and object in the lifeworld. Generally speaking, then, what the move from classico-modernism to phenomenology basically entails is an internalization of the relations among subject, object, and space.

Now, I have been stressing that the lifeworld is an *earthly* realm, a concrete sphere of embodied experience. This crucial feature is conveyed more fully by Merleau-Ponty than by Heidegger. Therefore, in introducing topological phenomenology, I will concern myself primarily with the work of Merleau-Ponty.

Although Merleau-Ponty's thinking was strongly influenced by both Heidegger and Husserl, Merleau-Ponty ranked the latter well above the former (Spiegelberg 1982, 537–38). And yet, especially toward the end of his career, it is not Husserl's essentially epistemological approach that Merleau-Ponty found most riveting, but the Heideggerian engagement with Being. This is particularly evident in his last work.

Merleau-Ponty's final philosophical position is adumbrated in "The Intertwining—The Chiasm," the enigmatic centerpiece of *The Visible and the Invisible* (1968). In this chapter, he articulated what had come to be his fundamental concept, an idea for which "there is no name in traditional philosophy" (139), in fact, "no name in any philosophy" (147): the *flesh of the world*. The flesh cannot be named in traditional philosophy because, with it, the most basic categories of the classical tradition are transgressed:

> The flesh is not matter, in the sense of corpuscles of being . . . is not mind, is not substance. To designate it, we should need the old term "element," in the sense it was used to speak of water, air, earth, and fire, that is, in the sense of a *general thing*, midway between the spatio-temporal individual and the idea, a sort of incarnate principle that brings a style of being wherever there is a fragment of being. The flesh is in this sense an "element" of Being. (1968, 139)

With the notion of the flesh, Merleau-Ponty posed his greatest challenge to the long-standing Cartesian division of subject and object. Neither a corporeal object nor a disembodied subject, the flesh is a "coiling over" of the body (146), a folding back of it upon itself from which the subject and its object first arise. The subjectivity produced in this "coiling over" is decidedly different from the transcendent subjectivity of idealism.

For Merleau-Ponty, to know is not simply to transcend a body that itself is unknowing. The act of knowing is not one of breaking away, of subject breaking free from object, of mind breaking out of the body's enclosing orbit into a boundless domain of its own. No such simple breaking with the body occurs when there is conscious awareness, since consciousness is not achieved by the mind's detachment from the body, but by the body itself—that is, by *the flesh*. Because the flesh is not closed, not a mute, unseeing, sealed objectivity, it need not be broken open by an agency that is alien to it. Rather, the flesh opens itself. It is the openness of this flesh to which the subject blinds itself when it places itself outside of and above the objects, which, in their turn, are made into *mere* flesh, closed bodies ("corpuscles of being"). But for Merleau-Ponty, object and subject, sensible body and sentient awareness, are of the same order, both arising from the paradoxical contortion in which the flesh turns out of and back upon itself. Thus Merleau-Ponty, stating that flesh is "the formative medium of the object and the subject" (1968, 147), describes it as a *"sensible for itself"* (135), a "coiling over of the visible upon the seeing body" (146). He suggests that the "primordial definition of sensibility," rather than entailing a "belongingness to one same 'consciousness,'" involves "the return of the visible upon itself, a carnal adherence of the sentient to the sensed and of the sensed to the sentient" (142). As a consequence of this, "he who sees cannot possess the visible unless he is possessed by it, unless he *is of it,* unless, by principle, according to what is required by the articulation of the look with the things, he is one of the visibles, capable, by a singular reversal, of seeing them—he who is one of them" (134–35). Merleau-Ponty goes on to portray the basic subject-object ambiguity found in the flesh as a "paradox of Being, not a paradox of man" (136). Or, in the language of the previous chapter, we can say that the flesh of the world constitutes an *ontological* paradox rather than merely an ontical one (again, whereas the ontical presupposes the reflective splitting of subject and object, the ontological bespeaks their prereflective intimacy).

2. EMBODYING THE FLESH THROUGH TOPOLOGY

In the "Working Notes" of *The Visible and the Invisible,* Merleau-Ponty indicated that the paradox of the flesh might best be approached via *topology.* The hint is given in an entry written in October 1959. Here Merleau-Ponty instructed himself as follows:

> Take topological space as a model of being. The Euclidean space is the model for perspectival being [and is consistent] . . . with the classical ontology. . . . The topological space, on the contrary . . . is the image of a being that . . . is at the same time older than everything and "of the first day" (Hegel). . . . [Topological space] is encountered not only at the level of the physical world, but again it is constitutive of life, and finally it founds the *wild* principle of Logos— —It is this wild or brute being that intervenes at all levels to overcome the problems of the classical ontology. (1968, 210–11)

In associating Euclidean space with "the classical ontology," Merleau-Ponty in effect was proclaiming it *ontical.* That is because, while classical "ontology" attempts to study the general nature of being, it neglects the "wild side"—the embodied dimension that constitutes the *Being* of beings. When Merleau-Ponty suggests that it is *topological* space that is truly ontological, he confirms the conclusion reached in chapter 1 that topology is indeed the appropriate means for exploring the fleshly lifeworld. The point is cogently amplified by phenomenological philosopher David Michael Levin, who "interprets fundamental ontology as a fundamental *topology*" (1985, 96).

Now, certain topological figures seem particularly well-suited for fleshing out the paradox of Being. In the previous chapter I referred to what is perhaps the simplest and most basic topological expression of paradox: the Moebius strip. Let us now examine this structure more closely by comparing it with a topological counterpart that is *not* paradoxical (Rosen 1994, 2004). A cylindrical ring (fig. 2.1(a)) is constructed by cutting out a narrow strip of paper and joining the ends. The surface of Moebius (fig. 2.1(b)) is produced simply by giving one end of such a strip a half twist (through an angle of 180°) before linking it with the other.

The cylindrical ring possesses the conventionally expected property of two-sidedness: at any point along its surface, two distinct sides can be identified. Now, in the Moebius case, it is true that if you place your index finger anywhere on the surface, you will be able to put your

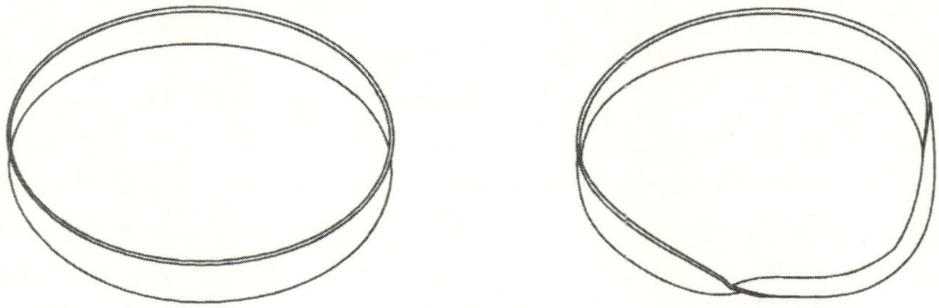

a b

Figure 2.1. Cylindrical ring (a) and Moebius strip (b)

thumb on a corresponding point on the opposite side. The Moebius strip does have two sides, like the cylinder. But this only holds for the local cross-section of the strip defined by thumb and forefinger. Taking the full length of the strip into account, we discover that points on opposite sides are intimately connected—they can be thought of as twisting or dissolving into each other, as being bound up internally. Accordingly, mathematicians define such pairs of points as *single* points, and the two sides of the Moebius strip as but *one* side. (If the Moebius property of one-sidedness is difficult to imagine in the abstract, it is very easy to demonstrate. For instance, when you draw a continuous line along the whole length of the strip, upon returning to your point of departure you will discover that your ink mark has covered *both* sides of the surface.)

It is important to recognize that the surface of Moebius is not one-sided in the homogeneous sense of a single side of the cylindrical ring. It is one-sided in the paradoxical sense, one-sided and also two-sided, for the local distinction between sides is not simply negated with expansion to the Moebius as a whole. In coming to interpenetrate each other, the sides do not merely lose their distinct identities. And yet, while the sides remain different, they also become one and the same.

It seems clear that, in signifying the paradox of Being via the Moebius surface, it is fleshed out more than words alone could do. For one thing, these words you read are but arbitrarily devised, conventionally agreed-upon tokens that refer to their content in a merely external manner, whereas the Moebius strip iconically embodies that paradoxical content. There is also the matter of the *dimensionality* of the signifier. The one-dimensional typographic marks appearing on the two-dimensional surface of this page obviously fall short of tangibly delivering the three-dimensional reality of the lifeworld they signify. In contrast, the Moebius strip is a two-dimensional surface embedded in three-dimensional space. Thus it can embody more concretely the paradoxical conjunction of opposites constitutive of the flesh.

Nevertheless, while the Moebius model manifests one-sidedness more tangibly than the written word, it *is* a model, an outward symbolization of the union of inside and out, rather than a full-fledged embodiment directly incorporating the inner depths of Being. What would be needed for the latter? Not a two-dimensional body enclosed as mere object in three-dimensional space, but a body of paradox that is itself three-dimensional.

There exists a higher-dimensional counterpart of the Moebius surface. By way of introduction, consider an interesting feature of the Moebius: its *asymmetry*.

Unlike the cylindrical ring, a Moebius surface has a definite orientation in space; it can be produced either in a left- or right-handed form (depending on the direction in which it is twisted). If both a left- and a right-oriented Moebius surface were constructed and then "glued together," superimposed on one another point for point, a topological structure called a *Klein bottle* would result (named after the German mathematician Felix Klein).

The Klein bottle (fig. 2.2) has the same property of asymmetric one-sidedness as the two-dimensional Moebius surface, but embodies an added dimension (see Rosen 1994, 2004). Mathematicians tell us that we cannot really produce a proper physical model of this curious bottle. That is, left- and right-facing Moebius bands cannot be superimposed on each other in three-dimensional space without *tearing* the surfaces. I am going to suggest that this inability to objectify the Klein bottle in three-dimensional Cartesian space actually derives from the fact that the bottle calls into play the dimension of *ontological flesh*.

There is a different but mathematically equivalent way to describe the making of a Klein bottle that, for our purposes, will be very instructive. Once again a comparison is called for.

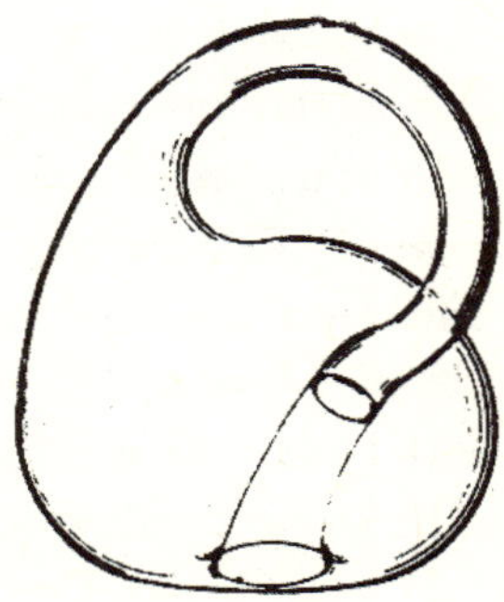

Figure 2.2. The Klein bottle (from Gardner 1979, 151)

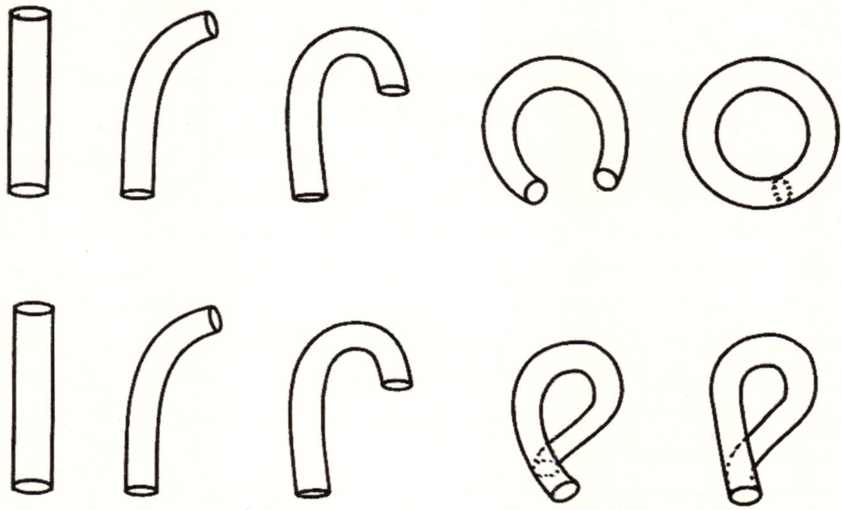

Figure 2.3. Construction of torus (upper row) and Klein bottle (lower row)

Both rows of figure 2.3 depict the progressive closing of a tubular surface that initially is open. In the upper row, the end circles of the tube are joined in the conventional way, brought together through the three-dimensional space outside the body of the tube to produce a doughnut-shaped form technically known as a *torus* (a higher-order analogue of the cylindrical ring). By contrast, the end circles in the lower row are superimposed from *inside* the body of the tube, an operation requiring the tube to pass *through* itself. This results in the formation of the Klein bottle. Indeed, if the structure so produced were cut in half, the halves would be Moebius bands of opposite handedness. But in three-dimensional space, no structure can penetrate itself without cutting a hole in its surface, an act that would render the model topologically imperfect. So, from a second standpoint, we see that the construction of a Klein bottle cannot effectively be carried out when one is limited to the three Cartesian dimensions that frame our experience of external (objective) reality.

Mathematicians are aware that a form that penetrates itself in a given number of dimensions can be produced without cutting a hole if an *added* dimension is available. The point is nicely illustrated by Rucker (1977). He asks us to imagine a species of "flatlanders" attempting to assemble a Moebius strip. Rucker shows that, since the "physical" (i.e., externally experienced) reality of these creatures would be limited to *two* dimensions, in trying to make an actual model of the Moebius, they would be forced to cut a hole in it. Of course, no such problem arises for us human beings, who have full access to three external dimensions. It is the making of the *Klein bottle* that is problematic for us, requiring as it would a fourth dimension. Try as we might, we find no fourth dimension "out there" in which to execute this operation.

I propose that the "fourth dimension" needed to complete the formation of the Klein bottle is the dimension of wild Being. This is not just another space for reflection extended before us; rather, it is folded within us, entailing the prereflective depths of the flesh. However, the standard mathematical approach to the Klein bottle is a modernist one. In accordance with the ontical formula, the Klein bottle is treated strictly as an *object-in-extensive-space.*

In conventional topology, the separation of the mathematical object from its containing space is reinforced by dividing container and contained along dimensional lines. To see how this is done, let us first consider some concrete facts of visual perception. Suppose we were viewing a box sitting on a table. As a cubic structure, the box possesses six sides. If perception were fully three-dimensional, we would be able to directly perceive all sides of the cube at once. It is clear we cannot do this. We can simultaneously view all four lines bounding a two-dimensional square. We can perceive at once both of the endpoints that bound the one-dimensional line segment. Yet when it comes to the six surfaces that bound the cube, we are limited to simultaneously viewing a maximum of *three* of them, along with their angular relationships. Still, while perspectival experience stops short of being wholly three-dimensional, neither is it merely two-dimensional. Were our perceptual repertoire strictly limited to two-dimensionality, we would be able to view but one surface of the box at a time. Our experience would thus lack any depth or perspective. Being restricted to the perception of a single plane, we would have no access whatsoever to the dimension perpendicular to that plane. That is certainly not the case. It is obvious that we *can* simultaneously view perpendicular planes of the box and their angular relationships. If we were observing a ball instead of a box, this angularity would be expressed as continuous curvature in the direction perpendicular to the plane of perception. Our ability to perceive angle and curvature attests to the practical truth that the experience of these objects is not strictly two-dimensional any more than fully three-dimensional. The concrete reality lies somewhere "in between."

How is this handled with the mathematical objects of modern topology? The objects we have been working with are conventionally classified as *surfaces*. The torus, cylindrical ring, and Moebius strip are considered to be two-dimensional surfaces embedded in three-dimensional space. But these structures are not simply flat, are they? Does not their curvature in the direction perpendicular to the plane demonstrate that their dimensionality is actually greater than two? The answer given is that the curvature of these topological surfaces is a "global" feature that results from the way they are *embedded* in three-dimensional space; this curvature does not alter the intrinsic dimensionality of the surfaces; *as* surfaces, they remain two-dimensional. Similarly, the Klein bottle is taken as a two-dimensional surface that would require a four-dimensional space for its containment. In this modernist analysis, the "fourth dimension" needed to complete the formation of the Klein bottle remains an extensive continuum, though this "higher space" is viewed as "imaginary"; the Klein bottle, for its part, is regarded as an "imaginary object" embedded in this space (whereas the cylindrical ring, Moebius surface, and torus are "real" mathematical objects in the sense that tangibly perceptible models of them may successfully be fashioned in three dimensions). By thus treating the mathematical object and its containing space as whole dimensions apart, their separation is enforced. And, whether the object must be approached through hyperdimensional abstraction or it is concretizable, the mathematician's attention is always directed outward toward an object, toward that which is cast before his or her subjectivity. Subjectivity itself is the detached position *from* which all objects are viewed—or, perhaps better, from which all is viewed as object. Never is subjectivity *as such* opened to view. In this way, the classical split between object and subject is upheld and the rule of the disembodied subject prevails.

Now, in his applied study of topology, the mathematician Stephen Barr advised that we should not be intimidated by the "higher mathematician. . . . We must not be put off because he is interested only in the higher abstractions: we have an equal right to be interested in the tangible" (1964, 20). The tangible fact about the Klein bottle that is glossed over in the higher abstractions of modernist mathematics is its *hole*. Because the standard approach has always presupposed extensive continuity, it cannot come to terms with the inherent *dis*continuity of the Klein bottle created by its self-intersection. Therefore, all too quickly, "higher" mathematics circumvents this concrete hole by an act of abstraction in which the Klein bottle is treated as a properly closed object embedded in a hyperdimensional continuum. Also implicit in this modernist approach is the detached subjectivity of the mathematician before whom the object is cast. I suggest that, by staying *with* the hole, we may bring into question the ontical conception of object-in-space-before-subject.

Let us look more closely at the hole in the Klein bottle. In a sense, this loss in continuity is *necessary*. One certainly could make a hole in the torus, or in any other object in three-dimensional space, but such discontinuities would not be necessary inasmuch as these objects could be properly assembled in space *without* rupturing them. It is clear that whether an object like the torus is cut open or left intact, the closure of the space containing that object will not be brought into question; in rendering such an object discontinuous, we do not affect the assumption that the space in which it is embedded is simply continuous. With the Klein bottle it is different. Its discontinuity does speak to the supposed continuity of three-dimensional space itself, for the necessity of the hole in the bottle indicates that space is unable to contain the bottle the way ordinary objects appear containable. Topologists tell us that if the Kleinian "object" is properly to be closed, assembled *without* a hole, an "added dimension" is required. Thus, for the Klein bottle to be accommodated, the idealized three-dimensional continuum must in some way be opened up, its continuity opened to challenge. Of course, we could attempt to sidestep the challenge, to skip over the hole by a continuity-maintaining act of abstraction, as in the standard mathematical analysis of the Klein bottle. Assuming we do not employ this stratagem, what conclusion are we led to regarding the "higher" dimension that is required for the completion of the Klein bottle? If it is not an extensive continuum, what sort of dimension is it?

My proposition is that the Klein bottle's "missing dimension" is the *ontological* dimension of human being; that it is not just another framework for reflection but a dimension that entails the prereflective depths of Being. Let me now attempt to make this clearer by refocusing on the threefold categorial disjunction implicit in the standard treatment of the

Klein bottle: contained object, containing space, uncontained subject. To reiterate: (1) The contained constitutes the category of the bounded or finite, that of the immanent contents we reflect upon, whatever they may be. These include empirical facts and their generalizations (which may be given in the form of equations, invariances, or symmetries). (2) The containing space is the contextual bounded*ness* serving as the means by which reflection occurs. (3) The uncontained or unbounded is the transcendent agent of reflection, viz. the subject. It is in adhering to this ontical trichotomy that the Klein bottle is conventionally deemed a "two-dimensional object" (i.e., a surface) embedded in "four-dimensional space." But the actual nature of the Klein bottle suggests otherwise. The concrete necessity of its hole indicates that, in reality, this bottle is not a mere object, is not simply enclosed in a continuum as can be assumed of ordinary objects, and is not open to the view of a subject that itself is detached, unviewed (uncontained). Rather than being contained in space, the Klein bottle may be said to *contain itself,* thereby superseding the dichotomy of container and contained. Rather than being reflected upon by a subject that itself remains out of reach, the self-containing Kleinian object may be said to *flow back into* the subject, thereby disclosing not a detached *cogito,* but the presencing of wild Being.

The Klein bottle's ontological blending of the contained object and its containing space confounds the idealized division of these terms into separate dimensions. The dimensional gap imposed by modernist mathematics is thus closed: the Kleinian object and its spatial context are of the *same* dimension. This surely does not mean that the Klein bottle is *two*-dimensional, as in its topological idealization. But neither is it simply three-dimensional. Rather, the Klein bottle uniquely accords with the phenomenological fact that actual human experience falls *between* the hypostatized dimensions of classico-modernism. Then are we to think of the Klein bottle's dimension number as some definite fraction lying on a scale between the numbers two and three? To think that would be to presuppose ontical continuity. The Klein bottle's dimension number is not "less than three" in the positive sense of constituting a finite value on a fixed numerical continuum (say, "2.5"). Kleinian dimensionality must be taken as "less than three" in a more qualitative, negative sense: it possesses a hole that discloses the incompleteness of the three-dimensional continuum. The "fractionality" of Kleinian dimension thus points to radical *dis*continuity, not to a less-than-three-dimensional continuum.

Yet the discontinuity in question does not merely replace continuity with its binary opposite. Rather than binary opposition, a *dialectic* of continuity and discontinuity is enacted with the Klein bottle. (Recall how postmodern topological efforts to simply negate classico-modernist continuity only wind up covertly reinforcing it—à la the Chinese finger puzzle!)

I have noted elsewhere the uncanny resemblance of the Klein bottle to the hermetic vessel of old alchemy (Rosen 1995, 137). The design of the enigmatic vessel is essentially that of the *uroboros,* the serpent that consumes itself by swallowing its own tail. To take itself in, the serpent must intersect itself, an operation requiring a hole (corresponding to the opening that is its mouth). The hole in the inside-out Klein bottle is of this sort. It is neither solely a hole in the spatial container, nor a hole in that which it contains, but the hole produced by the act of *self*-containment that integrates the container with its contents, in this way giving *(w)holeness.* By virtue of its strange one-sidedness, inside and outside do flow continuously together on the Klein bottle in a way that would be impossible on ordinary two-sided surfaces like the torus and cylindrical ring. Yet the boundary negotiated in the Kleinian movement is an *interior* one. Therefore, as on the Moebius strip, continuous passage from one side to the other also leaves us on the *same* side, for the distinction between opposing sides, their mutual discontinuity, is paradoxically upheld. This dialectic of continuity and discontinuity is doubtless a difficult notion to grasp, and I will seek to make it clearer in due course.

For now, let me attempt to offer a better picture of the Klein bottle via a schema for it provided by the communications theorist Paul Ryan (1993, 98). My adaptation of Ryan's diagram is given in figure 2.4.

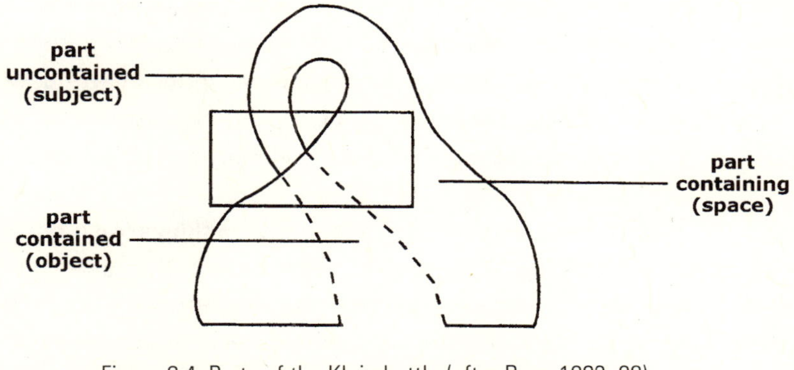

Figure 2.4. Parts of the Klein bottle (after Ryan 1993, 98)

Ryan identifies the three basic features of the Klein bottle as "part contained," "part uncontained," and "part containing." Here we see how the part contained opens out (at the bottom of the figure) to form the perimeter of the container, and how this, in turn, passes over into the uncontained aspect (in the upper portion of fig. 2.4). The three parts of this structure thus flow into one another in a continuous movement that flies in the face of the classico-modernist trichotomy. We can also see the aspect of *dis*continuity in the area of the diagram where a self-intersection is required (the region marked off by the rectangular box). Therefore, in its highly schematic way, the one-dimensional diagram lays out symbolically the basic terms involved in the "continuously discontinuous" circulation of three-dimensional Being.

Now, in employing topology to flesh out the dimension of Being, we have been following the advice of Merleau-Ponty: "Take topological space as a model of Being" (1968, 210). It is therefore not surprising that the self-containing Kleinian dimension we have arrived at should bear a close resemblance to Merleau-Ponty's own understanding of dimensionality, though he himself stopped short of spelling out this idea in topological terms. In his essay "Eye and Mind," Merleau-Ponty (1964) previews his ontology of the flesh by introducing the dimension of *depth*.

For Descartes, a dimension is an extensive continuum entailing "absolute positivity" (Merleau-Ponty 1964, 173). Descartes' assumption is that space simply is *there*, that it subsists as a positive presence possessing no folds or nuances; no shadows, shadings, or subtle gradations. Space is thus taken as the utterly explicit openness that constitutes a field of strictly external relations wherein unambiguous measurements can be made. Along with height and width, depth is but the third dimension of this hypostatized three-dimensional field. Merleau-Ponty contrasts the Cartesian view of depth with the true depth found in the lifeworld, where we discover in the dialectical action of perceptual experience a paradoxical interplay of the visible and invisible, of identity and difference:

> The enigma consists in the fact that I see things, each one in its place, precisely because they eclipse one another, and that they are rivals before my sight precisely because each one is in its own place. Their exteriority is known in their envelopment and their mutual dependence in their autonomy. Once depth is un-

> derstood in this way, we can no longer call it a third dimension. In the first place, if it were a dimension, it would be the *first* one; there are forms and definite planes only if it is stipulated how far from me their different parts are. But a *first* dimension that contains all the others is no longer a dimension, at least in the ordinary sense of a *certain relationship* according to which we make measurements. Depth thus understood is, rather, the experience of the reversibility of dimensions, of a global "locality"—everything in the same place at the same time, a locality from which height, width, and depth [the classical dimensions] are abstracted. (1964, 180)

Speaking in the same vein, Merleau-Ponty characterizes depth as "a single dimensionality, a polymorphous Being" from which the ontical Cartesian dimensions of linear extension derive, and "which justifies all [Cartesian dimensions] without being fully expressed by any" (1964, 174). The dimension of depth is "both natal space and matrix of every other existing space" (176).

Merleau-Ponty proceeds to explore the depth dimension via the artwork of Cézanne. Through the painter, he demonstrates that primal dimensionality is *self-containing.* For Cézanne works with a visual space that is not abstracted from its content but flows unbrokenly into it. Or, putting it the other way around, the contents of a Cézanne painting overspill their boundaries *as* contents so that, rather than merely being contained like objects in an empty box, they fully participate in the containment process. Like the serpent of old, Cézanne's paintings thus "swallow themselves." Inspired by these works, Merleau-Ponty comments that "we must seek space and its content *as* together" (1964, 180).

Merleau-Ponty also makes it clear that the primal dimension engages embodied subjectivity: the dimension of depth "goes toward things from, as starting point, this body to which I myself am fastened" (1964, 173). In commenting that "there are forms and definite planes only if it is stipulated how far from *me* their different parts are" (180; italics mine), Merleau-Ponty is conveying the same idea. A little later, he goes further:

> The painter's vision is not a view upon the *outside,* a merely "physical-optical" relation with the world. The world no longer stands before him through representation; rather, it is the painter

> to whom the things of the world give birth by a sort of concentration or coming-to-itself of the visible. Ultimately the painting relates to nothing at all among experienced things unless it is first of all "autofigurative." . . . The spectacle is first of all a spectacle of itself before it is a spectacle of something outside of it. (1964, 181)

In this passage, the painting of which Merleau-Ponty speaks, in drawing upon the originary dimension of depth, draws in upon itself. Painting of this kind is not merely a signification of objects but a concrete *self*-signification that surpasses the division of object and subject.

In sum, the phenomenological dimension of depth, as described by Merleau-Ponty, is (1) the "first" dimension, inasmuch as it is the source of the Cartesian dimensions, which are idealizations of it; (2) a uroborically self-containing dimension, not merely a container for contents that are taken as separate from it; and (3) a dimension that blends subject and object concretely, rather than serving as a static staging platform for the objectifications of a detached subject. In other words, depth constitutes the dimensional structure of wild Being, the prereflective action of the flesh obscured by the ontical reflectivity of classical thinking. In realizing depth, we surpass the concept of space as but an inert container and come to understand it as an aspect of an indivisible cycle of life-world action in which the "contained" and "uncontained"—object and subject—are integrally incorporated. And I would add that, when the depth dimension is elucidated topologically, it proves to be a *Kleinian* dimension.

The Kleinian circulation—from contained object to containing space back to uncontained subject—unites the reflective and prereflective in a concretely *self*-reflective embodiment of three-dimensional Being. The uroboric Klein bottle is a reflected-upon content that, in containing itself, flows backward into its own prereflective ground. That ground is the Klein bottle's "missing dimension." We may indeed say that *all* reflected-upon contents originate in the prereflective. But in the case of an ordinary content, we cannot move back into its ground without obstruction because this content lends itself to the appearance of being simply contained, closed into its spatial container in such a way that it is closed off from its prereflective source. Only a *self*-containing object of reflection can incorporate its prereflective origin without a break. To be sure, reflection continues in this self-reflective act. Like the movement from one side of the Moebius strip to the other that paradoxically keeps us on the *same* side, the retrograde Kleinian movement from reflection to prereflectivity at once maintains the reflective. But now, instead of one-sided domination by reflection, there is harmony with the prereflective.

3. CONCRETE REALIZATION OF TOPOLOGICAL FLESH

The time has come to acknowledge that the words I have written about the Kleinian embodiment of ontological paradox may themselves seem rather *dis*embodied. What can be done to make them more concrete?

As a first step, let me admit that, until now, I implicitly have been assuming the very classical posture I have sought to question. As noted above, in the classico-modernist paradigm of object-before-subject, only the object is open to view; the subject remains detached, out of reach, anonymous. And this is essentially how I have approached the "object" of primary concern in this book, viz. ontological flesh. In largely maintaining my own anonymity, in failing fully to situate *myself* in this text as the subject before whom Being is "cast," I have let Being appear as a free-floating abstraction. But it seems that if I am truly to challenge the ontical stance, I must make Being less remote by recognizing in explicit terms that, rather than standing before some anonymous subject, it stands before *this* subject. That is to say, Being is the content of *this text;* it appears before me, the one who writes these words, and before you who reads them. Surely I state the obvious, but it is precisely the obvious that we lose sight of when we are lost in abstraction. Thus brought down to earth, the question of realizing Being, of embodying three-dimensional (w)holeness through the Klein bottle, becomes one of whether this text we are working with, being Kleinian in character, can lead us back into the prereflective ground from which this reflection of ours originates, and can do so without a break, thereby surmounting the exclusive rule of reflection and (re)turning us to the lifeworld.

The classical text operates squarely within the reflective mode and raises no questions about doing so. Here the word or sign, whose signifier serves as surrogate for the subject, refers solely to what is other, making this signified object of reflection explicit, a well-bounded content closed into its context. The signifier/subject per se remains *im*plicit; it does not meaningfully refer to itself. Is classical reflection effectively challenged in modernist or postmodern deconstructionist writing? I suggest that it is not.

It is true that the modernist sign is self-referential. We can say that, in modernism, attention is withdrawn from the end-products of reflection and meaning is relocated in an abstraction of the process itself. We see this in intensely self-involved psychological works such as James Joyce's *Ulysses*. In fact, we may say that modernism as such is "psycho-logical": it seeks to apply *logos* to the *psyche,* i.e., to gain explicit knowledge of subjectivity. Here the sign is turned back upon itself so as to bring to light what formerly had been strictly implicit. In the language of Freudian psychoanalysis—that exemplar of modernism discussed in chapter 1—the goal is to "make the unconscious conscious." Stated most essentially, modernism wants to surpass classical signification by turning the classical subject into an object.

In carrying out its program, does modernism achieve its basic goal? What actually happens is that the classical approach is upheld at a higher level of abstraction. For, in making the old subject explicit, in rendering it a well-delineated content enclosed in its context, this object-*né*-subject implicitly must be given to, must appear before, a new, higher-order subject. (Deleuze and Guattari argue similarly that, in modernism, "a new type of unity triumphs in the subject . . . a higher unity"; see previous chapter.) The self-reflection of modernism is indeed akin to gazing at the reflection of one's eyes in a mirror: what *had* been the gaze of the subject now itself appears as an object gazed upon by a subject that is one step removed from the original. The important thing to recognize is that this transformation of terms leaves completely intact the classical *relationship* of object-in-space-before-subject. The move to modernism thus poses no fundamental challenge to the mode of signification that preceded it. Modernism's self-referential sign, by turning the self into an other, in fact maintains in abstraction the classical *split* between self and other, between the signifier and its signified object.

In postmodernism, the ultimate consequence of modernism is recognized and played out. The crux of postmodernism, I suggest, is the realization that modernism's objectification of the subject is but the first term of an infinite regress. It might seem that, in modernism, only the *classical* subject loses its privileged position as the unquestionable base of knowledge, as the transcendent, never-to-be-viewed perspective point from which all else is viewed. However, once this subject is viewed, made explicit, objectified, cast before the perspective point of a newly implied, higher-order subject, *no* subject can securely hold its position. Having established that the classical subject can be turned into an object, the new, modernist subject should be susceptible to the same fate. The objectification of that subject would bring a still newer order of subjectivity with the same susceptibility, and so on, ad infinitum. And each time the subject is undermined by being made into an object, what had been object to that subject is also undermined. Ultimately, then, we have neither subject nor object in any stable, abidingly meaningful form.

In the parlance of postmodern or post-structuralist literary theory, the fixed relationships between particular signifiers and their signified meanings give way to the restless, ever-shifting text, the ceaselessly self-alienating application of the sign to itself. This is what we found in Lacan's approach to language (examined in chapter 1). It is much the same for Jacques Derrida: "sign will always lead to sign, one substituting the other . . . as signifier and signified in turn" (Spivak 1976, xix). In Derrida's own words, language must be understood as a field "of *freeplay,* that is to say, a field of infinite substitutions" (see ibid.) in which identity fragments into sheer difference (*différance*). I am proposing that the specific way this takes place is by a recursive process of self-referential mirroring in which, time and again, the signifier/subject is displaced by being made into the signified object of a newly implicit subject. Therefore, if we view the self-reflection of modernism as a mirroring that maintains in abstraction the classical relation of object-in-space-before-subject, postmodernism would constitute an infinite repetition of this mirroring, one that maintains classical identity in such a way that, in the end, it also negates it.

The "field of infinite substitutions" constitutive of the postmodern text, the infinite regress of signs within signs within signs, is reminiscent of the old Ptolemaic *epicycles.* When the ancient Greek notion that heavenly bodies traverse perfectly circular orbits was called into question by new observations of the planets, instead of gracefully relinquishing the paradigm of the circle, this model was perpetuated for centuries in an artificial, ultimately unconvincing form. The orbits of planets were described in terms of epicycles, complex arrangements of circles within circles that gratuitously replicate the image of the circle. Today, the challenge to the status quo appears far greater than it was in Ptolemy's time. What is now being called into question is no mere image we can reflect upon but the reflective posture itself, that expressed in the relation of object-in-space-before-subject. Unable to let go of this deeply ingrained, ontical habit of comportment, postmodernism carries it forward in its infinite regress of signs. The sterility of this is not lost on the postmodernists. In their abstract self-reflections there is a distinct mood of disenchantment. Yet, because the postmodern writer can find meaning nowhere else, the rule of reflection lingers on, albeit in this negative, thoroughly self-subverting manner.

A phenomenological alternative to modernism and postmodernism alike is intimated in philosopher Eugene Gendlin's expansion upon Heidegger (mentioned in chapter 1). Gendlin (1993) offers us a text that can reflect upon its bodily, *pre*reflective source. "Speaking is a special case of . . . bodily living," says Gendlin. "Our bodies perform the implicit functions essential to language. . . . Our bodies imply our . . . linguistic meanings" (34). Moreover, when we speak or write, the prereflective source of this activity is not simply left behind; it continues to operate in the very midst of our linguistic functioning. Thus, for example, "the most sophisticated details of a linguistic situation can make our *bodies* uncomfortable" (34). Could we not reflect explicitly upon the prereflective source of our reflection? Let us attempt such an act of self-reflection here, with the very words on this page. If Gendlin is correct, our reading of these words arises from our bodies, and since this bodily source goes on functioning even as our words now turn back upon it, it seems we should be able to realize that source in a bodily way so that our words no longer appear as mere abstractions. This is what Gendlin means when he proclaims that "words can say how they work" (29): they work from the body, and, becoming cognizant of their own bodily underpinning, they can link back to it. As in Gendlin's approach, the words of the postmodern text do reflect upon themselves, but only as disembodied signs ultimately devoid of meaning. Gendlin points us beyond postmodernism. In Gendlin's form of self-reflection, the text is not merely conscious of itself *as* an abstract text, but calls attention to the concrete process from which it originates. Only by gaining access to this prereflective "subtext" can we supersede the old ontical trichotomy of object-in-space-before-subject and concretely approach the (w)holeness of the lifeworld. In Gendlin's terms, the prereflective is "pre-separated" (1991b, 116–17); that is, it "comes before," is more primordial than the divisions arising in classical thinking and perpetuated in modernism and postmodernity.

But a further step seems necessary if we are to close the gap between our reflection upon the prereflective and the prereflective itself. A *Kleinian* rendition of Gendlin's text is required, I suggest.

First let me emphasize that our post-postmodern text must be *paradoxical* in character. Again, what we are seeking to do is include in this reflection of ours the prereflective source of our activity, the ontological "subtext" that normally is kept implicit. Our text—whose signifiers stand in for our*selves* (I, who writes these words, and you, who reads them)—is to draw back in upon itself like Cézanne's auto-figurative painting, make reference to itself without alienating itself, as happens with modernist and postmodern texts; in so signifying itself, this text cannot merely turn itself into an other that is cast before a newly implied, more abstract self. Does this mean that the self that is signified must be the *same* self that is doing the signifying? Not exactly. If the self in question were *simply* the same, our reflection would collapse into mere self-identity. As long as we are engaged in reflection, are working with a text, are writing or speaking and not merely remaining silent, there can be no simple self-identity. Yet, even though the very act of reflecting upon the self turns it into what is other, it is possible for this other to flow right back into the source from which it arises, rather than appearing *merely* as an other cast before a new self. Thus, the self-reflection I am describing would give us neither self nor other, in the strictly oppositional sense of these terms. We would realize instead their paradoxical interpenetration. And this "paradox of Being" is what we require to supersede the supremacy of reflective predication. Signifier and signified would be more than reciprocally interdependent in such a self-reflective text. They would be identified, utterly one. Yet they also would be two. By virtue of the latter aspect, reflection would continue; by virtue of the former, the "preseparated," prereflective lifeworld dimension would be brought into play.

However, for the gap between reflection and the prereflective to be closed in this way, the paradoxical return of signification to itself requires a signifier that possesses *sufficient dimensionality.* It is here that the Klein bottle plays its crucial role. Our work with this structure clearly is no exercise in "pure mathematics" in which a mathematical object is signified by a definition or algebraic formula. For us, the Klein bottle is not merely a signified object; it is a *signifier,* one that, indeed, paradoxically signifies itself. It is the *dimensionality* of this signifier that permits us to close the gap between the reflective and prereflective.

In its dialectical fashion, the prereflective lifeworld is three-dimensional. Of course, this bare statement about the prereflective in itself is merely reflective; the three-dimensionality of the prereflective is the content of our predication, and that abstract content is simply contained as what we reflect upon in the closed linguistic space of our discourse; from here, we cannot pass smoothly to concrete prereflectivity itself. This limitation is related to the fact that the signifiers we use to convey our three-dimensional content themselves lack sufficient dimensionality. We saw in section 2 that, by themselves, these printed words, these arbitrarily devised typographic tokens, are inadequate to make a reality of ontological paradox—not only because their relation to their content is merely conventional, but also because the marks on this page are but one-dimensional. The topological concretization of paradox carried out in the previous section effected an increase in the dimension of the signifier. We went from mere words of paradox ("I am not-I") to its two-dimensional Moebius strip expression. With this step, our statement of paradox became more internally embodied. Yet we were still working with a mere symbol, a reflector whose prereflective content in fact remained abstract, being contained within our field of reflection. Only with the three-dimensional Kleinian reflector can we surpass the primacy of reflection, for only the Klein bottle signifies the prereflective dimension by signifying *itself.*

It is the uroboric movement of the Klein bottle back into itself that produces the hole in it. Since this opening is the mark of self-signification, with it, the classical gap between signified and signifier in fact should be *closed.* With the hole in the Kleinian text,[1] we should at once have *wholeness.* However, this will not happen if we continue to assume a reflective posture in our relation to our text; in that case, the (w)hole will appear as *merely a hole,* a breach in the text that renders it incomplete.

Clearly, then, we must change our posture, approach the Kleinian text in a different way. Specifically what way? How can we overcome the old, compelling tendency to turn whatever we signify into an object of reflection appearing "out there" before us, simply contained within its context and thus set apart from us? What must we do to allow the self-containing Kleinian signifier to signify itself by flowing right back into us?

I am proposing that the Kleinian text in truth cannot be simply contained. In containing itself, the Klein bottle should spill over the bounds of the context that would enclose it, flow backward to its own prereflective ground—*our* ground, we who read this text. Yes, as Gendlin would say, the bodily source of these words continues to operate as we read them. And, with the dimensional enhancement of these paradoxical words of the flesh that fleshes them out, makes them concrete, the gap between this reflection on the prereflective and its living source should indeed be closed. Thus, in properly completing our Kleinian discourse, this text we read would live within us as it stands before us, and would do so without interruption. But I ask again, how must we approach our text to complete it in this way?

Although the long-prevalent habit of classical reflection strongly disposes us to approach the Klein bottle as but an object of predication, this "object" does not lend itself to being predicated thus. The inherent character of the Klein bottle suggests that we adopt a *pre*reflective approach to it. This means, as Gendlin would say, that we are to obtain a "moody understanding" (1993, 30) or "felt sense" (1978) of our Kleinian text, a bodily cognizance that exceeds this text as a mere content we reflect upon; the "felt sense," of course, is the awareness of the prereflective subtext. It should be true that we could gain such a sense of *any* text, since all texts originate in the prereflective lifeworld. But when the text appears closed into its context and thus closed off from its prereflective source, the gap between the reflective and prereflective will persist. In the case of the Kleinian text, there can be no pretense that it is simply divided from its subtext, for, *as* a text, it is incomplete. To complete it, we must follow its own natural trajectory back into the living matrix that constitutes the origin of our reflection upon it. In this way, the Kleinian text comes alive, stands within us as well as before us. Our task is to surpass the long dominant habit of predication so that—in one paradoxical movement (a "continuously discontinuous" crossing, an "unbroken quantum leap")—we may pass from a reflective understanding of the prereflective ground of this text to the ground itself. The Klein bottle uniquely mediates this passage. The hole in this Kleinian text is the "perfect size and shape" for our inwardness, and, like a black hole in space (a break in the classical continuum!), it draws us toward it, extending to us a pregnant invitation to fill it with the whole of our selves.

Let me make the point in a different way. We are discovering that the Kleinian text requires a change not only in the *content* of thinking, but also in the underlying manner in which thinking operates: its posture, orientation, or gearing. In both classico-modernism and postmodernity, thinking is geared to move "forward." It is poised to move *from* prereflective Being *to* the reflective structure constituted by the old subject-object opposition. In this orientation, the *cogito,* though *starting out* from the lifeworld, is fundamentally inclined to move away from it. As a result, the lifeworld per se is eclipsed. What happens with the Kleinian reversal is a switching of gears whereby thinking is no longer simply moving straight ahead. This entails more than a mere change in the *direction* of thinking.

When proceeding from the lifeworld in the "forward gear," we may attempt to change direction by "turning around" upon this prereflective source—like turning a car around so that it now faces the direction away from which it was previously facing. In such a reversal, we are indeed inclined to *face* Being, to have it appear before us, over against our consciousness. With our thinking thus geared, we still try to *objectify* Being, but only succeed in obscuring it. We cannot know Being by simply turning around upon it any more than turning a car 180° to face the direction *from* which it *had* been facing allows us to capture that "from" as it was initially experienced; such a turning merely turns the old "from" into a new "to" that is seen *from* a new perspective. It is clear that the attempt to grasp the "from" of Being while maintaining the forward orientation is futile; to make such an effort is to "chase one's tail," to turn in a vicious circle.

It is for this reason that thinking must shift into a *different gear;* the prereflective realm must be approached in "reverse." Notice that, whereas turning around to face in the opposite direction turns us away from the original direction in which we faced, if we move in reverse, we continue to face in the same direction. A similar distinction can be made with respect to Being: In seeking to come "face-to-face" with it, we hope (in vain) to turn our backs on the subject-object dichotomy, whereas the movement *backward* into Being does not simply negate the reflective formula but brings to light the prereflective dimension of vital activity that first makes it possible.

The reversal of gears that is required for working effectively with the Kleinian text can be understood as a certain form of *proprioception.* We may contrast this mode of functioning with the better-known symbolic operations of perception and conception. Etymologically, to perceive is to "take hold of" or "take through" (from the Latin, *per,* through, and *capere,* to take), and to conceive is to "gather or take in." These activities are carried out in the "forward gear," the orientation in which we work with the ordinary, simply contained text. The term *proprioceive* is from the Latin, *proprius,* meaning "one's own." Literally, then, proprioception means "taking one's own," which can be read as a taking of self or "self-taking." It is true that the term's conventional meaning derives from physiology, where it signifies an organism's sensitivity to activity in its own muscles, joints, and tendons. But the physicist/philosopher David Bohm (1994) spoke of the need for "*proprioceptive thought*" (229), which he viewed as a meditative act wherein "consciousness . . . [becomes] aware of its own implicate activity, in which its content originates" (232). Years earlier, the social psychiatrist Trigant Burrow spoke similarly of the need for human beings to gain a proprioceptive awareness of the organismic basis of their divisive symbolic activity (see Galt 1995). What I propose here is that proprioception—broadly understood as a mode of "self-taking" inclusive of cognition—is the appropriate way to work with the Klein bottle, and that such a meditation is what the self-containing Klein bottle requires and invites. In thinking this Kleinian text, we must think proprioceptively, think our own thinking.

Note that, when I speak of "meditation," I am surely not referring to the *classical* kind. Whereas classical meditation generally aims at transcending the body, the self-reversal of thinking I have in mind would seek to move back into it. The goal would be *re*-embodiment, reconnection with the lifeworld. But would this not require a *disengagement* from thought and a return to the bodily *senses?* What we learn from Merleau-Ponty is that thinking in fact does not just entail sheer abstraction but possesses its own bodily grounding. For Merleau-Ponty, the dimension of language and thought operates as a "second flesh," a second order of embodiment: "It is as though the visibility that animates the sensible world were to emigrate, not outside of every body, but into another less heavy, more transparent body, as though it were to change flesh, abandoning the flesh of the [sensible] body for that of language" (1968, 153). It is this *bodily* character of language and thinking that is denied when thinking is geared to move exclusively "forward." And the proprioceptive self-reversal of thinking facilitated in the Kleinian text would amount to a withdrawal of the *cogito*'s projections that would disclose the actual grounding of Cartesian consciousness in the "second flesh."

Only through proprioception, where we would not simply be seeking to *break away* from symbolic reflection, could we truly surmount its pervasive influence. And while the reflective would no longer reign supreme, through Kleinian proprioception it actually would reach culmination, for no longer would it conceal the prereflective ground that sustains it so as to maintain the false impression of being complete unto itself. In *recognizing* its own incompleteness as a simply autonomous activity, in *acknowledging* its roots in the prereflective lifeworld, symbolic reflection gains *authentic* completion. The proprioceived Kleinian reflector brings reflection to fruition by bringing to light the aspect of it that previously had not been reflected upon. To use an old alchemical expression, the inside-out Kleinian vessel is *bene clausum* (well-sealed). It is not only entirely open to its own prereflective, "second flesh" source, but—*because* of this openness—also entirely closed, complete unto itself, fully three-dimensional.

But what about the *first* order of the flesh, that which Merleau-Ponty associated with the sensible body? We have begun to see that there is more than one realm of the flesh, more than one lifeworld dimension. In fact, we will soon discover that there are *four.* Over the next three chapters, the dynamic interrelationships among these dimensional spheres will be examined in topological detail. Then, in the three chapters that follow, the abstractions of chapters 3–5 will be given concrete expression. Ultimately, the aim is to realize each order of lifeworld signification as an existential *self*-signification. We focus on this question in the final chapter (chapter 8), which begins by returning to the Kleinian self-signification of cognitive flesh just enacted in the present chapter in order to work out certain aspects of the re-embodiment process we have not yet considered. The book then closes with an exploration of the possibilities for self-signification inherent in each of the other lifeworld spheres.

4. CONCLUSION

In making the (w)holeness of the second flesh a concrete reality, we are to read this text proprioceptively; read our own reading; read these words about passing beyond themselves (into their prereflective roots) in such a way that the passage actually takes place. Such a reading is mediated by fleshing out this text via the dimensional amplification provided by the Klein bottle. Now the body of our text does not consist merely of empty reflectors, intrinsically meaningless signifiers that can only point outside themselves to disembodied meaning; our text is the Klein bottle. It is in reading the (w)hole in this self-containing three-dimensional text that we should pass unbrokenly into its subtext. The Klein bottle's incompleteness when read ontically is at once an incompleteness in our reflecting upon it. In accepting the invitation to proprioceive the Kleinian text, we complete our reflective symbolic activity by circling back into its prereflective origin.

But old habits do persist. *Can* we read our own reading? Can we read the hole in the Klein bottle in such a way that we relax the compulsion to regard it as *merely* a hole, a gap in an ordinary object simply contained in space? Can we read the hole in the Klein bottle as an opening to "another dimension" and read that dimension as the prereflective source of our very own reading? Can we *enter* that dimension through proprioception? Generally, it is a matter of proceeding in such a way that conceptual mediation and existential immediacy come together. The Kleinian concept brings us to the limit of the conceptual. Because the true boundary of our symbolic activity must be paradoxical, a boundary that is *not* a boundary (lest we continue in the boundary-making mode of symbolic reflection), it is precisely at this inner horizon that we should be able to bridge the gap between the conceptual and the existential, between what we know and what we are.

PART II

Lower Dimensions of the Flesh

PREAMBLE

> Now as I beheld the living creatures, there appeared upon the earth by the living creatures one wheel with four faces. And the appearance of the wheels and the work of them was like the appearance of the sea: and the four had all one likeness. And their appearance and their work was as it were a wheel in the midst of a wheel . . . the spirit of life was in the wheels.
>
> Ezekiel 1:15 (Douay Version)

Abstraction per se is not a bad thing. We know that it has in fact been indispensable to us in the quest for self-understanding that is necessary for our self-fulfillment. But we have reached the point where advancing this process of individuation means surmounting the *one-sided rule* of abstraction. Only then can we bring to light the hitherto eclipsed lifeworld that grounds us, and reenter that world. We know as well that the supremacy of abstraction can be surpassed not simply by moving away from abstraction, but by going all the way through it to its interior horizon, the paradoxical boundary where it reaches culmination.

Having ended part I at the interior boundary of three-dimensional abstraction embodied by the Klein bottle, we shall now proceed to explore lifeworlds of different dimension. These other orders of the flesh turn out to be considerably more concrete than the Kleinian order of language and thinking. We will see that they involve the action spheres of feeling, sensuality, and embodied intuition. Yet, in the course of bringing them into relationship with the Kleinian dimension, even greater heights of abstraction must be confronted. That is because the boundary operating *between* orders of the flesh is also of an interior kind. As a consequence, entering into the non-conceptual lifeworld cannot just entail detaching ourselves from conceptual abstraction; rather, the non-conceptual must be realized conceptually. This is surely not to say that a conceptual account is *sufficient,* since genuinely engaging the fleshier dimensions does mean passing beyond conceptuality. But the boundary mediating this passage is such that we also remain *within* the conceptual sphere, achieving new levels of comprehension in the abstract knowledge we gain of the non-conceptual.

The remainder of this book is laid out according to the old alchemical dictum *solve et coagula,* "dissolve and coagulate." Like the verb "abstract" (from the Latin *abstrahere,* to draw from or separate), the verb "dissolve" (from the Latin *dissolvere*) indicates separation: to "dissolve" is "to disunite; to break up . . . to cause to separate into parts."[1] The alchemical substance to be transformed was thus initially broken into parts, sublimated so as to refine it. Then, in the second stage, the process was reversed and the material was made to congeal, its constituent elements being drawn together and solidified. This was the phase of "coagulation" (from the Latin *coagulatus,* past particle of *coagulare,* to curdle). What had been divided in order to refine it was now reconstituted. Over the next three chapters, the non-conceptual lifeworld dimensions are approached through a *solutio,* an abstract conceptualization of them that is necessary but hardly sufficient, since it fails to do justice to their concrete actuality. The three subsequent chapters are dedicated to *coagulating* these dimensions, transmuting them into living flesh.

While I believe the theoretical analyses offered in chapters 3–5 are indeed necessary for bringing conceptual understanding to fruition, I am fully aware of the challenging nature of the material herein and have attempted to facilitate comprehension of it by including illustrations and tables to supplement the written text. In fact, tangible models of many of the geometric structures I am going to describe are easy to construct and employing them should prove helpful. Nevertheless, there is no denying the complexity of these accounts, and more than one reading may be required. Upon completing your reading, perhaps you will find that your efforts have been rewarded by the insight you have gained into dimensions of Being that have been relegated to the shadows from time immemorial.

Finally, I want to acknowledge that some of the ideas we are going to explore may seem quite speculative, at least initially. Is it possible to establish the solidity of one's ground at the same time one is seeking to open *new* ground? If so, I have not found a consistent way of doing it. My hope is that, despite the conjectural nature of certain propositions set forth here, they will (1) nonetheless appear plausible, (2) gain in credibility when viewed in the context of the overall network of interrelationships that is laid down, and (3) be validated further, as theoretical abstraction is coagulated into fleshier form.

CHAPTER THREE

INTRODUCTION TO THE LOWER DIMENSIONS

1. BEING AND APPROPRIATION

Merleau-Ponty was not alone in intimating the existence of more than one non-ontical sphere of dimensional process. We find a similar indication in Heidegger's late lecture "Time and Being" (published in *On Time and Being,* 1962/1972). To begin, let us consider what Heidegger says about the ontological meaning of time.

Despite the prevailing Aristotelian tendency to view time as but a sequence of point-like "nows," in its ontological truth, time possesses a certain "fullness" or "thickness." According to Heidegger, "true time proves to be three-dimensional" (1962/1972, 15). Specifically, he tells us that, behind the appearance of an extensionless, point-like present moment are: (1) the *past*—which, though denied to us *as* present, nevertheless makes its presence felt in its own way; (2) the *future*—which is withheld from us in *its* own peculiar manner of making itself present; and (3) the present itself. Heidegger suggests that these three dimensions of time gather together and are held apart in every moment of lived experience. That is, they extend toward each other, yet, since they never merely collapse into one another to form a monotonic unity, the "space" of their intimate interplay remains open. The three dimensions of time determine each other; there is a "mutual giving to one another of future, past and present" (14):

> [The future,] being not yet present, at the same time gives and brings about what is no longer present, the past, and conversely what has been offers future to itself. The reciprocal relation of

> both . . . gives and brings about the present. . . . [Thus] future, past and present . . . belong together in the way they offer themselves to one another." (1962/1972, 13–14)

Heidegger next observes:

> With this presencing, there opens up what we call time-space. But with the word "time" we no longer mean the succession of a sequence of nows. Accordingly, time-space no longer means merely the distance between two now-points of calculated time. . . . Time-space now is the name for the openness which opens up in the mutual self-extending of futural approach, past and present. This openness exclusively and primarily provides the space in which space as we usually know it can unfold. The self-extending, the opening up, of future, past and present is itself *pre*spatial; only thus can it make room, that is, provide space. (1962/1972, 14; emphasis added)

What Heidegger is saying is that—while the "dimensionality of time, thought as the succession of the sequence of nows, is borrowed from the representation of three-dimensional space" (14)—when temporal dimensionality is thought in its ontological truth, as the threefold mutual giving, then it is three-dimensional space that derives from *it*. Expanding on this, Heidegger asserts:

> What is germane to the time-space of true time consists in the mutual reaching out and opening up of future, past and present. Accordingly, what we call dimension and dimensionality . . . belongs to true time and to it alone. Dimensionality consists in a reaching out that opens up, in which futural approaching brings about what has been, what has been brings about futural approaching, and the reciprocal relation of both brings about the opening up of openness. Thought in terms of this threefold giving, true time proves to be three-dimensional. Dimension, we repeat, is here thought not only as the area of possible measurement, but rather as reaching throughout, as giving and opening up. Only the latter enables us to represent and delimit an area of measurement. (1962/1972, 14–15)

In Heidegger's view, the "reaching throughout . . . giving and opening up" inherent in true time is the action of *Being.* Once again, this action is *prereflective;* it takes place *prior* to the ontical activity of the subject with respect to its object once that subject has become detached. Therefore, to give the ontological difference its due, we must steadfastly avoid confusing the action of Being with that of ontical reflection. The former is what first makes the latter possible.

Having associated Being with "true time," and spoken of the latter as the "three-dimensional" mutual giving of past, present, and future, Heidegger is prompted to ask, "from what *source* is the unity of the three dimensions of true time determined, the unity, that is, of its three interplaying ways of giving, each in virtue of its own presencing?" (1962/1972, 15; emphasis added). Heidegger answers the question in the following way: "The unity of time's three dimensions consists in the interplay of each toward each. This interplay proves to be the true extending, playing in the very heart of time, the fourth dimension, so to speak—not only so to speak, but in the nature of the matter" (15). In invoking a "*fourth* dimension," Heidegger is implying that the "dimension of interplay" at the heart of time must be taken as somehow constituting a dimensionality unto itself, rather than merely being a feature of three-dimensional time. Indeed, only thus can the "dimension of interplay" be seen as the *source* of three-dimensional time. Apparently, then, there is a dimension that exceeds the three dimensions of time and constitutes their "source."

Heidegger avoids the metaphysical implication that the "fourth-dimensional source" is a static realm or being by speaking of it in active terms, as a *giving:* "It gives time" (1962/1972, 16). But true time itself is a giving: it is that which "provides the space in which space as we usually know it can unfold" (14). Accordingly, true time is what gives us the ability "to represent and delimit an area of measurement" (15). Therefore, it seems we must regard the "fourth dimension" that gives true time as "the giving of a giving" (16).

Later in his lecture, Heidegger attempts to go further with his portrayal of the "fourth-dimensional It" that gives time and Being:

> In the sending of the destiny of Being, in the extending of time, there becomes manifest a dedication, a delivering over into what is their own, namely of Being as presence and of time as the realm of the open. What determines both, time and Being, in their own,

> that is, in their belonging together, we shall call: *Ereignis*, the event of Appropriation. (1962/1972, 19)

The ordinary meaning of the German word *Ereignis* is "event" or "occurrence." But Heidegger asks us to read this word in an unusually literalized fashion, to understand it as "making one's own." Thus we have the English translation of *Ereignis* as "Appropriation," from the Latin *ad*, to, and *proprius*, one's own. What this suggests is that the "fourth-dimensional giving" entailed in Appropriation is not an imposition upon three-dimensional Being and time of something that is foreign to them, but is a giving to them of what is *their own*. To the extent that "fourth-dimensional giving" is indeed fully committed to yielding over to Being what is truly Being's *own*, it seems the giving in question would not be so much an act of "fatherly condescension" that *determines* the "ownness" of Being, as a more nurturing gesture of support that encourages Being to give to *itself*. How does Being respond to this encouragement? Precisely by the presencing that gives an "area of measurement," gives space "as we usually know it," or, more generally, gives object-in-space-and-time-before-subject. Let us designate this giving of Being to itself as *self-Appropriation*.

Extending the metaphor, if "fourth-dimensional" Appropriation is not a "fatherly" act, neither is the self-Appropriation of Being. By "fatherly" action, I mean that of the deus ex machina, the agent acting from above or outside. This action functions exclusively in its own interests, with the assumption that actor and acted upon are essentially separate. In other words, "fatherly" action describes the operations of the detached subject upon its objects. To be sure, this reflective subjectivity is *what* Being gives to itself. But the *giving* is not reflective. In so giving, Being does not take for itself what is other, but gives to itself what is its own. Being's mode of action is *pre*reflective, as we know; it gives itself reflective subjectivity in a manner that precedes the division of subject and object.

The point may be clarified, at least metaphorically, by considering the relationship of mother to child. Through her nurturing support, the mother facilitates the child's autonomy, encourages it to develop its unique way of acting, which is distinct from her own motherly caregiving. But, in thus giving to her child, the mother is not just acting selflessly on behalf of another. Because the mother's very being is intimately bound up with that of her child, by giving to it, she is also giving to herself. I suggest that this is the sense in which the prereflective giving of reflective subjectivity is Being's gift to itself—an action that will be further elucidated when, in the next two chapters, we consider in explicit detail the diachronic aspect of Being, the stagewise process through which Being itself grows and develops. For now, I am proposing that Being acts *maternally,* and the "child" to which it gives birth is the subject. When we take into account Heidegger's sense of Appropriation, we have a tripartite relation. Operating in the manner of a "midwife," "fourth-dimensional" Appropriation encourages Being's three-dimensional "motherly" self-Appropriation, and this, in turn, results in the birth of the "father," that is, of reflective consciousness. Note that, in an earlier essay, Heidegger himself seems to lend support to the interpretation of Being as a "motherly" or "womanly" dimension when, in a circuitous fashion, he links Being with Μοῖρα, the goddess to whom "Gods and men are subordinated" (1946/1984, 55).

2. THE TOPODIMENSIONAL FAMILY

In the previous chapter, topology proved to be an essential tool in the fleshing out of ontological paradox. There the three-dimensional Klein bottle was used to articulate the dialectical action of the "second flesh," to detail concretely the workings of three-dimensional Being. But the fact that we must deal with *more than one* order of non-ontical dimensionality—whether viewed in terms of Merleau-Ponty's "first" and "second" flesh, or of Heidegger's Appropriation and Being—suggests that more than one dimension of topological embodiment is required; by itself, the Klein bottle will not suffice. We have found, of course, that the Klein bottle does have a differently dimensioned topological relative, viz. the Moebius strip. Can this structure be seen to embody a different lifeworld? We are now prepared to examine the whole family of topological structures to which the Klein bottle and Moebius strip belong. What I am going to demonstrate is that each member of the family in fact lends itself to non-ontical interpretation.

Mathematicians have investigated the transformations that result from bisecting topological surfaces. Suppose we were to cut the two-sided cylindrical ring of figure 2.1(a) down the middle, proceeding along its full length. Upon completing the cut, the ring would simply decompose into a pair of identical narrower rings each possessing the same topological structure as the original. A more interesting result is obtained in bisecting the one-sided Moebius strip. Rather than falling into two separate pieces as one might expect, the bisected surface retains its integrity but has now become the *two*-sided structure depicted in figure 3.1.

Figure 3.1. The lemniscate

Compare a single side of this two-sided surface with that of the two-sided cylindrical ring. Whereas revolution about the latter describes but one closed loop, traversal of the former gives us a doubly looped, figure-8 pattern known in mathematics as a *lemniscate;* turned on its side, the lemniscatory surface resembles the familiar sign for infinity, ∞. The two patterns of movement are schematically contrasted in figure 3.2.

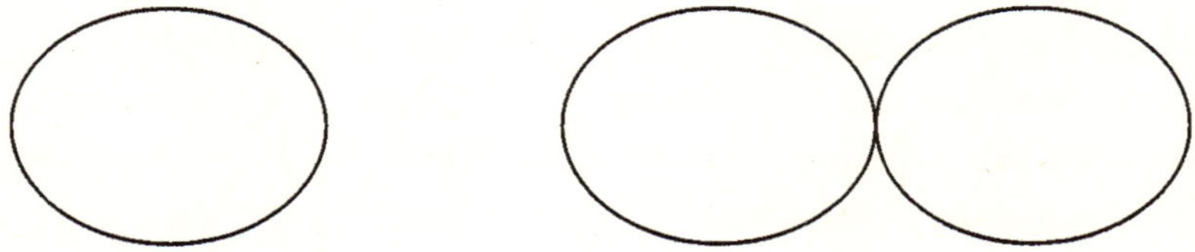

Figure 3.2. Schematic comparison of cylindrical (left) and lemniscatory (right) rotation

Now, we have found that the integrative quality of the Moebius surface lies in its paradoxical one-sidedness. The two sides of the Moebius flow unbrokenly into each other to form a single side, without either side actually losing its distinctness. When bisection of the Moebius strip transforms it into a *two*-sided structure, this integrity is lost. Yet we can now see that each of those sides, being lemniscatory in character, constitutes its own order of paradoxical "unity-in-diversity." That is, in a single side, we have a double cycle, these cycles being connected by a continuous movement through the central node of the figure-8. It is true that, in the lemniscate, we no longer have a complete *overlapping* of opposing elements, as we do in the Moebius. The lemniscate thus could be said to have less internal coherence than the Moebius. Be that as it may, a similar pattern of "transpolar flow" is evident in both structures.

Note that, whereas bisecting the simply symmetric cylindrical ring yields rings that are completely symmetric with respect to each other, bisecting the asymmetric Moebius strip produces lemniscatory sides that are related *enantiomorphically:* they complement each other in the manner of asymmetric mirror opposites. And just as the single lemniscate has its mirror counterpart on the other side of the bisected Moebius, the Moebius as a whole has its own enantiomorph. For we know from chapter 2 that, while there exists but one form of the cylindrical ring, the Moebius surface can be produced in either a left- or right-handed version. If both versions of the Moebius were constructed, then "glued together," superimposed on one another point for point, the result would be a Klein bottle.

The Klein bottle, Moebius strip, and lemniscate constitute a series of topological forms that are nested within one another. Bisecting the Klein bottle produces Moebius enantiomorphs; bisecting the Moebius yields mirror-opposed lemniscates. One more bisection is required to complete the series. Upon cutting the lemniscate, the surface neither retains its integrity nor simply falls into separate pieces. Instead, the single surface is transformed into two interlocking surfaces, each of which is itself lemniscatory (see fig. 3.3). The transformation brought about by this bisection is clearly the last one of any significance, since additional bisections, being bisections of *lemniscates,* can only produce the same results: interlocking lemniscates. The bisection series is completed, then, when we

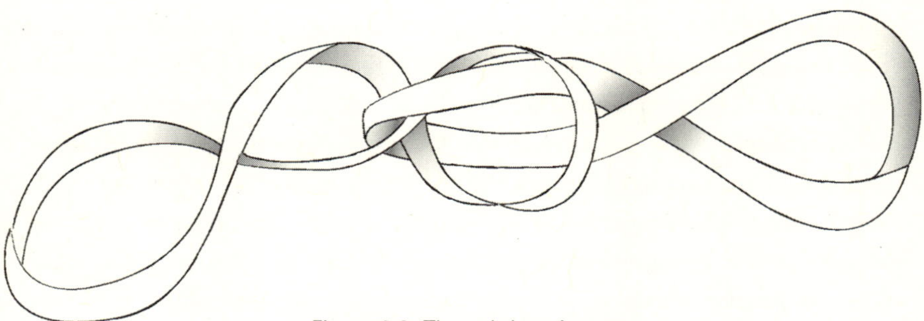

Figure 3.3. The sub-lemniscate

obtain interlocking lemniscates—a structure we shall henceforth refer to as the *sub-lemniscate.*

Now, if the Klein bottle embodies the "motherly" self-Appropriation of three-dimensional Being, and if we require a further topological embodiment to account for and flesh out the "midwifely" act of Appropriation that surpasses the three-dimensional sphere, can we not turn to our topological bisection series? Indeed, when we consider all the members of this closely knit topological family, it seems we are led to the idea that there are actually a total of four orders of Appropriation, rather than just the two that are indicated in Heidegger's "Time and Being." Naturally, a proposition of this sort will need to be unpacked and examined with considerable care.

In launching our topological investigation, let us not assume that the "midwifely" dimension or dimensions are *higher* than three-dimensional Kleinian Being. Despite Heidegger's metaphorical linking of Appropriation to the "fourth dimension" (1962/1972, 15), the topological bisection series actually suggests that what Heidegger calls Appropriation is *lower* than three-dimensional. The three-dimensional Kleinian member of the series contains the other members in the fashion of Chinese boxes: the two-dimensional Moebius strip is contained within the Klein bottle, the lemniscate within the Moebius strip, and the sub-lemniscate within the lemniscate. We may take this as an indication that the differently dimensioned midwives that assist Being's "self-birthing" are not mere abstractions that lie "beyond" Being. Rather than transcending three-dimensional Being, they are concretely nested within it. However, just as we were obliged to depart radically from the ontical interpretation of the Klein bottle in order to realize it as an embodiment of dimensional Being, it appears we must equally depart from the standard treatment of the sub-Kleinian members of the bisection series.

To reiterate a key principle, conventional mathematics is guided by the ontical intuition of object-in-space-before-subject. Accordingly, on the conventional view, all members of the bisection series are taken as two-dimensional topological objects embedded in a higher-dimensional continuum that is set up for the abstract operations of the detached mathematical analyst. What onto-phenomenological thinking has told us is that the Klein bottle is no such object; instead, it is the circulatory action that *fuses* object and analyst, and, in the process, brings hitherto idealized three-dimensional space to its true dialectical completion. The *necessary hole* in the Klein bottle was critical to our understanding of this. It was the initial observation that the Klein bottle cannot be completed as an object in three-dimensional space without being ruptured that invited an exploration eventually leading us to conclude that this paradoxical structure, in its intimate blending of object and subject, is an embodiment of three-dimensional Being.

Now, in considering the *sub*-Kleinian members of the bisection series, it is clear that they *can* be properly completed as objects in three-dimensional space. It is for this reason that, while a paradoxical structure like the Moebius strip can well *symbolize* the integration of the three-dimensional subject and its object, it cannot effectively *embody* said integration. That is precisely why we saw the need for "dimensional amplification" via the Klein bottle in the previous chapter. The proposition I now offer is that, although the sub-Kleinian members of our topological family indeed cannot embody the dialectic of *three*-dimensional Being, they provide embodiments of *lower-dimensional* orders of Being.

Whereas the standard interpretation of the bisection series says that each of its members is a two-dimensional object embedded in a higher-dimensional continuum, what I am proposing is that, ontologically, *none* of its members actually possess this status (not even the Moebius, which *is* two-dimensional, but which—at an ontological level—is no mere object). Since the necessary incompleteness of the Klein bottle in three-dimensional space is plain to us, it is relatively easy for us to see how this structure defies ontical objectification. Why can we not see necessary holes in the sub-Kleinian members of the series? It is because our seeing is three-dimensional. Within this frame of observation, lower-dimensional dialectics are imperceptible. To understand the role of the sub-Kleinian bodies of paradox vis-à-vis the lower dimensions, let us examine the exact manner in which lower-dimensional dialecticity is repressed within the three-dimensional framework.

In the foregoing chapter, we considered Merleau-Ponty's account of the dialectical opposition entailed in the perspectival interplay of the visible and invisible, wherein surfaces in space overlap and eclipse one another; all the surfaces of an object cannot be seen in a single glance. This is depicted in figure 3.4(c); the diagram suggests the concealment of some of the cube's surfaces.

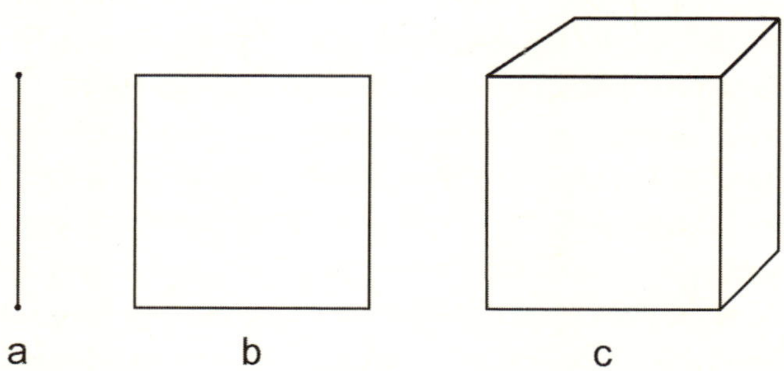

Figure 3.4. Endpoints of a line segment (a), lines bounding a surface area (b), and planes bounding a volume of space (c)

But while the bounding planes that enclose a volume of three-dimensional space are not all simultaneously perceptible, the endpoints bounding the line segment (fig. 3.4(a)) and the lines enclosing a surface area (fig. 3.4(b)) *can* all be perceived at once. The simultaneous perceptibility of the points and lines renders these geometric beings purely and simply positive, as Merleau-Ponty said of Cartesian space in general. Though we may *speak* of them as being opposed, their appearance together before us as just what they are lacks dialectical tension. Therefore, while *planes* give evidence of the "invisible," and while, in so doing, they implicate our subjectivity (Merleau-Ponty 1968, 140), points and lines apparently do conform to the Cartesian idea of dimensionality. They appear as but the "in-itself" (Merleau-Ponty 1964, 173), as sheer positivity. Incapable of expressing true opposition, points and lines, in their simultaneous perceptibility, seem to give us only *juxta*position. It is true that classical analysis treats *all* dimensional elements, planes included, as merely juxtapositional. Phenomenological investigation tells us that such treatment is not appropriate when it comes to the interplay of planes in space. But are we not seeing now that classical analysis is indeed warranted in the case of points and lines?

I propose that classical analysis does not suffice when we are inquiring into the deeper nature of points and lines. I suggest that while these geometric structures appear to be merely juxtapositional in the three-dimensional framework of observation, in actuality, they possess their own dialectical attributes. It is in the "repressed dialectics" of points and lines that we find the lower-dimensional orders of Appropriation that I have intimated.

To grasp the meaning of lower-dimensional dialecticity, let us begin by clarifying the classical distinction between three-dimensional space and the elements that bound it. Space is bounded by two-dimensional planes. Manifested in concrete form, the bounding plane serves to partition space into particular regions and neighborhoods. It is in its capacity as a concrete bounding element that the plane possesses opposing sides: the region of space on one side of the plane is set apart from that on the other. In the case of three-dimensional space itself, we cannot speak meaningfully of opposite sides since space does not function as a bounding element. Through its bounding elements, space plays the role that Plato attributed to it in the *Timaeus:* it is a *receptacle,* a medium or containing environment. For to contain an object in space, one must establish a boundary that sets off its interior region from what lies outside of it. Note that while mathematicians often abstractly refer to the two-dimensional plane as a "space," in ordinary concrete experience it is only the three-dimensional domain that serves in this capacity. What of the line and the point in this milieu? They may be regarded as subsidiary bounding elements. The line partitions the planar bounding element, divides it into two-dimensional areas, and the point partitions the linear bounding element, cuts it into one-dimensional lengths. With this as a background, we may now consider the possibility of a lower-dimensional "receptacle," i.e., a lower dimensionality that would actually function concretely as a space in its own right, not just as a bounding element for three-dimensional space.

Beginning with Edwin Abbott's imaginative excursion into "Flatland" (1884/1983), there have been many conjectures on what experience might be like for a lower-dimensional being. However, in much of this, the *dialectical* character of dimensionality is missed, and this has resulted in some inconsistencies. Take, for example, the speculations of the Russian philosopher P. D. Ouspensky in his theosophical work *Tertium Organum:*

> Let us . . . consider the two-dimensional world, and the being living on a plane. . . . In what manner will a being living on such a plane universe cognize his world? First of all we can affirm that he will not feel the plane upon which he lives. He will not do so because he will feel the objects, i.e., [the flat] figures which are on this plane. He will feel the lines which limit them. . . . The lines will differ from the plane in that they produce sensations; therefore they exist. The plane does not produce sensations; therefore it does not exist. (1970, 53)

Thus, according to Ouspensky's extrapolation of classical space, just as we three-dimensional beings do not at all sense the space in which we are embedded but only the two-dimensional surfaces of the objects in this space, two-dimensional beings would be restricted to sensing the one-dimensional edges of the objects in their space.

Ouspensky further notes:

> Sensing only lines, the plane being will not sense them as we do. *First of all, he will see no angle.* It is extremely easy for us to verify this by experiment. If we hold before our eyes two matches, inclined one to the other in a horizontal plane, then we shall see one line. To see the angle we shall have to *look from above.* The two-dimensional being cannot look from above and therefore cannot see the angle. (1970, 53–54)

These conclusions about two-dimensional experience appear to be directly contradicted by Ouspensky later in his text, when he speaks of two-dimensional beings as capable of perceiving *surfaces* (not just lines), of experiencing "simultaneously in two directions" (89), of viewing *circles,* figures that possess angles of 360° (92). Why the discrepancy? Why is it that, in one place, Ouspensky asserts that the perceptual capacity of a two-dimensional being is strictly one-dimensional whereas elsewhere he claims that it is two-dimensional? Perhaps Ouspensky's susceptibility to this error stems from the fact that, while his approach to dimension is classical, the truth of perception may be found in the non-classical realm that lies *between* dimensions.

Ouspensky's modus operandi is extrapolation by dimensional analogy. Thus, when he draws the conclusion that "the two-dimensional being has no idea of an angle" in the line it perceives (54), analogical reasoning evidently should tell him that we three-dimensional beings would be just as unable to perceive an angle in the *surface*. It is clear that we are incapable of seeing a solid cube or sphere "simultaneously from all sides" (87). Ouspensky makes the claim that we "can never see, even in the minute, any part of the outer world as it is, that is, *as we know it* [i.e., in three dimensions]. We can never see the desk or the wardrobe *all at once, from all sides and inside*" (88). Evident in these two sentences is Ouspensky's presupposition of a categorical division between three- and two-dimensional sense perception, one that leads to the conclusion that if we cannot see *all* sides of a solid object at once, then we are limited to viewing but a *single* side of it. According to Ouspensky, while we do *know* of the existence of the other sides of any solid object we encounter, this knowledge is not directly sensed; it derives from conceptual inferences we draw about the object. But is it true that our sense experience permits us to apprehend just a single surface of a solid at a time; that we cannot simultaneously sense a surface that is perpendicular to it; that we have no sense of the angle joining two or more surfaces of the solid?

Obviously it is not. The phenomenological fact of "fractionally" three-dimensional perception discussed in the previous chapter is that—although we surely cannot view all six outer surfaces of a solid cube in simple simultaneity—we can see up to *three* of them in this way. Thus, we *can* view angles joining surfaces, though, to be sure, we cannot apprehend them as completely as we can the angles that connect the lines of a plane figure. These are the facts of perspective that Ouspensky glosses over in his classical approach to dimension. "We cannot imagine the cube . . . seen, not in perspective, but simultaneously from all sides," says Ouspensky (87). In failing to qualify this further, Ouspensky allows the inaccurate impression that perspectival experience entails a *single*-sided view of the cube, one that would preclude the sensation of angle.

In Ouspensky's construal of dimension, the dialectical character of lived experience is neglected. At polar extremes—where we would experience either all sides of the cube in simple simultaneity or just a single side, but nothing in between—dialectical tension is absent. The vitality of the dialectic is found only by taking into account the intermediary zone, where there is *opposition,* not just simple juxtaposition. The sine qua non of the perspectival three-dimensional lifeworld is the opposition of the sides of surfaces. And it should follow by analogy that, in a *two*-dimensional lifespace, there would be a dialectic of *opposing edges of lines.* This is the lower-dimensional action that is occluded in three-dimensional space, where the edges of lines appear to be merely juxtapositional. Note that a lower order of *subjectivity* should be implicit in this action, since dialectical action is not merely objective but entails the interplay of object and subject. Thus, just as the dialectic of the plane in the three-dimensional lifeworld gives evidence of Merleau-Ponty's "invisible," and, in so doing, implicates our subjectivity, the dialectic of the line in the two-dimensional world should entail a subjectivity of its own. Indeed, a different order of *Being* should be involved here.

If the Klein bottle embodies the dialectical self-Appropriation of three-dimensional Being; and if there is in fact a second order of Appropriation, one of different dimensionality than the first; then I am proposing that this second order is embodied by the lower-dimensional counterpart of the Klein bottle, the structure that results from bisecting the Klein bottle, i.e., the *Moebius* structure.

It is true that, in three-dimensional space, the Moebius appears as a well-formed topological object, a strip or surface that can readily be completed without the necessity of self-intersection. But remember what Rucker (1977) said about "Flatlanders" attempting to assemble a Moebius strip in a *two*-dimensional space: they would in fact be forced to cut a hole in it, just as we three-dimensional beings must do with the Klein bottle. So it seems that, in two-dimensional space, the Moebius structure indeed would have the special, self-intersective hole necessary for playing out the dialectic at that level. Said dialectic clearly would not involve the interplay of opposing sides of a surface, since there could be no such opposition in two-space. But note that, while a Moebius strip embedded in three-dimensional space does take the form of a one-sided *surface,* it also has the property of being one-*edged.*

Unlike the closed surfaces of the torus and Klein bottle, the cylindrical ring and the Moebius strip possess edges[1] (see fig. 2.1). We already know that expansion upon the Moebius strip from a local cross-section to its full length brings about the merger of its opposing sides. It is also true that the *edges* of the Moebius become integrated. To confirm this, run your index finger continuously along an edge of the Moebius until the whole length of the surface has been traversed. Upon returning to your point of departure you will discover that you have covered *both* edges of the surface. In contrast, tracing an edge of the cylindrical ring maintains the simple distinction between the edges.

Of course, in three-dimensional space, the opposition of edges lacks the dialectical force of opposing *sides*. Though we may describe the boundary lines constituting the edges of the Moebius strip as locally "opposed" to one another, the fact that we can view them simultaneously (unlike locally opposed sides of the surface) places them in the same simply positive context, thus rendering their relationship merely juxtapositional, not truly oppositional. And if there is no genuine perspectival opposition, the Moebius *integration* of edges must lack full-fledged *trans*perspectival dialecticity. But this would not be the case for the Moebius structure of Flatland. In this *two*-dimensional space, the fusion of edges would possess the same dialectical character as the Kleinian fusion of perspectives in three-dimensional space.

In two-dimensional space, the Moebius structure would not be an open surface but a *line,* one that is paradoxically both open and closed, just as the Klein bottle is an open and closed surface. We are aware, however, that this Kleinian "surface" is no simple two-dimensional object in three-dimensional space but is a structure that brings three-dimensionality to its dialectical completion. By the same token, the Moebius "line" would bring *two*-dimensionality to its (w)holeness. To better appreciate the threefold distinction among the Moebius line, the Moebius surface, and the Klein bottle, let us consider the simpler case of three non-paradoxical counterparts: the circle, the cylindrical ring, and the torus (fig. 3.5).

We know that the cylindrical ring (fig. 3.5(b)), taken as an object in three-dimensional space, is an open surface possessing two edges or bounding circles. This surface can be topologically transformed into a closed structure by elongating it, stretching it to form a tube, then bringing the circular edges of this tube together. The second part of this operation is

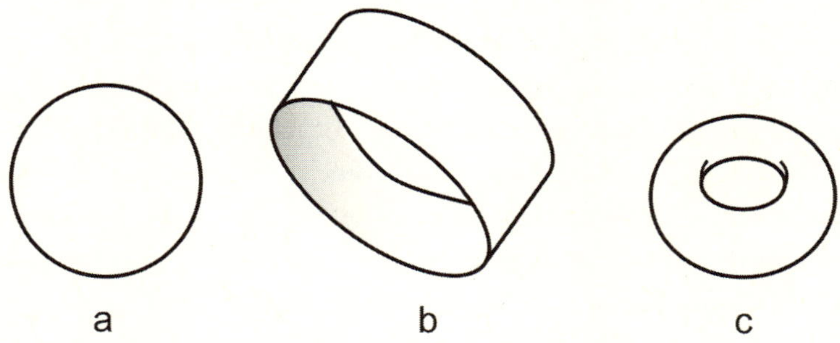

Figure 3.5. Circle (a), cylindrical ring (b), and torus (c)

identical to that depicted in the upper row of figure 2.3. The closed surface that results is the torus (fig. 3.5(c)). Now, if we proceed in the other direction, narrowing the cylindrical ring instead of elongating it, the circular edges draw closer and closer together. In the limit, we obtain but a single circle, the two-dimensional open surface having been reduced to a one-dimensional closed line (fig. 3.5(a)).

Consider, now, parallel operations upon the Moebius counterpart of the cylindrical ring. Like the cylinder (figs. 3.5(b) and 2.1(a)), the Moebius strip (fig. 2.1(b)) is an open surface, one that locally possesses a pair of edges. If the Moebius were stretched in a manner similar to the elongation of the cylindrical ring, and if its edges were glued together, we would obtain the Klein bottle (fig. 2.2). Of course, joining the edges of the Moebius strip is not so easy, given the odd one-edgedness of this structure (edges twist together to form a single edge). In fact, we are unable to execute the operation in three-dimensional space without tearing the surface, an action that topology does not permit. This limitation is familiar to us. What we are seeing is that the topological operation of identifying opposing edges of the Moebius strip is equivalent to that which we considered earlier, and which is shown in the lower row of figure 2.3: the end circles of an elongated tube are joined from *inside* the tube's body, a procedure requiring the tube to pass through itself and thus to produce an impermissible breach when attempting to assemble the Klein bottle as an object in three-dimensional space. Moreover, identification of the opposing edges of a single Moebius strip is also equivalent to gluing together two Moebius strips of opposite orientation; we saw previously that the latter operation yields the Klein bottle as well.

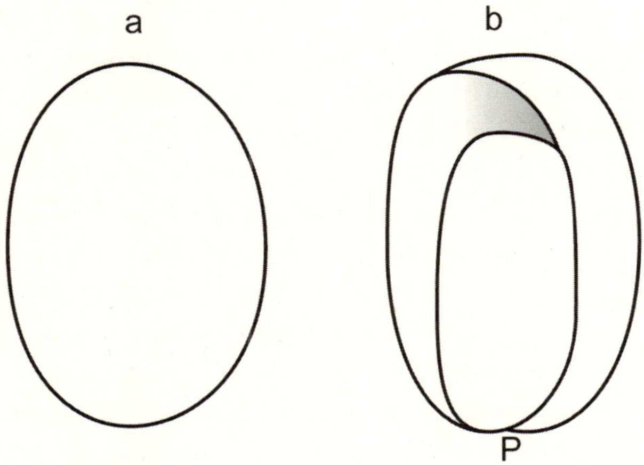

Figure 3.6. Edgewise views of cylindrical ring (a) and Moebius strip (b)

Now suppose, instead of elongating the Moebius strip, we made it narrower. What would happen in the limit of this operation? Would we obtain a simple circle as with the cylindrical ring, a simply closed line of a single dimension? To see what would actually result from this operation, let us compare the *perceptual* reduction of the cylindrical and Moebius strips in three-dimensional space.

Figure 3.6(a) illustrates the fact that we three-dimensional observers can rotate a cylindrical ring in such a manner that only one of its edges is visible. In this way, the ring is perceptually reduced from a two-dimensional surface to a one-dimensional circle. It is clear from inspection of the Moebius that no such reduction is possible. The one-dimensional, strictly edgewise view obtainable over the full length of the cylindrical ring can be realized in the Moebius case only at a *cross-section* of the strip. Note, moreover, that in viewing the Moebius in the edgewise fashion we do not actually see the cross-sectional line itself, but just the endpoint of this line that is nearest to us, illustrated in figure 3.6(b) by point P. However we position the Moebius, however we rotate it in three-dimensional space, at any given moment no more than a single point will be visible to us on the Moebius's edge at which extension in two dimensions will have vanished. It is only at this singular point that the perspectival opposition of sides is, in effect, perceptually neutralized (with no outer surface visible here, neither can we speak of an inner surface on the opposing side). Thus, the attempt to reduce the Moebius strip perceptually to a simple circle—a one-dimensional, purely positive presence in three-dimensional space—is frustrated by an unavoidable "dialectical surplus."

In the dimensional reduction thus far considered, only the properties of topological objects in three-dimensional space have been affected, not the characteristics of space itself, or those of the observer. Suppose, instead of merely reducing the dimension of our objects, we were able to change the whole dimensional frame of observation, lower the dimensionality of *object-in-space-before-subject.* We would then obtain the Flatland state of affairs, that wherein space itself is two-dimensional. In this environment, the cylindrical surface would indeed become a circular line, but a line with peculiar properties when compared with our three-dimensional perception of lines. Just as we three-dimensional observers cannot fully view both sides of a surface at once but must experience them in perspectival opposition, the Flatland observer would not be able to view both edges of the circle at once. Contra Ouspensky, this would not mean that the Flatlander would simply be limited to perceiving but a *single* edge of the circle, that it would be incapable of detecting the circle's angularity, its curvature in the second dimension. Rather, the Flatlander would experience the edges of the circular line in a perspectivally oppositional fashion akin to the way we three-dimensional observers experience the sides of the toroidal surface. This dialectic of edges is certainly not detectable in the three-dimensional environment wherein lines are viewed in a strictly juxtapositional manner.

Now, in the foregoing dimensional reduction, the open-edged cylindrical surface is transformed into a circular Flatland line whose property of closure is akin to that of the torus, the higher-order counterpart of the cylinder in three-dimensional space. Similarly, the dimensional reduction of the open-edged Moebius band would yield a Flatland line whose structure would correspond to the band's higher-order counterpart, viz. the Klein bottle: unlike the open Moebius surface, the Moebius *line* would be both open and closed. Assuming, then, that the Flatland Moebius takes on the role played by the Klein bottle in three-dimensional space, would a Flatland structure exist whose role would correspond to that of the Moebius strip in three-space? The topological bisection series tells us that it would be the *one-dimensional lemniscate.*

Recall our study of the two-dimensional lemniscatory surface (fig. 3.1). Though the opposing lobes of this dual structure are linked by a continuous movement through its central node, the lemniscate lacks the integrative quality of the one-sided Moebius band, whose opposing sides completely overlap one another. The dynamics of dimensional reduction suggest that, in Flatland, the lemniscate should lose its duality and assume the integrality of its higher-order Moebius counterpart. As the Moebius strip is a one-sided surface with an exposed edge, the Flatland lemniscate would be a one-edged line with an exposed *point.* And just as *closing* the exposed edge of the Moebius strip forms the Kleinian surface, the Flatlander would form the Moebius line by closing the exposed point of the linear lemniscate. Of course, in each milieu, a self-intersection would be required to complete the closure. We have seen that the necessary hole in the Klein bottle bespeaks its unobjectifiability in three-dimensional space. This observation eventually led us to recognize the one-sided Klein bottle as the perspectivally integrative dialectical action that invites backward, proprioceptive[2] movement, thereby disclosing the internal circulation of subject and object constitutive of three-dimensional Being. A similar conclusion is called for with respect to the Moebius line of Flatland.

Above I noted the inability to eliminate the dialecticity of the Moebius surface in three-dimensional space (fig. 3.6). The "dialectical surplus" of opposing Moebius sides in three dimensions bespeaks the intrinsic, *edgewise* dialecticity of the Moebius in *two*-dimensional space: this one-edged line would be unobjectifiable in said space. The necessary gap in the Moebius line would invite a lower-dimensional proprioception that would disclose the internal circulation of subject and object constitutive of two-dimensional Being. To reiterate, the Moebius *surface* is indeed but an object in three-dimensional space, a structure that merely *symbolizes* the dialectic of three-dimensional Being; this dialectic can only be truly embodied via the Klein bottle. What I am suggesting, however, is that when the Moebius is transposed into its own element, when it is given expression in the dimensional milieu of Flatland, which is the life-world of two-dimensional space and concomitant two-dimensional subjectivity, we then have the Moebius *line,* the unobjectifiable structure that fully embodies the dialectic of two-dimensional Being.

But we must take note of an essential *difference* between Kleinian and Moebial orders of the flesh. The Kleinian interplay of subject and object is more complex than its Moebius counterpart. In the three-dimensional

Kleinian container, whatever the content of experience, all perception is framed by three degrees of dimensional differentiation, three bounding elements—planes, lines, and points. In contrast, Moebius experience would be limited to just two such terms: lines and points. This would seem to mean that the perception of objects in the "flat" Moebius lifeworld would be less differentiated, that the subject-object dialectic would be weaker in this environment. The perspectivity here, that of edges of the line, should be less sharply defined than the Kleinian perspectivity of sides that is familiar to us. Moreover, the dialecticity of lower-dimensional lifeworlds should become even weaker as we go further down.

The next lowest order of Being is that of the lemniscate (table 3.1, below, summarizes the several orders of topodimensional Being). In three-dimensional space, the lemniscate is objectively manifested via an open, two-sided surface (fig. 3.1). Above we surmised that, in the *two*-dimensional milieu, the lemniscate would function as an open, one-edged line analogous to the open one-sidedness of the Moebius surface in three dimensions. A further dimension down, however, the lemniscate would enter its own element, the realm of one-dimensional space. Just as the open Moebius surface of three-dimensional space becomes the open-and-closed, self-intersective Moebius line of two dimensions, the open lemniscatory line of two-dimensional space would become an open-and-closed, self-intersecting *point* of a one-dimensional lifeworld. If the Kleinian structure constitutes the uroboric circulation of subject and object by which the three-dimensional lifeworld is completed in earnest; and if the self-intersecting Moebius structure is the subject-object action cycle that completes the two-dimensional lifeworld; then the self-intersecting lemniscate would be the ontological action that would bring to fruition the one-dimensional lifeworld. Still, with each step down to lower dimensionality, the dialectical tension of the dimensional action would be less, the perspectivity weaker; that is, the opposition and integration of the perspectives of objects, and of subject and object themselves, would be less sharply defined. The relative weakness of dialectical opposition in the lemniscatory lifeworld is indicated by the fact that here, only one element of differentiation would be operative, that of the point.

The last member of the bisection series is the sub-lemniscate. In three-dimensional space, this structure appears as the interlocking lemniscatory surfaces illustrated in figure 3.3. How would the sub-lemniscate be

Table 3.1. Topodimensional orders of Being

TOPOLOGICAL SERIES				
LIFEWORLD DIMENSION	OPEN/CLOSED INTEGRATIVE CYCLE	OPEN INTEGRATIVE CYCLE	OPEN DUAL CYCLE	INTERLOCKING DUAL CYCLES
3d	Kleinian surface	Moebius surface	Lemniscatory surface	Sub-lemniscatory surfaces
2d	Moebius line	Lemniscatory line	Sub-lemniscatory line	
1d	Lemniscatory point	Sub-lemniscatory point		
0d				

manifested in the two-dimensional environment? Table 3.1 illustrates the general idea that, in moving into lower-dimensional lifeworlds, lower-order members of the bisection series assume the roles of their higher-order counterparts. Therefore, in the two-dimensional lifeworld, just as the Moebius structure would play the Klein bottle's role (it would be an open-and-closed integrative circulation), and as the lemniscate would play the role the Moebius played in three-dimensional space (it would be an open integrative circulation), the sub-lemniscate would function as the lemniscate did in three dimensions: it would be an open circulation of dual character, a two-edged line, with each edge tracing a lemniscatory pattern: ∞. Continuing the reduction, the sub-lemniscate of the one-dimensional lifeworld evidently would operate as an "open integrative point," thereby assuming the role that was played by the lemniscatory line in the two-dimensional lifeworld. Then, in the zero-dimensional realm, the sub-lemniscate would finally come into its own.

Dialecticity would be completely absent in the zero-dimensional sphere. This is borne out by a clear-cut difference between the zero-dimensional situation and those of higher dimensions. The three orders of lifeworld dialecticity are expressible in terms of the dimensional interplay of container, contained, and uncontained. To begin to appreciate the uniqueness of zero-dimensionality, let us for the moment regard higher-dimensional containment relations in simplified ontical terms rather than concerning ourselves with the paradoxical *self*-containment that underlies them.

We can then say that, at each level, the containment of objects is enacted by the lower-dimensional bounding elements of the containing space: the planes bounding three-dimensional space, the lines bounding two-dimensional space, and the points bounding one-dimensional space. These relations suggest the general idea that the $n-1$-dimensional bounding element of n-dimensional space itself serves as the containing space in relation to a lower-dimensional bounding element. For example, the two-dimensional plane that bounds objects in three-dimensional space would function as space itself in the lower-dimensional lifeworld wherein the highest-dimensional bounding element is the line.

What we encounter in making the transition to the zero-dimensional realm is the complete absence of bounding elements and thus, the absence of space. The zero-dimensional point that bounds or differentiates one-dimensional space does not translate as a spatial container possessing its own, lower dimensional bounding element, since there is no dimensional element below zero. This is reflected in the fact that—unlike the plane or the line—the point is utterly indivisible; it cannot be partitioned or subdivided in any way. It is clear that, without the bounding elements necessary for spatial containment, there can be no objects-in-space. So the dialectic of container and contained could not be enacted in the zero-dimensional sphere.

There is, of course, a third term of the dialectic inseparable from the other two: the *un*contained, the subject. Obviously, if there is no spatial container and no contained object, neither could there be the detached vantage point of reflection upon the object that constitutes the subject. Since the zero-dimensional situation would be characterized by a complete lack of containment, there could be no ontological *self*-containment here either, no dialectic of Being. It is this total absence of dialectical structure that requires us to leave blank the zero-dimensional row of table 3.1.

Now, if the zero-dimensional state of affairs is unique in that it alone does not constitute an order of dialectical Being, can we say what it does constitute? What additional light can be shed on it? This issue is important enough to our overall grasp of inter-dimensional relations as to warrant a further digression from the topological account.

What does it mean to declare that there is no Being in the zero-dimensional sphere? What sort of negation is this? Evidently, it cannot entail a merely relative kind of negativity. That is, the negativity of the zero-dimensional cannot be expressed merely in relation to, and as the absence of, the "positive" Being of higher dimensions. Why not? Because Being itself is not simply positive. Rather than constituting a simple whole, it entails *(w)holeness,* the paradoxical blending of positivity and negativity (the contained and uncontained, object and subject). To negate Being then, must mean negating the positive in such a way that the negative is also negated. Instead of flatly declaring that Being does not exist in the zero-dimensional realm, we must say that it neither exists nor does it not exist!

The Kyoto philosopher Tanabe Hajime (1986) associated this enigmatic principle of neither/nor with Zen: the "logic of Zen . . . is [that of] *neither/nor*—as in the phrase 'Neither do I say that it is life, nor do I say that it is death'" (56). According to Tanabe, it is the double negation that permits us to surpass the relative concept of "nothing" and arrive at the notion of "absolute nothingness" (7–12). Here negation is not defined strictly in relation to affirmation, constituting its lack; rather, "Nothingness means *transformation*" (22, emphasis added; see discussion below).

The notion of "nothingness" is indeed pervasive in Eastern thinking and it is tempting to try to link it with Western existential phenomenology, via writings such as those of Sartre (1943/1956) and Heidegger (1929/1975). In this regard, Heidegger translator Walter Kaufmann proposes an interesting correlation in a footnote to Heidegger's essay "What Is Metaphysics?" Heidegger has commented that, whereas "*Dasein* ['there-being'] naturally relates to what-is . . . *Da-sein qua Da-sein* always proceeds from Nothing" (1929/1975, 251). Kaufmann's (1975) editorial response is to refer the reader to the *Tao Te Ching* of Taoism: "for though all creatures under heaven are the products of Being, Being itself is the product of Not-being" (381n3). However, it is important not to confuse "Dasein qua Dasein" with *Being qua Being.* In relating authentically to itself, Dasein ceases to relate exclusively to "what-is," to finite particular things. Thus it "proceeds from Nothing," i.e., from the "no-thing-ness" of Being, for, as we know, Being is not a thing, not an isolated object upon which a detached subject reflects, but is the *pre*reflective action that first gives rise to object-in-space-before-subject. Is the passage from the *Tao Te Ching* also alluding to the relationship of Dasein to itself that reveals the primordiality of Being? Does it not in fact refer to *Being's* self-relationship ("Being itself"), which discloses the "Not-being" that is even more primordial than Being? This deeper primordiality is not addressed in Heidegger's early essay "What Is Metaphysics?" The "Nothing" therein named has only to do with Being, not with the Taoist Not-Being of which "Being itself is the product."

Still, if we are seeking a parallel to non-Western nothingness in Heidegger's writing, we may actually find it in his later work, where he has abandoned his preoccupation with Dasein's relation to Being, and is now more concerned with the origin of Being per se. In the previous section, we examined Heidegger's later thinking on the question of what produces Being. He names the "fourth-dimensional" source of three-dimensional Being as *Appropriation* (1962/1972, 19): that which gives to Being what is Being's own. What I now propose is that the source of Being to which Heidegger refers is actually *zero*-dimensional, and that we may correlate it with the non-Western idea of nothingness, as governed by the Zen logic of neither/nor. Note that Heidegger never addressed himself to the question of *intervening* orders of Appropriation. We will return to them once we have further clarified the zero-dimensional order.

Tanabe Hajime's particular approach to nothingness lends itself well to phenomenological thinking, for Tanabe strongly emphasizes the *processual* character of nothingness. Throughout his principal work, *Philosophy as Metanoetics* (1986), he associates nothingness with action, transformation, mediation. Absolute nothingness is grasped as "absolute mediation" or "absolute transformation" (18), as "pure movement" or as an absolute "flux of activity" (74). In this context, Tanabe distinguishes ordinary movement or change from *action:*

> Action means that absolute nothingness emerges to work in such a way that being is converted into nothingness and nothingness into being. The notion of a self-identical immovable posits a substratum that sustains the movement or qualitative change in a thing. Here, on the contrary, "action" points to pure movement without any immovable substratum, an incessant conversion in which being is transformed into nothingness and nothingness into being. What is continual in this process (that is, what is not annihilated) is discontinuity (or annihilation) itself and transformation into nothingness.

> Nor does the self-consciousness of this fact mean an awareness of a conversion or change occurring in some substratum of the self that continues to exist without being annihilated. If that were the case, we could speak of movement and change, but not of action. There is simply no self-identical something lying beyond and outside of the transforming process of action in order to bring a self-identical unity to action. Were such a unity to ground action, action would not be spontaneous and free; we could not call it action. Thus it is not self-identity but the negation of self-identity that brings unity to action. (1986, 74–75)

Because the action of absolute nothingness is not centered in a being, identity, or self, Tanabe portrays it as the action of *tariki,* of "Other-power." In this dimension of pure process "that is neither life nor death" (7),

> the Other is absolute precisely because it is nothingness, that is, nothingness in the sense of absolute transformation. It is because of its genuine passivity and lack of acting selfhood that it is termed absolute Other-power. Other-power is *absolute* Other-power only because it acts through the mediation of the self-power of the relative that confronts it as other. (18)

In other words, absolute nothingness cannot act through its *own* power since there is no "own" here, no core identity that would give it positive definition. Similarly, Tanabe describes the action of nothingness as "an 'action of no-action,'" that is, "an activity without an acting self in which action ceases to be merely the doing of the self" (81).

But does "the self" simply mean the *subject* here? Was Tanabe speaking of a mere being, or was he alluding to Being? Evidently, by the term "self," Tanabe meant the former, the merely ontical. For he associates "the doing of the self" with that which is relative, with what is simply, non-paradoxically self-identical. As a result of this limitation in Tanabe's account, the threefold distinction among being, Being, and absolute nothingness—among the ontical, the ontological, and that which surpasses ontology—collapses to the twofold opposition of being and nothingness. Consequently, we miss the fact that Being constitutes its own distinct order of non-relative action, of genuine process. Indeed, I suggest that absolute nothingness could never be mediated by a self that is merely relative, as Tanabe seems to think. That is because a simply relative being could not survive its confrontation with the absolute; instead of being able to mediate or contain absolute nothingness, such a being would be overwhelmed and devastated by it. I propose, then, that it is not the action of being that mediates absolute nothingness but the action of *Being,* that only an *absolute* self would be equal to the task of such mediation. In this dialectical account, the attainment of "absolute selfhood" is of course no mere positive accomplishment, given that it entails self-*negation.* Again, it is not simple wholeness that would be achieved in realizing the absolute, but *(w)holeness.* The negative aspect of the absolute self is precisely what enables it to mediate the negativity of absolute nothingness.

In the "Other-power" of absolute nothingness, we see the kind of thoroughly selfless action that Heidegger termed *Appropriation.* I pointed out in section 1 that this other-dimensional action, in giving to three-dimensional Being what is truly *Being's* own, cannot be an act of "fatherly condescension" that *determines* the "ownness" of Being. In Tanabe's language, patriarchal action of the latter kind would not come from Other-power (*tariki*), but would be centered in *jiriki,* "self-power" (1986, 7–9). It is Appropriation that would constitute the action of Other. Zero-dimensional Appropriation—as utterly selfless Non-Being, as absolute nothingness—can do no more than mediate *Being's* self-development, rather than imposing anything of its own upon Being: there simply is no "own" from which such imposition could originate. Because Appropriation possesses no *center* of action, it must use Being's center; Being is thus the "axis" of Appropriation, to use Tanabe's term (in particular, he speaks of "the center of the self" as "the axis of absolute transformation"; 1986, 10–11). It is in this sense that Appropriation is mediated by Being. Nevertheless, in its Tao-like, "passive" fashion, Appropriation does act, and the action in question does not merely derive from Being, any more than the Taoist action of Not-Being so derives. This is the "action of no-action" through which Appropriation mediates Being. In this way, Being and Appropriation mediate one another (as Tanabe said of relative self and absolute nothingness).

It is only when we understand Appropriation as selfless, as radically (non-relatively) Other, that we can effectively avoid the condescension that might be implicit in the ministrations of a "higher self." Only as Other—where self does not get in the way—can Appropriation truly extend the nurturing gesture of support that encourages Being's *self*-Appropriation, its capacity to give to itself. But let us be clear about the *tripartite* relationship that is involved here. What I proposed in the last section is that, if other-dimensional Appropriation does not act in a "fatherly" way, neither does the self-Appropriation of three-dimensional Being. The action of the "father" is the reflective operation of the detached subject upon objects in space wherein the actor takes for itself what is other. In contrast, Being's self-Appropriation is a prereflective giving to itself of what is its own. But the matter becomes more subtle when we ask, What is this "own" that Being gives itself? Acting prereflectively, Being gives to itself the sharpness of reflective consciousness! Thus, even though Being does not simply act in the detached fashion of the reflective subject, reflective subjectivity is nonetheless its own, is the gift that Being bestows upon itself. In section 1, I attempted to clarify this analogically by considering the relationship of mother to child.

To repeat: through her nurturing support, the mother facilitates the child's autonomy, encourages it to develop its own way of acting distinct from her motherly caregiving. And yet, in thus giving to her child, the mother is not merely acting on behalf of another. The mother's identity is intimately bound up with that of her child, so that, by giving to it, she is giving to herself, promoting her *own* development. This is the sense in which the prereflective giving of the capacity for reflective subjectivity is Being's gift to itself: by its "motherly" action, Being gives itself the clarity it requires in order to achieve its individuation (see next chapter). The proposition, then, is that three-dimensional Being, in its self-Appropriation, acts *maternally.* And what encourages this action is the extra-dimensional Appropriation of which Heidegger spoke, the totally selfless, Tao-like Not-Being that plays the role of *midwife,* giving to Being what is Being's own.

In topological terms, the upshot of the foregoing is that sub-lemniscatory zero-dimensionality, as absolute nothingness, as selfless Appropriation, functions in the role of midwife to the maternal self-Appropriation of three-dimensional Kleinian Being. But what about the *intervening* orders of topodimensional action, those associated with the lemniscatory and Moebius spheres of the flesh? Our bisection series suggests a more complex set of relationships than was adumbrated by Heidegger. For him, only two non-ontical dimensions had to be accounted for: three-dimensional Being and the "fourth-dimensional It" that engages in Appropriation, gives to Being what is Being's own. Whereas the former, by definition, is ontological, the latter is *me*ontological (concerned with Not-Being, from the Greek, *mē,* not, and *ontos,* being). What the bisection series indicates is that *four* non-ontical dimensionalities must be taken into account. Here, while zero-dimensional nothingness is strictly limited to the role of midwife and three-dimensional Being to the role of mother, the intermediary dimensional actions evidently would function in *both* capacities.

Take the case of the one-dimensional lemniscate. To be sure it is an order of Being; it possesses a modicum of subject-object differentiation, of dialecticity; here there is an aspect of individuality, a minimal core of self. This means it will engage in self-Appropriation, and, for the maternal activity entailed in this, it will require the midwifely support of the selfless zero-dimensional sub-lemniscate. However, considered in relation to two-dimensional Moebius action, one-dimensional lemniscatory action is *less* self-centered, since the core of identity of the latter is more weakly developed than that of the former. Bearing in mind the intimate family relationship of the Moebius and lemniscate implicit in the bisection series, it appears we must say that—just as the lemniscatory self receives midwifely encouragement from the totally selfless sub-lemniscate—the Moebius order of self-Appropriation would receive midwifely support from the Appropriative action of the less self-centered lemniscate. In like manner, Moebial Being—in addition to *receiving* support for its self-Appropriation—would provide midwifely support to the most highly differentiated, well-developed order of self-Appropriation, that of three-dimensional Kleinian Being.

All relationships among the members of the topological family are spelled out in the *topodimensional matrix* (table 3.2). The word "matrix" is consanguineous with "mother"; from the Latin, it means "womb," "place of origin." Millington and Millington (1966) note that a matrix is an "array of numbers or letters, called elements, in rows and columns, used as a form of tabulation in problems where the relations between the elements are of fundamental importance" (142). This is what we have in the topodimensional matrix of table 3.2.

Table 3.2. Interrelational matrix of topodimensional bodies
SLB is zero-dimensional sub-lemniscatory body; LB is one-dimensional lemniscatory body; MB is two-dimensional Moebial body; KB is three-dimensional Kleinian body.

0d (SLB)	0d/1d (SLB/LB)	0d/2d (SLB/MB)	0d/3d (SLB/KB)
1d/0d (LB/SLB)	1d (LB)	1d/2d (LB/MB)	1d/3d (LB/KB)
2d/0d (MB/SLB)	2d/1d (MB/LB)	2d (MB)	2d/3d (MB/KB)
3d/0d (KB/SLB)	3d/1d (KB/LB)	3d/2d (KB/MB)	3d (KB)

In matrix algebra, a square matrix like that given in table 3.2 possesses a *principal diagonal,* the one extending from the upper left-hand corner to the lower right. It is here that the "eigenvalues" of the array are displayed, i.e., the "self-values" or primary elements. All other elements in the matrix play the role of specifying the relationships among the principal elements. In the topodimensional matrix, the four eigenvalues correspond to the four basic orders of Appropriation.

Consider the two principal elements of highest dimensionality: the two-dimensional Moebial body (MB) and the three-dimensional Kleinian body (KB). In the matrix, these elements are linked by the combining elements given in the two corresponding non-principal cells, 3d/2d and 2d/3d.[3] Said coupling cells are related to each other *enantiomorphically;* they are mirror images of one another. The cells in question signify the non-ontical counterparts of the ontically observable, oppositely oriented Moebius strips which, when glued together, produce the Klein bottle. The ontological Moebius element, taken strictly as an eigenvalue, is the open/closed, one-edged structure of the two-dimensional lifeworld that topologically embodies motherly self-Appropriation. But when we shift our view of the Moebius, consider it in relation to higher, Kleinian dimensionality, a kind of "doubling" takes place in which the single Moebius structure becomes a pair of asymmetric, mirror opposed twins. Whereas, ontically speaking, these are open Moebius surfaces contained as objects in three-dimensional space, what we now require is the non-ontical realization of enantiomorphs.

Before proceeding, I must acknowledge a limitation of table 3.2. Though we are essentially dealing with topodimensional birthing processes, events unfolding over time, the table gives us only a synchronic view. We do not see the "doubling" of the Moebius, the merging of enantiomorphs to form the Klein bottle, and so forth; all matrix elements are displayed as simply co-present. In chapter 5, the matrix will be "set in motion" to provide a dynamic rendition of topodimensional nativity (see table 5.1). For the preliminary account presently given, we are obliged to read our synchronic table in a diachronic fashion.

Let us say that, with the doubling of the Moebius, with the creation of enantiomorphs that support the genesis of the Kleinian body, one enantiomorph functions as a nascent form of the Kleinian mother engaged in self-birthing, while the other operates in the capacity of a selfless Moebius midwife that aids the mother's self-Appropriation. The midwife can thus be said to contain the mother as she gives birth to herself. The dimensional aspect of the collaboration is seen in the asymmetric relationship between enantiomorphs: 2d/3d and 3d/2d. Although each enantiomorph is a hybrid term spanning both dimensions, the 2d/3d enantiomorph uniquely expresses the role of the incipient mother; her "fractionality" reflects the fact that she has not yet fully reached her potential as an integral three-dimensional structure. Whereas the "fractional" motherly enantiomorph falls short of her native Kleinian dimensionality, the reciprocal 3d/2d enantiomorph *exceeds* her own two-dimensional Moebius sphere of action so as to function as selfless midwife to the mother.

The midwife-mother relationship culminates with the *fusion* of enantiomorphs. We can say that, in completing their merger, enantiomorphs are "annihilated"; they cease to exist *as* free-standing enantiomorphs. The midwifely enantiomorph, having played out its role, is absorbed into the three-dimensional Kleinian body; the other enantiomorph, no longer a "fractional" dimensional hybrid, is brought to term as the fully three-dimensional Kleinian mother herself. Invoking the *alchemical* metaphor, this ontological event—whose ontical counterpart is the gluing together of oppositely oriented Moebius strips—coincides with the "hermetic sealing of the uroboric vessel" (see chapter 2) that promotes the individuation of Kleinian Being, its self-containment. The Kleinian mother thus gives birth to herself with the encouragement of the lower-dimensional Moebius midwife, whose selfless action adds indispensable support.

We see from table 3.2 that the three-dimensional Kleinian mother does not receive midwifely support exclusively from the two-dimensional

Moebius midwife but from two other midwives as well: those of the one-dimensional lemniscatory body (LB) and the zero-dimensional sub-lemniscatory body (SLB). The linkage of the LB and KB is brought about through the 3d/1d and 1d/3d enantiomorphically related coupling cells, and the SLB and KB are joined via the 3d/0d and 0d/3d enantiomorphs. With respect to the LB, the table discloses a *second* pair of enantiomorphs, 2d/1d and 1d/2d. It is through this lower-dimensional coupling that the one-dimensional lemniscate provides midwifely containment for the self-Appropriation of the two-dimensional Moebius. In terms of the series of topological structures ontically observable to us in three-dimensional space, the LB-MB relationship is expressed by the fusion of lemniscatory surfaces that yields the Moebius surface.[4] Ontologically, the LB's 2d/1d support of the MB is its *primary* midwifely action, just as the 3d/2d enantiomorph of the MB provides primary support for the KB. The second midwifely LB enantiomorph, 3d/1d, constitutes a secondary source of containment for the three-dimensional Kleinian mother. Here the enantiomorphic liaison between midwife and "fractional" mother (1d/3d) does not directly culminate in the production of a fully developed Klein bottle. The fusion of 3d/1d and 1d/3d enantiomorphs leads instead to the 3d/2d and 2d/3d coupling, the higher-dimensional enantiomorphic pair whose subsequent merger does bring the Klein bottle to fruition.

We arrive at similar conclusions regarding the third order of midwifely support for Kleinian self-Appropriation, that which is provided by the sub-lemniscatory enantiomorph, 3d/0d. The SLB possesses *three* midwifely enantiomorphs. Through the first of these, 1d/0d, the 0d/1d motherly self-Appropriation of the lemniscate is supportively contained (the ontical counterpart of the SLB-LB relationship is given in the gluing together of sub-lemniscatory surfaces [fig. 3.3] to yield the lemniscate [fig. 3.1]). The second sub-lemniscatory midwife, 2d/0d, supports Moebial self-Appropriation in a secondary way, its fusion with 0d/2d leading to the 2d/1d–1d/2d enantiomorphic relationship that climaxes in the self-realization of the two-dimensional Moebius mother. Finally, the third midwifely SLB enantiomorph, 3d/0d, supports Kleinian motherly development in a tertiary manner: 3d/0d–0d/3d fusion → 3d/1d–1d/3d fusion → 3d/2d–2d/3d fusion.

3. SUMMATION

In this chapter, we have obtained our first glimpse of lifeworld dimensions beyond the three-dimensional world that is familiar to us. Our point of departure was Heidegger's distinction of two non-ontical action spheres: three-dimensional Being and its enigmatic "fourth-dimensional" source, viz. Appropriation. After exploring the implicitly maternal nature of these process dimensions, we proceeded to elaborate upon Heidegger's account. Working topologically and by inter-dimensional analogy, a total of four non-ontical dimensions were identified, with the three hitherto eclipsed lifeworlds being recognized as *lower* than three-dimensional. Upon noting that the lower dimensions possess less dialectical tension and complexity than their higher-dimensional counterparts, particular attention was given to the lowest non-ontical order, the zero-dimensional. This mysterious sphere of primordial action was related to the non-Western concept of "absolute nothingness," on the one hand, and to Heidegger's dimension of Appropriation, on the other. The chapter concluded with a matrix (table 3.2) providing a preliminary overview of all topological dimensions and their modes of interrelationship.

CHAPTER FOUR

DIMENSIONAL ONTOGENY

In this chapter and the next, we will see how the dimensions of ontological flesh engage in dynamic processes of individuation in which they give birth to themselves by bringing themselves to (w)holeness. Although this diachronic or developmental aspect of Being was certainly touched on above, it was not spelled out in a comprehensive manner. It is to this task that we now turn. In the present chapter, I will offer a general account of the basic stages of topodimensional evolution, limiting myself to the more familiar three-dimensional realm. Then, in chapter 5, we will proceed to investigate the specific patterns of development of all dimensional spheres and study the ways in which these different lifeworlds are woven together.

1. THE THREE BASIC STAGES OF ONTOGENY

In Plato's metaphysics, reality is sharply divided into changeless Being and beings that become. For his student Aristotle, Being is a "mover" that itself is unmoved. Down through the centuries into the present time, this static view of Being has been profoundly influential. This is the view that was challenged in Merleau-Ponty's later works. "Wild or brute [B]eing" (1968, 211) is no immutable spirit transforming the earth from on high. Because Being itself is of earth, because it is fleshly and embodied, it participates in transformational processes in such a way that it is affected by them. Earthly matter is in process, and living matter develops, passing through stages of biogenesis. Being, in like manner, possesses the character of a living organism—though not a *finite particular* organism, to be sure. Rather, Being is a generic organicity, a whole dimension of life; neither object nor subject alone, it is a sub-objective lifeworld. Nonetheless, that world develops. To speak as we have of Being's motherly self-Appropriation, of her "midwife-assisted self-giving," is to intimate that Being indeed engages in a bodily self-transformation: *She gives birth to herself.* Before proceeding to consider this developmental process in a systematic way, let me add a word of clarification on the question of gender.

Though I have used feminine descriptors to characterize Being, hardly can "she" be regarded feminine *as opposed to* masculine, since to do so would be to define "her" one-sidedly, in the dualistic terms of patriarchy. To avoid gender splits and stereotypes, my portrayal of individuation as a "motherly" or "maternal" process is to be understood in a dialectical sense of maternity, not in the stereotypically "feminine" sense. This does not mean that Being is gender *neutral.* It is not androgynous but *hermaphroditic*—to adopt another metaphor often found in the alchemical literature. In fact, the specific sense in which Being is a "hermaphrodite" has already been adumbrated: rather than being "all woman," she contains within her the seed of a masculinity or fatherly subjectivity whose development is inextricably entwined with her own.

1.1. Individuation as ontogeny

The word "ontogeny" (from the Greek, *ontos,* being, and *genesis,* generation) is commonly defined as the development of an individual being, with "being" normally understood in the ontical sense. To be "in-dividual" is to be undivided, to possess a unitary core, a coherent and stable center of identity. From very early on, human growth and development is guided by the fundamental tendency toward individuation.

In *Volatile Bodies* (1994), Elizabeth Grosz examines the original condition for the emergence of individuality:

> For the subject to take up a position as a subject, it must be able to be situated in the space occupied by its body. This anchoring of subjectivity in its body is the condition of coherent identity, and, moreover, the condition under which the subject *has a perspective* on the world, and becomes a source for vision, a point from which vision emanates and to which light is focused. (47)

The emergent subject is "the focal point organizing space. The representation of space is . . . a correlate of one's ability to locate oneself as the point of reference of space: the space represented is a complement of the kind of subject who occupies it" (47). Fully developed, the space in question is "the perspectival space that has dominated perception at least since the Renaissance" (48); i.e., it is the continuum. In sum: "A stabilized body image, . . . a consistent and abiding sense of self and bodily boundaries, requires and entails understanding one's position vis-à-vis others, one's place at the apex or organizing point in the perception of space" (48). So our earliest, most basic sense of ourselves as stable and autonomous individuals is inseparably linked to our experience with space. (The idealized Cartesian subject might fancy itself a disembodied spirit that is freely transcendent of space, but the underlying reality is that—however high this subject may appear to fly—it remains "tethered to its ground." The subject's relation to space necessarily is maintained, as the old formula implies: object-in-space-before-subject.)

Grosz's analysis builds on Freud and Lacan. Freud's (1914/1957) concept of "primary narcissism" implies that there is a primary constancy grounding all others: the image of one's body, which constitutes the earliest manifestation of the ego. Because one's imagined body serves as the frame of reference from which all observations are made, an object seen to change with changes in perspective can be taken as the same object only insofar as the observer's body is implicitly sensed as remaining the same. Object constancy thus depends on subject constancy, or, in Freudian terms, on the constancy of the ego.

In Lacan's (1953) elaboration on Freud, the "mirror stage" is crucial to the initial appearance of an invariant body image. The child begins to develop a stable sense of its identity, to form an image of its own body as a whole, when the mother's body can be mirrored back to it as its primary object or alter ego. This possibility arises at around six months of age and depends, in turn, on the psychological separation of the infantile body from that of the mother. Grosz (1994), in drawing on Lacan, thus observes that it is in this mirror stage that "the division between subject and object (even the subject's capacity to take itself as an object) becomes possible for the first time" (32). And, of course, this is the stage in which *space* first appears. It is through the medium of space that confirmation of one's basal identity is mirrored back to one. In encountering expected variations in the appearance of the mother's face with changes in spatial perspective upon her, the emergent individual is reassured that he is indeed *one.*

Grosz cites Lacan's observation that, in the mirror stage, "'the total form of the body by which the subject anticipates in a mirage the maturation of his powers is given to him only as a *Gestalt*'" (1994, 42). Her conclusion is that the "mirror image provides an anticipatory ideal of unity to which the ego will always aspire" (42)—even though this "model of bodily integrity" is something "the subject's experience can never confirm" (43). However, while "the stability of the unified body image . . . is always precarious" (43), it is just this imagined unity of the ego or subject that anchors human identity and permits the process of human development to go forward.

1.2. Individuation as Ontogeny

If the mirror stage is the period during which object-in-space-before-subject first emerges, it marks the beginning of ontogeny. But it is not the beginning of Ontogeny. In capitalizing the latter term, I mean to indicate that we are now speaking *ontologically,* dealing with *dimensional* development, with the individuation of Being itself.

To understand the developmental process in Ontogenetic terms, it will not suffice to go back to the young child's first experiences with objects in space. We must go still further backward into infancy. At the very outset of development, there is no "infantile observer," no immature Cartesian subject who may be epistemically fused with its objective environment in that it can *perceive* no differences between itself and others, but is nonetheless *ontologically detached* from that environment. Ontologically—Ontogenetically—what we are dealing with initially is the opening phase in the development of Being itself, a stage in which a subject has simply not yet been separated from an object, that wherein the projection of object-in-space-before-subject has not yet occurred. In the primal situation, then, it is not just a matter of the subject being unable to *know* the object as distinct from itself; rather, subject and object are *ontologically* undifferentiated here; they do not yet constitute well-formed modalities of Being.

Let us say that, in infancy, embryonic subjectivity "rests in the womb" of a Being that itself is embryonic, having not yet given birth to itself.

Then three-dimensional Being responds to the encouragement it receives from the Appropriative urgings of the lower-dimensional midwives. The self-Appropriative labor of Being thus begins, and, when the period of gestation is completed, Being can give to itself what is its own by sending forth from its womb the ontical orientation of object-in-space-before-subject. Note again that, because the "child" is motherly Being's gift to *herself,* the child's development is not only, or even primarily, what is at stake here. Of foremost concern is the mother's own development. By giving itself subject-object differentiation, Being *clarifies* itself.

It may be useful to distinguish ontical clarification from the self-clarification of Being. In the former case, what is clarified is a specific matter, a finite particular being assumed to be already projected before a reflecting subject. In contrast, the self-clarification of Being that constitutes its self-Appropriation is the *prereflective self-orienting action* that first makes possible clarification of the ontical kind. In the initial phase of Ontogeny, what is unclear is not some *particular* object, but the distinction between subjectivity and objectivity as such. We could say that, in this stage, Being is disoriented; it lacks direction. Then, in the ensuing stage, Being commences to clarify the relationship between subject and object, to differentiate them, by conferring a *direction* upon their interaction. With the projection of object-in-space-before-subject, awareness is now oriented to function in a single direction: *from* the subjective ground of reflection *to* the objects that are cast before the reflecting subject. Of course, this unidirectional from-to gearing accords with the "forward" propensity of ontical thinking that was discussed in chapter 2. Obscured by the subject-to-object movement of reflection is the prereflective self-Appropriation of Being that *first opens up* the field of objectification, and, in the same stroke, first gives a subject who detachedly reflects upon the objects in this field. Initially disoriented Being thus affords itself direction by transforming itself into intentional subjectivity, subjectivity-in-reflection-upon-its-field-of-objectification. In the process of clarifying itself in this manner, Being *conceals* itself. It facilitates its own development by moving away from itself. In this second stage of Ontogeny, Being gives itself reflective clarity by means of a prereflective process that is itself excluded from the giving.

It is important to recognize that Being's self-giving is not deliberate or even conscious in this stage of its development. If it seems strange that

Being could clarify itself and conceal itself without being consciously aware of doing so, then we must be clearer about the difference between prereflective maternal action and the *father's* way of acting. It is the latter that entails conscious acts of deliberation carried out by a reflective subject. Intentional subjectivity is what Being *gives,* but the giver itself is no such subject. Perhaps we can characterize the prereflective action of Being as "instinctive": "impelled by an inner or animating agency; hence, imbued; filled; charged; as, a poem *instinct* with passion."[1] The modus operandi here is "natural, voluntary, spontaneous . . . impulsive."[2] The point to keep in mind is that wild Being is no coolly deliberative *cogito.*

What of the *third* stage of Ontogeny? The possibility of entering it begins when the *cogito* has gone as far as it can on its own. In historico-cultural terms, this coincides with the "end of philosophy" (Heidegger 1964/1977, 373–92), that momentous occurrence we considered in the preface. "The end of philosophy proves to be the triumph of the manipulable arrangement of a scientific-technological world" (377). Here Heidegger is alluding to the nineteenth-century culmination of philosophy in which it is transformed into the modern sciences. In thus achieving its climax in the lofty abstractions of positivistic science, philosophy takes the objectification of the world to its "most extreme possibility" (375). And it is precisely in arriving at this "last possibility" for thinking that now, after centuries of being relegated to shadow, the "*first* possibility for thinking" (377) can finally be considered. For only by reaching the very pinnacle of abstract reflection that completes philosophy do we reach philosophy's *interior boundary,* the place where the supremacy of reflection can genuinely be surpassed and prereflective Being—the original *source* of reflection (that "from which the thinking of philosophy would have to start," 377)—can be brought to light. The realization of Being is the third and final stage of Ontogeny.

It is worth noting that the reversal of beginning and end featured in Heidegger's 1964 essay is also found in his 1946 lecture on the oldest fragment of Western thought:

> The antiquity pervading the Anaximander fragment belongs to the dawn of early times in the land of evening [*Abend-Land,* i.e. the West]. But what if that which is early outdistanced everything late; if the very earliest far surpassed the very latest? What once

> occurred in the dawn of our destiny would then come, as what once occurred, at the last . . . that is, at the departure [or end] of the long-hidden destiny of Being. The Being of beings is gathered . . . in the ultimacy of its destiny. . . . If we think within the eschatology of Being, then we must someday anticipate the former dawn in the dawn to come; today we must learn to ponder this former dawn through what is imminent. (1946/1984, 18)

What is today imminent is the *end of philosophy.* This "departure" thus calls us back to the beginning. That is, it calls us back to *Being.* So the "task of thinking" at "the end of philosophy" (Heidegger 1964/1977, 373ff.) is to *think Being.*

If, in the second stage of Ontogeny, Being gives itself reflective clarity in such a way that its prereflective giving is eclipsed, what the third stage basically calls for is a *conscious acknowledgment* of the giving. This requires the "switching of gears" that reverses the prevailing orientation of thinking. Instead of moving out of and away from Being, we must now move backward *into* it, thereby drawing back in the reflective light that prereflectively had been emitted, withdrawing its projection. In our present day—having spent more than a century avoiding the task of thinking via the excursion into modernism that has now reached its postmodern dead end—the task remains.

To put it a little differently, the task of thinking in the third stage of Ontogeny is to think *proprioceptively.* We already know the etymology of the word "Appropriation": it means "to one's own," from the Latin *ad,* to, and *proprius,* one's own. Whereas stage two of Ontogeny involves an act of self-Appropriation in which Being gives to itself what is its own in the posture of moving away from itself, in stage three Being is called back to itself so as to *receive* the gift. Being must therefore engage in Proprio-*ception,* the taking or accepting of its own (I now capitalize "Proprioception" so as to emphasize its ontological status). With this realization of the "first possibility for thinking," the circle would be closed, the cycle of giving completed in earnest.

Regarding Proprioception as an acknowledgment of the Appropriation that has taken place, a reciprocation of the gift that has been given by "saying thanks" for it, we are led back to Heidegger's 1954 meditation on the sort of thinking called for in the call to Being. As mentioned in the preface, Heidegger noted the etymological consanguinity of "thinking" and "thanking":

> What gives us food for thought ever and again is the most thought-provoking. We take the gift it gives by giving thought to what is most thought-provoking. In doing so, we keep thinking what is most thought-provoking. We recall it in thought. Thus we recall in thought that to which we owe thanks for the endowment of our nature—thinking. As we give thought to what is most thought-provoking, we give thanks. (1954/1968, 145–46)

What is it that is "most thought-provoking"? To what do "we owe thanks for the endowment of our nature"? It is clear that, for Heidegger, it is Being. Therefore, in thinking Being, we think Proprioceptively; we thankfully receive what has been Appropriated, what has been given over to what is our own.

2. TOPODIMENSIONAL ANALYSIS OF THE BASIC STAGES

The three basic stages of Ontogeny have now been identified. They may be summarized as follows: (1) disorientation, the embryonic stage of Being wherein subject and object are largely undifferentiated; (2) the stage of self-Appropriation in which subject and object are differentiated via a "forward"-oriented action exclusively directed from the former to the latter; in thus clarifying itself, Being moves out of and away from itself, and into the ontical; (3) the climactic "reversal of gears" in which Being returns to itself via the Proprioceptive circulation wherein subject and object flow unbrokenly into one another. Of course, in previous chapters, the ontological dimension has been fleshed out by means of *topology*. Since our preliminary description of Ontogeny does not take this into account, the task presently at hand is to provide a topodimensional rendition of the stages.

According to classical intuition, the spatial continuum is bounded by an infinity of juxtaposed planes, elements that are themselves unextended in three dimensions and thus possess no internal structure of their own.

As noted in chapter 1, the elements of classical space are not internally substantial, concretely bounded entities but constitute a condition of abstract boundedness as such. It is within the infinite boundedness of space that *particular* boundaries are formed, boundaries that enclose what is concrete and substantial. The concreteness of what appears within such planar boundaries is the particularity of the object. To reiterate, the object is that which is bounded, and space, via its infinite reservoir of planar bounding elements, provides the context that enables the object thus to be differentiated.

Bearing in mind our present concern with topodimensional Ontogeny—or shall we say, *Topogeny*—let us sharpen our focus on the nature of dimensional boundedness. Do the dialectically opposed surfaces that bound an object in space derive from and reduce to the inert juxtapositional relation of structureless planes bounding space per se, as is implicit in classical thinking? We have seen that the answer is no. With the help of Merleau-Ponty, we have come to understand that, on the contrary, the dialectic goes "all the way down." As a consequence, the boundedness of space must itself undergo dialectical development.

It will prove helpful to state this proposition in terms of the issue of *symmetry.* The idea of symmetry is associated with invariance, with the classical principle of changelessness. Mathematically, an object is symmetric to the extent that it remains the same when some specific transformation is introduced (e.g., a sphere possesses perfect rotational symmetry because its appearance does not change however it may be turned upon its axis). We may say that, in this kind of symmetry, invariance is relative in nature. For it is not that the object is not transformed at all (remember how, in conventional topology, we can even change coffee cups into doughnuts), but that concrete changes in the object are offset by a substrate of changelessness (the coffee cup and doughnut will "look the same" from the abstract standpoint of their continuous inter-convertibility). However, while the symmetry of an object in space is in this sense relative, it is dependent on symmetry of a stronger sort: that of space per se. Whatever structural variations may arise in the objects that are embedded in space, the spatial continuum itself—assumed to be composed of infinitely many structureless bounding elements all inertly packed together—remains *absolutely* invariant. So, on the classical view, space is complete unto itself and totally impervious to change. Exactly why is the absolute symmetry of the spatial container assumed to be the precondition for all relative symmetries in the objects contained? It is because the bounding or containment of change that classical object symmetries entail requires the forming of particular boundaries from the infinite boundedness that is space.

In stark contrast to the closed juxtapositional stasis of the idealized symmetric continuum is the openness of the actual lifeworld. It is in participating in the lifeworld that we encounter the dialectical phenomena of "orientation" and "polarity" of which Merleau-Ponty spoke (1964, 173); that is, lifeworld experience is not blandly homogeneous but distinctly directed, oriented, inherently susceptible to the development of polar opposition. It is here that "there are no events without someone to whom they happen and whose finite perspective is the basis of their individuality" (Merleau-Ponty 1962, 411). Of course, on the classical reading, the perspectival finitude of the lifeworld, its incompleteness or asymmetry, is merely a surface appearance concealing an underlying condition of symmetry. When Merleau-Ponty *questioned* the classical idea that lifeworld asymmetries are just "derived phenomena"; when he challenged the notion that "space remains absolutely in itself, everywhere equal to itself, homogeneous" (1964, 173); when, for classical space, he offered the more primordial dimension of *depth*—he was exploring the possibility that non-symmetric dialecticity indeed goes all the way down into the roots of space. We have seen that the classical idealization hinges on denying phenomenological fact. It is a fact of perspectival perception in three-dimensional space that we cannot see all six surfaces of a cube at once but are limited to viewing a maximum of three. The idealization glosses over this asymmetry by offering us abstract assurance that, despite the *appearance* of the cube's incompleteness, the whole cube is "really there" at the moment we are viewing it; that its opposing perspectives are at bottom *juxta*posed in simple simultaneity; and that the space in which the cube is embedded remains a seamless continuum whose absolute symmetry is upheld. To Merleau-Ponty, however, phenomenological reality must take precedence over abstract ideals.

But does phenomenological reality dictate that space is *merely* asymmetric, that it is simply *dis*continuous, that its bounding elements are simply open, incomplete? What I have proposed is that wild Being, in its process of self-Appropriation, undergoes *development*, that it transforms itself through several stages of Ontogeny. Given that Being is *dimensional*, the stages of Ontogeny must entail dimensional self-transformation. It will therefore not suffice to replace the classical idealization of the simply closed continuum, of simply symmetric boundedness, with the merely asymmetric notion of the incomplete or open boundary. We require instead an understanding of the full course of dimensional development whereby the boundedness of space—including its classical idealization—is dialectically self-generated. Transposing these terms, we can speak of the *generation of self-boundedness* or self-containment. This way of describing dimensional development highlights the *uroboric* nature of the ontological body: it is a dimensional Being whose individuation entails the realization of its self-containing character. I submit, then, that ontological dimensionality requires us to think in terms of a process of transformation by which the dialectical development of spatial self-boundedness occurs. With this in mind, we shall now inquire into the topodimensional characteristics of the general stages of Ontogeny set forth in the previous section. For the remainder of the chapter, the account will focus on the lifeworld most familiar to us: the Kleinian world of three-dimensional space. Then, in the following chapter, we will go on to consider the genesis of the lower-dimensional lifeworlds.

Let us be clear that the classical idealization of space is not an errant deviation from onto-dimensional development but is an integral part of that dialectical process. In particular, it is an aspect of the *intermediary* stage of Being's self-Appropriation, that in which Being moves away from itself and into the ontical. This is a stage of covert dimensional asymmetry. Here dimensional boundedness is actually incomplete, though it is *projected* as complete by positing the ideal of perfect symmetry. In the previous chapter, I related the *genuine* fulfillment of dialectical process to the completion of the enantiomorphic fusion by which lower-dimensional midwives selflessly support the self-development of the higher-dimensional Kleinian body. Seen in relation to the developmental analysis of the present chapter, this climactic fusion event is now associated with the third and final stage of Ontogeny. In this way, enantiomorphic fusion serves as our index of topodimensional development. With the full merger of dimensional enantiomorphs, the Kleinian vessel is sealed; Kleinian Being realizes self-boundedness or self-containment; achieves individuation.

Does this mean that, in the intermediary stage, enantiomorphic fusion is but *partial?* In a general sense that is true, yet stage two fusion is not *simply* incomplete but is distinguished by the fact that its incompleteness is concealed and the appearance of completeness is projected. Therefore, rather than saying that stage two in the generation of dimensional self-boundedness entails a *partial* fusion of enantiomorphs, a partial sealing of the dimensional vessel, it appears more accurate to describe this as a *quasi*-fusion or closure: the vessel *seems* to be sealed but it actually is not. In fact, there is a topological model of dialectical process that displays this feature of quasi-closure in a precise fashion: the Moebius strip.

In the initial comparison of the Moebius strip and lemniscate, I made the observation that the integrative one-sidedness of the former is paralleled by the trans-cyclical flow of the latter: just as there is a flowing together of the sides of the Moebius in which neither side loses its distinctness, the lemniscate entails a flowing together of distinct cycles, viz. the cycles that constitute the figure-8 or infinity sign (see fig. 3.1). What I did not spell out in that earlier analysis is that merging the sides of the Moebius strip also involves the fluid traversal of two cycles. This can be illustrated most simply by returning to our comparison of the Moebius strip and cylindrical ring (fig. 2.1).

Consider a circulation about the cylindrical ring (fig. 2.1(a)). Positioned on the inner surface of this simply two-sided structure, we move 360° around to complete one revolution, returning to our point of departure. Naturally, our passage around the inside of the ring never takes us to the outside. Throughout the journey, we remain on the side on which we began.

It is not like this with movement around the one-sided Moebius surface (fig. 2.1(b)). While a 360° revolution does seem to bring us back to our point of origin, at the same time, it is *not* our point of origin, since we are now on the *other side* of the surface. Notice the way ordinary cylindrical revolution maintains the simple dichotomy between point of origin and displacement from that point: by 180° of movement on the cylindrical ring, we are furthest removed from where we began, and by 360°, we are back where we started, the displacement being simply and completely reversed. In contrast, with a 360° turn around the Moebius, there is a circling back to the point of origin that at once is the point *most remote* from said origin, since—when sidedness is taken into account—it is the point on the other side of the surface that is diametrically opposed to the actual starting point. What happens if we continue our movement beyond this 360° point of quasi-return? After an additional 360°, we find that now we have truly returned, since we have come back to our genuine point of departure on the side of the surface from which our movement commenced.

The double Moebius cycle thus essentially involves a quasi-return to origin, followed by an additional circuit that repeats the process, now completing it in earnest via a second return. In this way, Moebius action models the dialectical process in which there is first a quasi-closure, a quasi-sealing of the dimensional vessel, and then an authentic ("hermetic") closure or act of containment. We can also see how the initial closure of the Moebius circuit symbolizes the quasi-fusion of enantiomorphs indicative of the intermediary stage of the dialectic, whereas the second closing of the Moebius circuit corresponds to the realization of full-fledged enantiomorphic fusion associated with the final stage of Ontogeny. The sides of any topological surface are related to each other as mirror opposites. In the case of the simply two-sided cylindrical ring, a full 360° turn naturally leaves us on the same side on which we started; opposites thus go unreconciled. In contrast, the Moebius dialectic brings mirror-opposed sides together. In the initial Moebius circulation, this happens only to the extent of completing the passage from one side to the other. At this point, both sides have not been traversed in their entirety; in fact, only *half* of the one-sided surface has been covered. It is the *complete* encompassment of enantiomorphically related sides that is required to fulfill Moebius one-sidedness, and, for this, there must be the additional Moebius circuit.

Note a correlated feature of Ontogeny that is modeled by Moebius action: the stage-three "switching of gears" from "forward" to "backward"; from classico-modernist and postmodern thinking to onto-phenomenological thinking; from self-Appropriation to Proprioception. In fact, such a reversal is symbolically evidenced even in the simpler action of the lemniscate. Moving continuously from one loop of the figure-8 to the other, clockwise circulation becomes counterclockwise (or vice versa). The Moebius counterpart of this does better justice to the complexity of the transformation.

To demonstrate the property of gear reversal, we incorporate a test body into the Moebius model: an asymmetric profile. Figure 4.1(a) shows a left-facing profile revolving backward in a counterclockwise direction around a cylindrical ring. It is clear that action on the simple ring will continue indefinitely in this manner, with the orientation of the profile never changing (though the half-face is turned upside down). In figure 4.1(b), we see the profile moving counterclockwise about the Moebius strip. Entering the twist, the left-facing form is changed into a right-facing form, the transformation being completed after 360° have been traversed. The transformation of left into right thus coincides with the occurrence of one full cycle of Moebius action. This change in orientation can be seen to reflect a change in clock sense, for what is counterclockwise to a left-facing profile will be clockwise to one that faces right. (Though the profile continues to move counterclockwise along the strip from our external perspective, its internal clock sense is reversed, as indicated by the dashed arrows.) Thus we may say that Moebius action involves a transformation of clock sense, a reversal of the counterclockwise or backward orientation that results in a forward or clockwise inclination. It is obvious that cycle one is not simply clockwise from beginning to end, this being followed by a shift in gears that gives a second, uniformly counterclockwise cycle. Rather, the clockwise orientation of cycle one is fully realized only at the end of the cycle as the culmination of an ongoing transformation from an initial counterclockwise orientation. Then, in entering cycle two, the direction of the gear shift has itself shifted so that the momentum is now from clockwise to counterclockwise, forward to backward. The fact that change in clock sense is *inherent* to Moebius action means that no *separate* switching of gears is required to bring it about. The gears are shifting throughout each cycle, with the quasi-return to origin marking the completion of the shift in one direction, and the readiness to begin shifting in the other. In this way, cycle one reaches its climax in the forward orientation symbolic of self-Appropriation, and, with commencement of cycle two, we begin the retrograde or backward movement that signifies Proprioception.

I hasten to emphasize, however, that these ontical observations of Moebius transformation provide us with no more than an analogy of the *ontological* transformations we actually seek to understand. For we know that, in the three-dimensional context, the Moebius strip is naught but an object-in-space. Though the Moebius can *symbolize* ontological transformation here, it lacks the special, self-intersective property required for playing out the dialectic of opposing sides of surfaces. Therefore, to

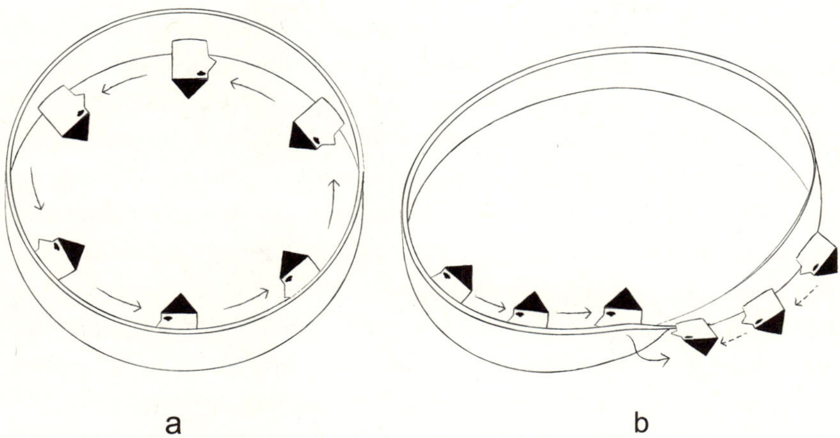

Figure 4.1. Revolution of asymmetric figure on cylindrical ring (a) and on Moebius strip (b)

embody Ontogeny in the three-dimensional milieu, it is the one-sidedness of the *Klein bottle* to which we must turn.

Like its Moebius relative, the Klein bottle should consist of a double cycle. That is because, as with the Moebius, two cycles should be required to fully express the dialectical property of one-sidedness, whereas the simply two-sided, non-dialectical counterpart of the Klein bottle, i.e., the torus (fig. 2.3), entails but a single cycle of 360°. With this in mind, imagine a circulation that commences from a point on the inside of the Klein bottle and proceeds for 360° about the length of the structure (see fig. 2.2). It seems that this revolution should produce the quasi-return that brings us back to our point of departure, but on the outside (similar to what happens with 360° of Moebius action). Continuing movement along the same trajectory for an additional 360°, we should then regain our true point of departure on the inside of the bottle. However, unlike what happens with Moebius revolution in three-dimensional space, the continuity of side-to-side Kleinian circulation is broken by the occurrence of the self-intersection: to pass from inside to outside and back again, the surface of the bottle must be breached. It is this interruption of ontical action that invites an *ontological* understanding of the double cyclicity of the one-sided Klein bottle. Nevertheless, we have seen that conventional topology has its way of repressing the challenge to continuity posed by Kleinian one-sidedness. It may be helpful to clarify a little further the conventional approach.

I noted in chapter 2 that ontical topology treats the curvature of a surface as a global feature resulting from the way this mathematical object is embedded in space. Whatever may be the global structure of such an object, the spatial continuum itself is assumed to consist of local bounding elements that are structurelessly juxtaposed. To ontical thinking, the topological merging of the sides of a surface is a perfectly permissible operation, as long as it can be understood as a merely global effect, a transformation in which local (point-to-point) continuity is not compromised. By the same token, a surface's "non-orientability" is acceptable as a strictly global feature. Although the fusing of sides brings a "switching of gears," a change in left-right or clock orientation involving asymmetric opposition, this poses no problem for the ontical approach if the operation can be seen as having no effect on the self-symmetric bounding elements of the juxtapositional continuum.

Of course, in the case of the Klein bottle, it is not actually true that the topological transformation can properly be regarded as a global event that conserves the local continuity of three-dimensional space. Since Kleinian transformation requires negotiating the *hole* it produces in itself, we in fact cannot trace a simply continuous path on the Klein bottle from one side to the other. In the classical approach, this crucial discontinuity is avoided by the questionable act of abstraction we already have examined: the "black hole" in three-dimensional space that prevents continuous Kleinian transformation is circumvented by gratuitously invoking an imaginary four-dimensional continuum in which the Klein bottle is purportedly embedded. It is when we resist the inclination to proceed in the classical fashion that we are led to the ontological realization that the Kleinian phenomena of "orientation" and "polarity"—the dialectical interplay of left and right, backward and forward, inside and outside—do not merely involve the global dynamics of an object embedded in static space but mirror the core dynamics of space itself, or, better, of dimensional Being. Moreover, these transformations are consistent with the *development* of Being described above.

Once again, the stages of Ontogeny adumbrated in the previous section are: (1) the initial non-orientation of Being wherein subject and object are largely undifferentiated; (2) the "forward"-oriented self-Appropriation of Being in which ontical differentiation occurs; (3) the "backward"-oriented Proprioception of Being in which subject and object are paradoxically integrated. The task at hand is to grasp all three stages topodimensionally, in terms of the generation of self-boundedness or self-containment, which we are now prepared to see as the dialectical transformation of the *local* relationship between "sides" of the Klein bottle. In the topodimensional account presently to be given, the earlier analysis of Ontogeny is carried forward and refined.

When fully comprehended, the special hole in the Kleinian container indeed subverts the classical interpretation of containment. Understood ontically, a container is a finite bounding structure, a particular object that encloses within itself a smaller object, thereby segregating it from objects that lie outside. Classically, this finite container must not be confused with the alleged infinite boundedness of the *spatial* container; nor should the condition of not being contained within the finite vessel be mistaken for the non-containment of the Cartesian subject, its purported total freedom from the constraints of space. We have seen that, with the self-containing Kleinian vessel, the categorial separation of contained object, containing space, and uncontained subject is surpassed. In penetrating itself, the Kleinian container brings these three terms into intimate involvement with one another (see fig. 2.4). The Kleinian bounding vessel can be said to constitute space itself (rather than a mere object in space), but a space that is a far cry from the structureless symmetric juxtapositionality of the classical continuum. In the dialectical exchange that is mediated by Kleinian space, opposing sides of the vessel are not finite particular objects. Rather, the "sides" are constituted by objectness and subjectness per se. It is when the opposing elements of the Kleinian container are understood in this manner that they are no longer considered merely to be global features of a topological object-in-space; instead, the "sides" of the vessel are grasped "locally," i.e., as features intrinsic to dimensionality as such. And the opposition and ultimate fusion of these "sides" *develop* through the stages of Kleinian Ontogeny we have preliminarily discussed.

To facilitate comprehension of the developmental process, picture again the ontical operation in which the Klein bottle would be formed by the progressive merging of mirror-image Moebius strips. Initially, Moebius enantiomorphs are completely separated; then, as they begin coming together, Kleinian closure is approached. Yet, prior to the actual moment of fusion, we have only opposing Moebius strips, not yet the Klein bottle. Upon completing their conjunction, asymmetric enantiomorphs would be "annihilated" in favor of the finished bottle. That is, opposing Moebius enantiomorphs would be introjected, absorbed into the body of the completed bottle, opposition now taking the form of the enantiomorphically related *sides* of the bottle. Of course, the Klein bottle's necessary hole precludes any such ontical sealing of the vessel. For proper closure, the process must be *ontological.*

Through the fusion of dimensional enantiomorphs, the two-dimensional Moebius midwife surrenders itself so as to facilitate the self-development of three-dimensional Kleinian Being. The task is to seal the Kleinian container in a thoroughgoing way, so that it bounds itself completely. We know there is a paradox here. In hermetically sealing itself, in closing itself to the fullest extent, the one-sided Kleinian container is also entirely *open, un*bounded (the "liquid contents" of a hermetically sealed Klein bottle would be more tightly secured than in an ordinary bottle, yet, at the same time, they would freely spill out!). Grasped ontologically, the enantiomorphs that participate in this process are subject and object. It is they that must be paradoxically integrated as the opposing "sides" of the one-sided vessel.

But let us begin at the beginning. From the topodimensional standpoint, let us consider the *inaugural* phase of the dialectic, the stage of inchoate flux predating the first manifestations of object and subject constancies, the first crystallization of object-in-space-before-subject. Here perspective does not yet exist. We do not yet have that unidirectional propensity to move *from* subject *to* object that is the hallmark of ontical consciousness. Instead Being lacks orientation; it is "parked in neutral" with its "gears" not yet engaged; it has no well-defined direction.

Topodimensionally, this is the state of affairs in which the Kleinian vessel is unformed. Dimensional enantiomorphs have not yet commenced their fusion. Note that these primordial enantiomorphs are not actually related as three-dimensional subject and object per se, but as mother and midwife, the inter-dimensional precursors of subject and object. If we were to apply ontical logic to this embryonic situation, we would be led to the conclusion that the lack of integration of enantiomorphs means that they stand opposed, are differentiated from one another. What is not together must be apart; what is not the same must be different, says either/or thinking. But it would be a mistake to impose a mode of thinking that only emerges in stage two of Ontogeny upon the circumstances prevailing in stage one. For the latter we require the kind of thinking that Tanabe spoke of, viz. the logic of *neither/nor.* True, in the pre-ontical stage, no fusion of enantiomorphs has as yet taken place so that dimensional elements are unintegrated. But it is also true that there has been no differentiation of them (at least no *positive* differentiation; motherly and midwifely enantiomorphs are distinguished on the subtle basis of the midwife's "negativity," its utter selflessness). Therefore, primordially, dimensional elements neither share common ground nor do they stand opposed. Neither identity nor difference can be found here. Thus, the *neither/nor* logic we have applied to the zero-dimensional sub-lemniscate evidently applies to the primordial stage of three-dimensional Kleinian Being as well. We may express this in terms of the *boundedness of space.* Because enantiomorphic elements have not yet begun to come together, the dimensional vessel has no closure or boundary, as I emphasized above. Applying the logic of neither/nor, what I now want to emphasize is that elements have not yet begun to be separated either, so that the vessel has no openness. The primordial vessel, then, is open or unbounded to no greater extent than it is closed or bounded.

With this account of the initial stage of Ontogeny, we can better appreciate the limitation of the ontical model of double cyclicity given above. According to the model, topological action commences from a point on one side of the Moebius surface, the ontical assumption being that, at this local juncture, sides are indeed well differentiated, are perspectivally opposed. (The underlying presupposition is that, although a paradoxical integration of sides may be displayed in the global structure of a topological object, when any object is considered locally it mirrors concretely the local differentiation of space itself.) Revolution through the first cycle then brings us to the other side of the surface, with orientation being reversed in the process, say, from counterclockwise to clockwise (or left to right). Finally, passage through the second 360° cycle brings us back to our point of departure on the original side, and restores the initial orientation.

The ontological interpretation of the cycles of action is markedly different. To express this in three dimensions, a Klein bottle is of course required. At the outset, we do not have the condition of local two-sidedness so compelling to perspectival intuition, but, rather, the *pre*-perspectival state of affairs in which opposition between Kleinian sides does not yet exist. Moreover, ontologically, "opposing sides" of the Klein bottle do not involve the sides of a topological object-in-space. It is object and subject themselves (or their primordial precursors, mother and midwife) that are neither differentiated nor integrated in the opening stage of Ontogeny.

What happens with entry into stage two, the stage expressed by commencement of the first cycle of Kleinian dimensional action? Contra the ontical model, we do not merely have a continuous movement from one locally well-differentiated side of the topological surface to the other. Instead we have the *differentiation and integration* of "sides," that is, of subject-object enantiomorphs. The process unfolds in such a way that a distinct orientation arises, a unidirectional forward movement *from* subject *to* object, as we have seen. So it is not the mere *switching* of orientation that occurs in cycle one as suggested by the ontical model, but the *establishment* of a definite orientation from an initial condition in which orientation is lacking. (Bear in mind the discrepancy between cycle number and stage number: cycles one and two of topodimensional action correspond respectively to stages two and three of Ontogeny, with the first stage entailing no directed cyclical action.)

Above I described the second stage of Ontogeny as involving a quasi-closure or integration of the dimensional vessel. What is now evident is that this quasi-integration of enantiomorphs is just as much a quasi-*differentiation*. For, in moving from stage one to stage two, we do not just go from openness to closure, but from an initial situation in which there is *neither* openness *nor* closure, to one in which there is *both*. These features can be studied in more detail via our topological models.

Consider once more the torus (fig. 3.5(c)). This higher-order counterpart of the cylindrical ring possesses two distinct planes of revolution, longitudinal (L) and transverse (T) (see fig. 4.2). The planes of action of the torus are orthogonal, simply independent of one another. Just as the earth revolves around the sun, and, at the same time, rotates upon its own axis, we can imagine perpendicular toroidal circulations to be occurring simultaneously. Yet any combination of these actions is strictly linear: their respective contributions add together without an internal change in either.

Let us take the longitudinal action plane of the torus as our plane of reference and regard it as a topological object with opposing sides. A

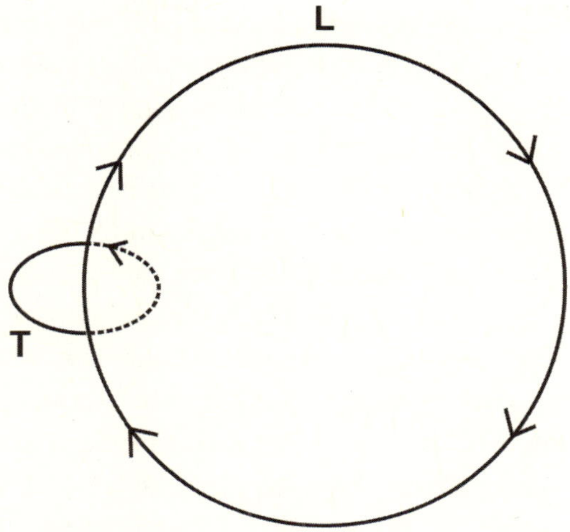

Figure 4.2. Orthogonal circulations of the torus: longitudinal (L) and transverse (T)

360° orbit about one side of this plane would be restricted to that side and would describe a circle of simple self-return. For its part, the 360° transverse circulation, the revolution occurring at right angles to the reference plane, would constitute an orbit of "simple departure" insofar as it would bring us to the diametrically opposite side of the reference plane. If we regard the longitudinal "circle of return" enacted on the given side of the reference plane as signifying closure or boundedness, self-identity or integrity (integration), the transverse "circle of departure" would then express opening, the transgression of boundaries, difference or differentiation. The strictly dichotomous relation of toroidal action components thus symbolically expresses the dualism of difference and identity. The logic of *either/or* is embodied here, for even though difference and identity operations may be thought of as occurring "at the same time," any *single* operation must involve *either* difference *or* identity, given the external relation between these opposites. Thus, in the single operation, there will either be the self-return taking place on the same side of the reference plane, or the transverse displacement to the other side, but never both. Moreover, this dualism of identity and difference is accompanied by a dualism of completeness and incompleteness. That is, whether the action is longitudinal or transverse, it will be either simply complete or simply incomplete. When 360° are spanned, the action is complete; for anything less, it is incomplete.

Returning to the Moebius model, we find a contrasting action pattern, one in which longitudinal and transverse components are thoroughly blended. Yet can we not identify two distinct circulatory components of Moebius action? Consider again the movement of the asymmetric profile upon the Moebius strip (fig. 4.1(b)). The profile orbits the length of the Moebius and also rotates upon its own axis in the direction perpendicular to the plane of longitudinal revolution. The longitudinal component of Moebius circulation is what brings the profile back to its starting point after 360°; this is the aspect of return, that which signifies closure, integration, boundedness. The transverse Moebius component, which involves only 180° of action, carries the profile to the *other side* of the surface; in so doing, it constitutes the aspect of departure that symbolizes opening, differentiation, and the surmounting of boundaries. However, unlike the action components of the torus, the circulations of the Moebius are interwoven in such a way that the return is at once a departure, and the departure a return! It is because the first cycle of Moebius action combines return and departure (identity and difference) in an integral way that a second cycle is required to fulfill these operations.

Under the continuing influence of classical logic, we might persist in asking whether there is or is not a return to origin after one cycle of Moebius action. Yes or no, we may inquire. The answer, however, is yes *and* no. The circle of identity is indeed closed with 360° of Moebius action. But because this circulation upon the same side of the longitudinal plane is inextricably coupled with the transverse displacement to the other side, the return to origin must be enacted again. It is for this reason that the cycle-one return is a *quasi*-return. In cycle two, the return movement is completed in earnest by reversing the departure that took place in cycle one: by a second transverse action, the profile now reverts to the side of the Moebius from which it began.

The first Moebius cycle constitutes a quasi-departure as well as a quasi-return. With 180° of rotation, we do have more than a mere *semi*-departure bringing us halfway round to the other side of the longitudinal action plane. Rather, the 180° transverse component of Moebius action brings us *all* the way around to the other side. And yet, since this transverse circulation is integrally combined with the longitudinal circulation that keeps the profile on the same side of the plane, further action is required in cycle two for departure to fully and truly be realized. Coupled with the transverse rotational component of cycle two that reverses the transverse action of cycle one by returning the profile to the side of the surface from which it had embarked, there is a *longitudinal* element that, in effect, offsets the longitudinal action of cycle one that had then limited the act of departure. Whereas the first cycle of longitudinal action had had the effect of maintaining the profile's position on the original side, in the second cycle longitudinal action serves to maintain the profile's position on the side to which it was displaced by cycle-one transverse action. In this way, the quasi-departure of cycle one becomes full-fledged departure in cycle two. The paradox, of course, is that the departure *is* the return, and vice versa; the true differentiation of sides is at once their true integration.

Transposed to the Klein bottle for an ontological reading, the quasi-differentiation and quasi-integration characteristic of cycle one constitute the state of affairs prevailing in the second stage of Ontogeny, where it is *subject* and *object* that are thus differentiated and integrated. In the quasi-complete fashion, dimensional enantiomorphs fuse, come together in an act of self-bounding that weds the emergent subject to the immanent world of objects; by the same paradoxical stroke, enantiomorphs move apart, so that the subject gains a measure of autonomy, of transcendence. Thus there arises the dialectic of constraint and freedom, containment and unboundedness. The two are not just joined at the surface but compounded in depth, blended so thoroughly that they cannot be teased apart.

Stage two is further distinguished by the fact that its dialectical entwinement and quasi-completeness are not recognized as such. Instead there is the ontical tendency to project simple separation and completeness, as well as simple attachment and incompleteness. The appearance is therefore created of the wholly self-sufficient, fully transcendent subject, on the one hand, and the utterly dependent object, on the other. Adding to this the third term of the classical triad, i.e., *space,* we have once again the old classical formula: object-in-space-before-subject. Like the subject, space is projected as complete within itself and unchanging, constituting as it does the infinite boundedness of the self-symmetrical continuum. Only the objects that are enclosed within this spatial container are said to change, but these transitory structures are ultimately regarded as secondary phenomena, as not essential in themselves. In the language of ontical topology, they are merely *global* entities, their dialectical properties having no effect on the local character of space per se. This classical ideal of dichotomy and stasis is what we project upon the dynamic Kleinian blending of subject, object, and space when we assume the ontical posture in stage two. The second stage of Being's development is thus the stage in which dimensional development itself tends strongly to be denied. Since the very notion that Being changes is exceedingly difficult to grasp here (Being is misconstrued as an "unmoved mover," if recognized at all), it will surely be no less difficult to comprehend the idea that Ontogeny is only *quasi*-complete in this second stage (first cycle). To acknowledge that the integration and differentiation of subject and object are incomplete would be to admit the lingering influence of the inchoate state of affairs that held sway in stage *one,* the stage in which *neither* identity *nor* difference were clearly in evidence. It is particularly this remnant of "infantile flux" that classical thinking cannot come to grips with, since it belies the "clean transcendence" of the phallic subject. The "womanish" vestige of primal chaos (known in alchemy as *prima materia:* prime matter or mater) is what the idealizations of the classical mind have always sought to obscure.

Our continuing predilection for ontical projection can explain the difficulty we have with one-sided topological structures like the Moebius strip and the Klein bottle. They seem so puzzling and circuitously complex to us, compared with the "straight-forward" properties of their two-sided counterparts. When faced with the quasi-closure of the Moebius strip and Klein bottle, we are strongly inclined to mistake it for the simply self-symmetrical closure of the cylindrical ring and the torus. For the latter pair of topological objects well embody the classical ideals of dichotomy and stasis, of simple completeness. The simplifying idealization imposed on paradoxical one-sidedness can be regarded as a kind of "tunnel vision" or narrowing of viewpoint, as illustrated by our topological models. In the case of the Moebius strip, it is when we restrict our view of this surface to a local cross-section that we eliminate its curious global features and thereby obtain the two-sidedness that defines the cylindrical surface in its entirety. In like manner, narrowing our view of the one-sided Klein bottle yields toroidal two-sidedness. Classical two-sidedness therefore can be considered as but an idealized limit case of the more general Moebius/Kleinian truth we find so difficult to countenance.

But *what about* the stage-two limitative tendency to project an appearance of simple completeness upon a process that in fact is incomplete? Is there a specific *topodimensional* feature that reflects this inclination?

We have seen that the self-Appropriative passage into the second stage of Ontogeny brings with it the emergence of a definite *orientation* from an initial situation in which direction is lacking. The direction of action thus established is the movement of classical consciousness out of and away from Being, the transition to the ontical that brings us forward, *from* subject *to* object. It is through this propensity that changelessness and completeness are idealistically projected. Obscured in the process is the underlying fact that this is just a stage in Ontogeny, one in which the development of Kleinian Being is actually only *quasi*-complete. Evidently, the "forward gearing" of dimensional action characteristic of stage two does find specific topological expression. In the ontical model of the Moebius strip, I used an asymmetric profile to bring out the *vectorial* aspect of the action: the shifting of orientation from left to right, counterclockwise to clockwise, backward to forward, in cycle one, and back again, in cycle two. No such "switching of gears" takes place with simply symmetric cylindrical or toroidal action. Of course, ontologically, cycle one does not just entail a linear change in orientation but the *establishment* of a definite orientation to begin with. And this is what we have with the Klein bottle when it is apprehended in the ontological way.

If, in stage two (cycle one), the circular dimensional action is *directed,* and if the self-Appropriative momentum of this stage spins us in the "forward" direction, what happens when we pass into stage three, the second cycle of Kleinian action? Ontical observation of circulation on the Moebius strip indicates that the cycle-one transition from backward to forward is reversed in cycle two. Exactly what does this mean in ontological terms?

In the second cycle of Kleinian action, the projection of simple completeness enacted in cycle one is *counter*acted. What arises here is a *retrograde* circulation, a movement against the "forward" orientation that had prevailed in cycle one. The "gears have shifted" in cycle two so that, instead of moving away from the inchoate origin in concealment of it, we move backward into it. This movement back into the undifferentiated, unintegrated source is not merely regressive; it does not merely cancel the forward thrust toward differentiation and integration, any more than the cycle-one move toward differentiation and integration cancels out the unindividuated source (though that source does become repressed). For in neither case are we dealing with the point-to-point, linear kind of movement that takes place in the infinitely divisible continuum. It is when movement is linear that, in arriving at point two, point one is simply left behind. In contrast to this action in classical space, the action *of* space, i.e., *dimensional* action, can be said to be "holistic" or "integral," since it entails an *indivisible* aspect not unlike the quantized spinning found in the subatomic realm (the nature of dimensional action will be further specified in section 2 of the following chapter). Therefore, just as the forward orientation of cycle one does not simply leave behind the original lack of orientation, the backward action of cycle two does not merely nullify the forward action.

Suppose you were handling a textured piece of fabric, one whose fibers were arranged in a certain direction. Is it not by running your fingers *against* the grain of the material that its direction becomes more clearly discernible? Similarly, in the cycle-two movement against the grain of cycle one, awareness is gained of the very process of forward-directed projection that transpires in that first cycle. In this way, the projection is retracted, consciously taken back, in the midst of its ongoing occurrence. It is in so acknowledging the Appropriative activity of cycle one that we engage in *Proprioception.* Here, rather than simply going *with* the grain—i.e., naively buying into the ideal of perfect integration and differentiation—we move in the retrograde manner that allows us to bear witness to how this idealization is first projected from the inchoate origin that had been thoroughly concealed when movement was geared exclusively forward the "first time around." The projection of difference and identity, of autonomy and integrity, is the gift that Kleinian Being gives to itself in self-Appropriation. And the "reversal of gears" that takes place in the second Kleinian circulation is the Proprioceptive *acknowledgment* of this giving that completes it in earnest.

Of course, the self-Appropriation of three-dimensional Being is a self-birthing that is aided by the Appropriative gestures of lower-dimensional midwives. In cycle one, the circle of individuation achieves its "pre-mature closure" not only by obscuring the Kleinian nature of dimensionality, but also by denying the midwifely dimensions. This inclination is reversed in cycle two. The retrograde action that now operates brings genuine maturation, true closure and opening, by bringing to light all uroboric dimensionings in acknowledgment of their gifts. Kleinian (w)holeness thus is realized, its Ontogeny fulfilled, not in phallic transcendence of the lower-dimensional action spheres, but in intimate relation to them. We then have (w)holes relating to (w)holes in a pattern of inter-circulation that constitutes a harmony of topodimensional spheres.

And this brings us back to the dimensional action matrix (table 3.2), with its four orders of circulation and their interrelationships. Having now completed our topodimensional analysis of the general course of Ontogeny by elucidating its double cyclicity, we will resume our study of the several specific orders of dimensionality and their consanguineous ties to one another. What I will show is that when these linkages are included in the developmental account, the cyclical action stages identified above resolve into *sub*stages, with variations in substage patterning from one topodimensional body to another serving to index specific differences among the bodies.

But we are not quite ready to reenter the action matrix in detail. I first want to take some time to reprise the general stages of topodimensional Ontogeny in terms of our working metaphor of midwife, mother, and child—anchored by a further articulation of the parameters of topodimensional transformation we have been discussing. In the initial stage of the birthing process, the midwife, acting selflessly, extends a meontological "gesture of support" to the pregnant ontological mother. She offers to contain the mother in a way that will encourage her to act on her own. At this initial point, however, the mother is not yet ready to accept the midwife's generosity, for she is not yet prepared to exert herself, begin her labor, commence the birthing process. When the mother does respond to the midwife, the midwife will contain her in a manner that entails her *self*-containment. The self-closure is at once an opening, an act of self-liberation brought about through the liberation of her child: in giving birth to it, she gives birth to herself. But, at the outset, the midwife's selfless encouragement of motherly closure and opening, having not yet been taken to heart by the mother, cannot be transposed into motherly self-encouragement. The mother is *pregnant,* of course, pregnant with the possibility of such self-giving. So it could not be said that the closure and opening simply are absent. But her self-Appropriation has yet to be made a reality.

In topodimensional terms, the liaison of midwife and pregnant mother found in Kleinian Ontogeny is expressed by the enantiomorphically related coupling elements of table 3.2 that give the interrelationships of the Kleinian dimensional body (KB) with the other dimensionalities: the Moebial, lemniscatory, and sub-lemniscatory bodies (MB, LB, and SLB, respectively). We are reminded here that *three* lower-dimensional midwives actually attend the self-birthing of the three-dimensional Kleinian mother, with each midwife to be taken over by Kleinian Being as its self-Appropriation progresses (the fine points of this developmental process will be examined in the next chapter, where we explore the substages of self-Appropriation). Before any of this happens—before any enantiomorphs begin to fuse—the rule of neither/nor prevails in the nascent three-dimensional sphere. At the outset of Kleinian Ontogeny, there is neither closure nor openness, boundedness nor unboundedness, symmetry nor asymmetry, object-contained-in-space nor uncontained subject (though there is the pregnant possibility of these).

In the second general stage of Kleinian birthing, the mother goes into labor and the pregnancy is consummated. As the midwife had supported the mother, the mother now supports her child. She lovingly embraces it, dotes on it, and, in so doing, her own needs appear on the surface to be subordinated. The selflessness of the mother thus seems to parallel that of the midwife. In fact, in the earliest practice of midwifery, the midwife was hardly a professional or a specialist but was a member of the family, most often the grandmother. So, at least superficially, it appears that the grandmother's selfless attention to the mother is now given by the mother to the child. But there is of course the crucial difference between the roles of midwife and mother, for the self-denial of the latter is enacted in the interest of self-realization. In ministering to her child, the mother fulfills herself, though this may not be immediately obvious.

What is the *ontological* significance of the mother's ostensive self-sacrifice, her apparent denial of her own nature in deference to her child's? In stage two of Ontogeny, the "male child" that is born from the instinctive maternal action of prereflective Being is reflective consciousness, the detached subject reflecting upon objects-in-space. In the forward-oriented circulation of stage two, motherly prereflectivity is eclipsed and the "child" gains ascendancy. He appears to go off on his own here, to soar into the stratosphere, leaving his mother down below. But however far the child may stray, however high he may fly above the motherground, the underlying fact—one that at all costs must be concealed in stage two—is that the cord is still attached: the child remains *her* child.[3] Why the ascent? Why the odyssey? Why the departure from the house of Being, the circuitous venture into the ontical? We already have considered the answer. By losing herself as she does, by repressing her prereflectivity and launching her child into the sphere of reflection, the mother is moving toward self-clarification. The gift of clarity is what Being grants to itself in self-Appropriation.

Nevertheless, if the giving is to be completed in earnest, it does have to be acknowledged. Thus, having lost herself in the granting of clarity, of light, Being now must (re)discover herself; the projection of light must be retracted, reeled backward to its origin. Expressed in terms of the nativity metaphor, a "second pregnancy" is required for the full maturation of the mother. Did the mother not go into labor following the first occasion? Was the child not born? Did it not develop? Yes, but, despite appearances, the process of individuation begun in the womb was never actually carried to completion. The circle of individuality only *seems* to close in stage two; the closure is *idealized,* consisting as it does of a boundary that defines the inner core of identity in simple segregation from what lies outside. Therefore, since the first nativity entails a "premature closure," a second one is required. This is what happens in the climactic stage of Organismic process. There is a re-naissance, a "second coming," at it were. But this is no mere repetition of the first kind of onto-natal action. For the "maternal child" now moves in the retrograde gear, moves *backward* through the birth canal, returning to the moment and point of her conception. In this way, through the Proprioceptive action of cycle two, the mother gains self-recognition; wild Being returns to herself in genuine self-completion. Stated in the simplest graphic terms, the rebirth that occurs in stage three of Ontogeny surpasses the single, idealized circle of individuality, O, to bring out the double circulation of the uroboric individual: ∞. Only when it is sealed a second time can the motherly vessel be sealed in earnest.

What are the topodimensional correlates of the second and third stages of Ontogeny? In stage two, with the onset of dimensional self-Appropriation, three-dimensional Being projects itself as object-in-space-before-subject. Thus there is the quasi-closure of the continuum, the purported infinite boundedness that constitutes perfect spatial symmetry. And—in categorial separation from the positivity of space and the objects it contains—is the negativity of the subject, of that which is unbounded, asymmetric, utterly open or transcendent. Therefore, whereas the dimensionality of stage one entails neither boundedness nor unboundedness, neither symmetry nor asymmetry, in stage two we have both, but have them in idealized, irreconcilable opposition to one another. As a consequence, the stage-one logic of neither/nor is supplanted by the logic of either/or.

*Topo*logically, the either/or of stage two is signified by an idealized structure like the torus, whose two directions of circulation appear simply and completely orthogonal to one another (fig. 4.2). Above we found that the classical orthogonality of the two-sided torus is but a limit case of the more general "hybrid" truth of the one-sided Klein bottle: it is when we restrict our view of the Klein bottle to a local cross-section (neglecting its full length) that it appears indistinguishable from the torus. Then we may associate the quasi-completeness of the second stage of dimensional generation—the quasi-boundedness and quasi-opening, the quasi-symmetry and quasi-asymmetry, the quasi-individuation—with a narrowed-down view of uroboric Kleinian reality, one that gives it the merely circular appearance of a torus. Note once again that this occlusion of Kleinian dimensionality constitutes a stage in the development of Kleinian Being itself: to find herself, the Kleinian mother first must lose herself.

Through the retrograde movement of stage three, the "dimensional blinders" are removed and we are able to recognize three-dimensional action for what it is: the uroboric circulation of Kleinian Being. As a result of this disclosure, the dualistic juxtaposition of symmetry and asymmetry at play in stage two of dimensional development is replaced by a dialectical blending of these opposites that I have elsewhere named *synsymmetry* (Rosen 1994, 17ff.). The kind of dialectic I am speaking of is nicely elucidated by Merleau-Ponty. Let us digress briefly to hear his account.

In the chapter of *The Visible and the Invisible* on "Interrogation and Dialectic," Merleau-Ponty advances a critique of Sartre's magnum opus, *Being and Nothingness*. Sartre applies the ancient formula, "*being is, nothingness*[4] *is not*" (Merleau-Ponty 1968, 88); in so doing, he presupposes the classical division of these two terms. Being is *en soi;* it is the pure positivity of the object. In categorial opposition to this is the *pour soi,* the sheer negativity of the subject. "It is with this intuition of Being as absolute plenitude and absolute positivity, and with a view of nothingness purified of all the being we mix into it, that Sartre expects to account for our primordial access to the things" (52). However, says Merleau-Ponty, "from the moment that I conceive of myself as negativity and the world as positivity, there is no longer any interaction" (52). In his attempt to achieve interaction, Sartre does juxtapose being and nothingness, but the two do not blend organically and the outcome is but an ambivalent oscillation between mutually contradictory poles that "belies the inherence of being in nothingness and nothingness in being" (73). Whichever pole we consider, whether "the void of nothingness or the absolute fullness of being," what we ignore are "density, depth, the plurality of planes, the background worlds" (68).

For his part, Merleau-Ponty insists that being is not pure positivity; that nothingness is native to it, is incorporated within its very core. Thus, neither being nor its negation exist in unalloyed immediacy; instead they mediate each other internally and the result is *depth:*

> The negations, the perspective deformations, the possibilities, which I have learned to consider as extrinsic denominations, I must now reintegrate into Being—which therefore is staggered out in depth, conceals itself at the same time that it discloses itself. . . . If we succeed in describing the access to the things themselves, it will only be through this opacity and this depth, which never cease: there is no thing fully observable, no inspection of the thing that would be without gaps and that would be total. . . . Conversely, the imaginary is not an absolute inobservable: it finds in the body analogues of itself that incarnate it. (1968, 77)

The mutual mediation of being and nothingness reflected in the phenomenon of depth leads Merleau-Ponty to conclude that "for the intuition of being and the negintuition of nothingness must be substituted a *dialectic*" (1968, 89). According to Merleau-Ponty:

> Dialectical thought is that which admits reciprocal actions or interactions. . . . Dialectical thought is that which admits that each term is itself only by proceeding toward the opposed term,

> becomes what it is through the movement, that it is one and the same thing for each to pass into the other or to become itself, to leave itself or to retire into itself. . . . Each term is its own mediation. . . . Is not the dialectic . . . in every case the reversal of relationships, their solidarity throughout the reversal, the intelligible movement which is not a sum of positions . . . but which distributes them over several planes, integrates them into a being in depth? . . . In sum [dialectical thought is] capable of differentiating and of integrating into one sole universe the double or even multiple meanings, as Heraclitus has already showed us opposite directions coinciding in the circular movement. . . . The circular movement is neither the simple sum of the opposed movements nor a third movement added to them, but their *common meaning,* the two component movements visible as one sole movement. (89–91)

In Topogenetic terms, the "circular movement" of the dialectic indicative of Merleau-Pontean depth is the serpentine Kleinian action brought to realization in cycle two, the third stage of Ontogeny. We have seen that this climactic action is retrograde in nature, is a Proprioceptive rebirthing of Being. Here the logic of either/or gives way to that of *both/and:* depth, dimensional Being, is *synsymmetric,* both symmetric and asymmetric—not in Sartre's sense of the mere juxtaposition of contradictories, but in Merleau-Ponty's sense of a dialectical blending in which opposing terms fulfill each other as they permeate each other. In the paradoxical hybridity of synsymmetric depth, we have both *more* symmetry and *more* asymmetry than the classical idealizations of these allow. That is, in its very openness, the dimensional vessel is actually bounded more completely than is the classical continuum, with its alleged "infinite boundedness": the synsymmetric vessel is *hermetically sealed.* And in its closure, the Kleinian vessel is more open, for this plenum is suffused in its entirety by void. That is to say, the vessel is a *(w)hole.* In short, the same uroboric self-circulation that bounds Kleinian dimensionality, that binds it indissolubly to the lived world, at once sets it free.

Now, while the dialectical circulation of immanence and transcendence in the three-dimensional lifeworld is indeed well expressed by Merleau-

Ponty, what is not as clear from his account are the *other* dimensional circulations, the lower-dimensional actions that are in fact *less* dialectical. In the closing stage of Ontogeny, where backward action permits us to recognize three-dimensional Being's motherly self-Appropriation, our attention is also called to the set of natal actions that were even more deeply repressed in stage two: the midwifely assistance given to the mother by the lower orders of Appropriation. It turns out that, in addition to the supportive grandmotherly role they play, two of these lower dimensionalities undergo their own processes of motherly self-development. In the present chapter we have focused primarily on three-dimensional Being so as to gain a better grasp of the overall course of Ontogeny. But now we are finally ready to return to the broader investigation begun in the previous chapter, that in which all orders of dimensional action are included.

CHAPTER FIVE

CO-EVOLVING LIFEWORLDS

1. SETTING THE MATRIX IN MOTION

We have found that Being is:

1. *Dimensional action.* It is not a mere object contained in space, not a static spatial container, and not a transcendent subject. It is the paradoxical transpermeation of contained, container, and uncontained constituting a lifeworld.
2. *Organic.* It changes, grows, undergoes dialectical development. It is transformed through several stages of Ontogeny.
3. *Polydimensional.* There are several spheres of Being.
4. *Topological.* The four dimensions of the flesh are embodied by the members of a topological "matriarchy," a family of maternal action spheres that corresponds to the bisection series: Klein bottle, Moebius strip, lemniscate, sub-lemniscate. Accordingly, we may alternatively characterize the development of Being as *Topogeny.*

In consideration of the foregoing points, the task now at hand becomes that of describing the underlying patterns of *dimensional Topogeny:* the transformations of the several spheres of topological Being as they evolve in relation to each other. In this reckoning, we must also take into account the important asymmetry among the dimensions that was discussed in chapter 3: lower-dimensional circulations entail less dialectical tension than do higher-dimensional ones. There is less differentiation and integration of subject and object in lower dimensionalities; less individuation; less Ontogeny. In the case of the lowest dimensionality—the zero-dimensional sphere corresponding to the sub-lemniscatory member of the bisection series—there is no individuation whatsoever. The zero-dimensional action that serves as "great-great-grandmother" to the three-dimensional Kleinian mother is selfless indeed. It is the only action with nothing to call its own: no subject, no object, no dialectical tension, no space or time, no development. In fact, we know that zero-dimensionality is not an order of Being at all but the meontological sphere of *absolute nothingness* (Tanabe 1986). Generally then, the topodimensional interrelationships that we must describe are among circulations with differing degrees of dialecticity and developmental potential.

Consider again table 3.2, the interrelational matrix of topodimensional bodies. Taken by itself, this table affords but a static picture of dimensional associations, one that is "averaged over," i.e., abstracted from, the actual facts of dimensional development. Therefore, to fill in the concrete details of how the several dimensional bodies evolve in relation to one another, we must set the matrix in motion. This is achieved in table 5.1(a).

Table 5.1(a) displays the full course of development of all orders of Topogeny. The dimensional ratios express relationships within and between orders of Being, and between Being and nothingness. The terms on the principal diagonals of the matrices (the diagonals extending from upper left to lower right) are "eigenvalues": they show as unitary ratios (3d/3d, 2d/2d, 1d/1d) the relationships of the topological dimensions to themselves (the zero-dimensional case is an exception; since the action here is utterly *selfless,* no eigenvalue is given and the cell is left blank). The ratios appearing in all other cells are members of the enantiomorphic pairings that account for the linkages *between* dimensional actions (cf. table 3.2 and accompanying discussion).

The general design of table 5.1(a) is circular. Moving upward from the matrix at the bottom, we have the stages of Appropriation, indexed by the S numbers appearing to the left of the matrices. The turn to Proprioception is then enacted by moving back down through those same matrices, with stage numbers now displayed to the right. The terms in parentheses accompanying each stage number indicate the topodimensional body or bodies to which that number applies; since the zero-dimensional SLB (sub-lemniscatory body) does not itself develop but serves only in a

Table 5.1(a). Stages of dimensional Topogeny

LB is one-dimensional lemniscatory body; MB is two-dimensional Moebial body; KB is three-dimensional Kleinian body.

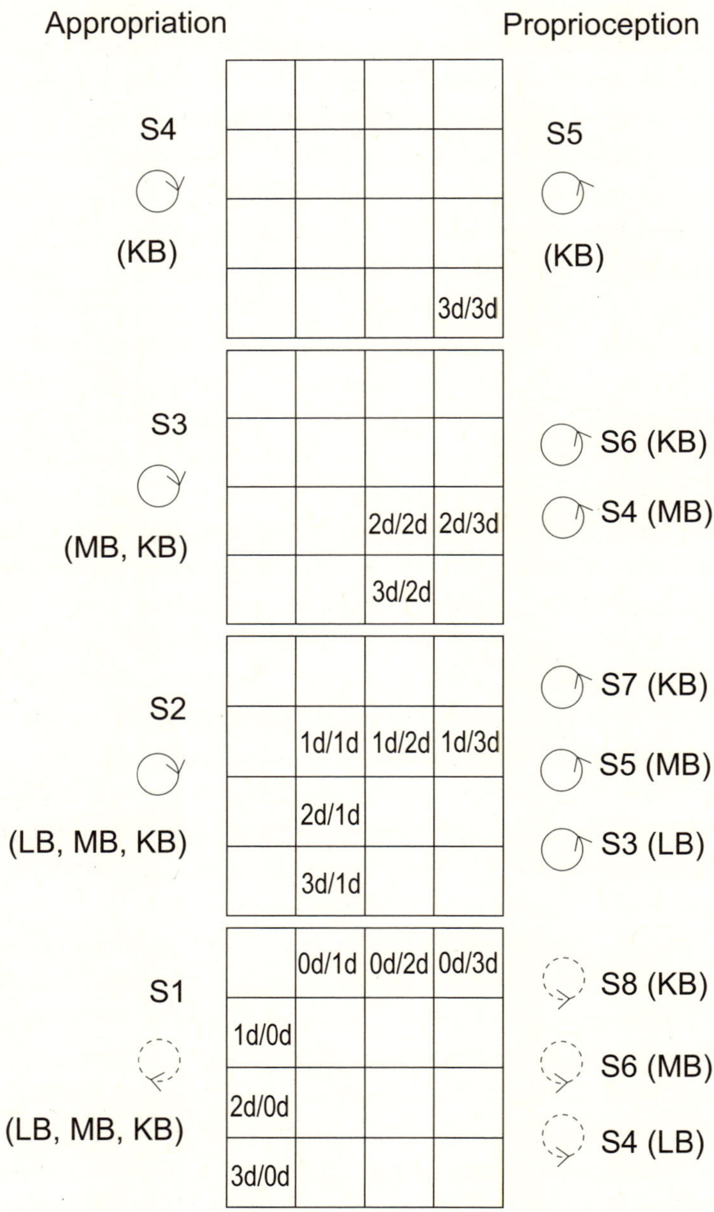

midwifely capacity, it does not appear in parentheses. If the sequence of stages for each topodimensional body is considered separately from that of the other bodies, we see that table 5.1(a) in fact does not describe a single developmental circle but circles nested within circles, so that the overall pattern is actually that of a *spiral.*

To preserve the thoroughly interwoven, nonlinear character of dimensional interrelatedness, table 5.1(a) displays the several windings of the dimensional spiral as *overlapping* one another. However, this makes the table fairly difficult to read. To facilitate understanding, I offer table 5.1(b) as a visual aid. Here the circulations of the dimensional spiral have been parsed, teased apart for easier identification.

Upon inspecting table 5.1(b), we can readily observe the simple arithmetic expansion (1, 2, 3, 4) in the number of matrices that are involved as we go from lower- to higher-dimensional windings of the dimensional spiral. The increase reflects the progressive increment in the number of stages through which Topogeny occurs. Note that something else changes in the process. From one winding of the spiral to another, there is a *logarithmic* increase in the number of cells found within each square matrix: 1, 4, 9, 16. This growth in matrix size signifies a progression in dimensional complexity, with larger matrices possessing higher degrees of dialecticity. We saw in chapter 3 that the strength of the dimensional dialectic, its intricacy and sharpness of definition, depends directly on the number of dimensional bounding elements that operate within the given lifeworld. These basic elements serve to differentiate and integrate the lifeworld. Thus, the three-dimensional Kleinian lifeworld, possessing three such bounding elements (point, line, and plane), is more dialectically refined than its lower-dimensional counterparts with their fewer bounding elements. In the matrices, the bounding or containing elements correspond to the pairs of enantiomorphically related cells that sit astride the principal diagonal. These "midwife-mother couplings," when appropriated by the dimensional mother via enantiomorphic fusion in the course of her stage-to-stage development, serve the purpose of dimensional self-containment, of dimensional differentiation and integration. Larger matrices can accommodate greater numbers of different enantiomorphic pairings.

The first winding of the dimensional spiral is that of the sub-lemniscatory (SLB) matrix, which consists of but a single cell. In effect, the "radius" of

Table 5.1(b). Dimensional spiral, parsed into separate windings

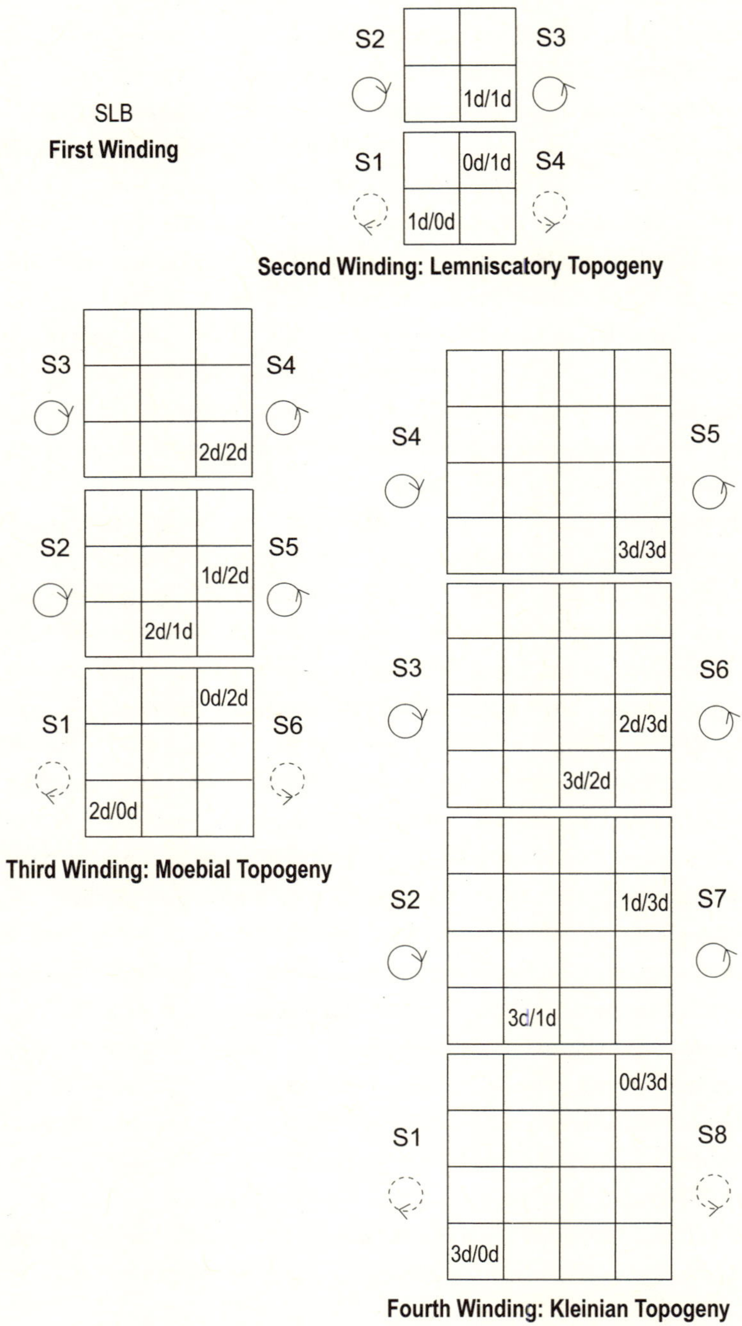

this circulation is "zero," for it entails no transformation whatsoever.[1] No stages of development can be found for the SLB, nor are any necessary, since this zero-dimensional action functions purely in the capacity of providing selfless midwifely support (via its enantiomorphs: 1d/0d, 2d/0d, and 3d/0d) for the self-generation of the higher-dimensional actions.

Advancing to the second winding of the spiral, the matrix expands to the 2 × 2 structures associated with the Topogeny of the one-dimensional lemniscatory body (LB). Again, while table 5.1(b) displays this winding as separate from the others (enabling us to study it more easily), the fact is that no such simple detachment of circulations exists. But neither do the windings simply coincide, as they seem to do in table 5.1(a). Rather, each developmental circulation possesses its own distinct timing, and this precludes it from being merely simultaneous with any other.

The hands of a clock provide an elementary analogue of the different temporal "scales" associated with the different windings of the dimensional spiral. With a full cycle of the second hand, there is but a unit movement of the minute hand, and a full cycle of the minute hand corresponds to a unit movement of the hour hand. Thus we have faster cycles embedded within slower ones. Of course, in the ontological counterpart of this, the cycles are not simply commensurable. The circulations of Being do not constitute subdivisions of a well-established unitary time metric (comprising seconds, minutes, hours, days, etc.), but entail different degrees of temporality itself. As Heidegger noted, three-dimensional Being is the dimension of "true time" (1962/1972, 15); also, it is the dimension of *depth,* the *spatial* aspect of Being articulated by Merleau-Ponty (1964). And, just as the lower dimensions possess less perspectival opposition and thus less spatiality, they possess less temporality as well; the interplay of temporal elements (past, present, future) is weaker. For the very lowest dimension, there is no temporality or spatiality at all. The zero-dimensional sphere of "absolute nothingness" is a sphere of absolute timelessness and spacelessness. (I must acknowledge that quasi-quantitative language—"less" temporality, "weaker" dialecticity, etc.—will not suffice if we wish to do justice to the truly radical differences among dimensional lifeworlds. Said differences will be explored more concretely in the final chapters of this book.)

We are following the second winding of the dimensional spiral, the "fastest" circulation among those topodimensional actions that are subject to development. In its opening stage, we have the "prenatal" matrix wherein the zero-dimensional sub-lemniscatory midwife or grandmother, 1d/0d, offers her meontological support to the "pregnant," "fractionally" one-dimensional lemniscatory mother, 0d/1d. (It was established in chapter 3 that, while each member of any given pair of enantiomorphs is a hybrid term spanning two different dimensions, the fractional member is an embryonic form of the higher dimension, whereas the native dimensionality of the reciprocal member is the lower dimension.) Matrical action is but nascently oriented here, as indicated by the dashed lines that compose the clockwise-directed circle. The mother is indeed pregnant with the possibility of forward projection, of quasi-individuation, and the midwife is "encouraging" her. But she has not yet accepted the encouragement, has not yet gone into labor and given birth to herself. Note the implication of ascribing circulatory activity to each individual matrix (as denoted by the vectorial circles appearing beside each in the table): Dimensional action does not just lie in the stage transitions from one matrix to another, for *each* matrix is itself a sphere of action. Like planets rotating upon their own axes as they revolve around the sun, the dimensional action matrices are spinning internally as they progress through their stages of growth.

Moving upward now to the stage two (S2) matrix of the lemniscatory winding, we find that the pregnancy has been consummated. In her act of self-Appropriation, the lemniscatory mother appropriates as her own the containment that had been proffered by the sub-lemniscatory midwife. This entails the quasi-fusion of enantiomorphs in which they are "annihilated," with the 1d/0d enantiomorph being absorbed into the one-dimensional lemniscatory body in such a way that lower-dimensional midwifely containment or boundedness takes on the character of a point-based boundedness internal to the lemniscatory dimension. By this self-transformation, the 0d/1d enantiomorph is brought to term as 1d/1d, the quasi-integral lemniscatory mother. Of course, the primordial situation is not simply eradicated; rather, it is repressed, relegated to the background (or "underground"). The S2 occlusion of 0d/1d and 1d/0d enantiomorphs is denoted by reduction of the matrix to the single entry, 1d/1d. The filling in of the clockwise circle of rotation indicates that oriented projective activity—the unidirectional movement *from* the lemniscatory subject *to* its object—is now occurring in earnest. Whereas the lemniscate's matrical spinning is but incipiently oriented in S1, it assumes a definite direction in S2 and the direction is "forward." In the "clockwise forespin" of this phase, not only are 0d/1d and 1d/0d enantiomorphs occluded, but also the 1d/1d lemniscatory mother herself, though not quite as completely. For this is the phase in which the mother defers to her "male child," the lemniscatory subject. It is here that one-dimensional Being is concealed and the basic ontical posture assumed, that of *object-in-space-before-subject.* (The S2 occlusion of the 1d/1d mother is not shown in table 5.1. Matrix reductions of the table account only for the deeper occlusions of the lower-dimensional enantiomorphs.)

However, I note again that, in this one-dimensional winding of the spiral, dialecticity is relatively weak. The lemniscatory subject is simpler, more rudimentary than his higher-dimensional counterparts. Lemniscatory differentiation and integration, lemniscatory individuation, lemniscatory space and time, are all comparatively undeveloped. In moving to the Moebial and Kleinian windings of the dimensional spiral, dialecticity increases in the logarithmic fashion, with matrix size expanding from 2^2 to 3^2 to 4^2. But before said passage to higher dimensionality can take place, the forward-spinning lemniscate must "switch gears," go into reverse so as to Proprioceptively acknowledge the Appropriation that hitherto had been obscured. It is by "going against the grain," by counteracting the forward action of Being, that *awareness* is gained of this action, thereby bringing it to fruition. Quasi-individuation then becomes genuine individuation.

We know that the "reversal of gears" is embodied in the double cyclicity of topological action, as symbolized most simply by the emblem of infinity, ∞. And what we now find, in table 5.1, is a stagewise specification of the crucial transition from forward to backward action cycles for each order of Topogeny. Every action matrix is to be interpreted in two ways: initially as spinning forward or clockwise, subsequently as spinning backward or counterclockwise. With regard to the second winding of table 5.1(b), we see that, in passing from stage two to three, lemniscatory action has shifted in this way, the movement "against the grain" being expressed by the reversal of the arrowhead from clockwise to counterclockwise.

How does the table elucidate the "second pregnancy" that was intimated in the previous chapter? It might be thought that retrograde lemniscatory action is only strongly manifested in S4, this stage being preceded by an S3 preliminary phase in which the "backspin" is merely incipient—just as the manifestation of S2 "forespin" is preceded by S1 nascency. This is consistent with the nativity metaphor wherein the mother cannot give birth a second time without becoming *pregnant* a second time. It would appear, then, that, for the Proprioceptive orientation to be fully realized in the final stage of lemniscatory development (LB S4), a period of gestation would be required in S3, as it was for Appropriation in S1. However, this does not take into account that, in the second nativity, the movement through the birth canal *proceeds in reverse*. The transition from S3 to S4 is not from embryonic action to well-oriented action, as in S1→S2, but the other way around. In the Proprioception carried out in S3, the decisively forward action of S2 that had obscured lemniscatory Being is counteracted by an equally strong retrograde action (indicated in table 5.1 by the solidly drawn rotation circle); with this, the lemniscatory mother gains cognizance of herself. Still, this *is* only a first Proprioception, one that does not include the concrete apprehension of her primal relation to the sub-lemniscatory midwife. It is in making the transition to S4 that the mother moves backward through the birth canal to the "time before her birth," thereby properly acknowledging the midwife. In S4, the embryonic clockwise movement of S1 is counteracted by an embryonic counterclockwise movement (denoted by the dashed rotation circle). With this Proprioceptive return to her "intrauterine origin," the lemniscatory mother has rebirthed herself in the "backward" fashion, and the limited individuations of S2 and S3 are surpassed by one that brings uroboric (w)holeness to full-fledged concrescence.

To grasp the significance of S4, we must realize that while the S2 reduction of the 2 × 2 matrix is reversed here, this is no simple reversion to S1. For, unlike what happens in S1, the lemniscatory mother is now moving *Proprioceptively.* This movement against the S1 grain enables her to witness concretely for the first time the S1 nascency from which the S2 projection originates. The matrix reduction of S2 signifies the fusion of enantiomorphs that entails the introjection of the sub-lemniscatory midwife into the lemniscate. But, again, this is a *quasi*-fusion. 1d/0d and 0d/1d enantiomorphs are in fact not *completely* integrated in stage two, which means that the midwife is not completely introjected. Instead of the lemniscate truly containing the sub-lemniscate, fully appropriating this selfless midwife for the mother's own development and thereby gaining bona fide autonomy, the lemniscatory mother actually remains dependent upon the midwife. Thus the midwife, operating "behind the scenes" in stage two, must continue her ministrations. Of course, the mother also operates covertly in S2, given that she has abandoned herself to the objectifying consciousness of her "fatherly child." In S3, the mother's concealment is counteracted by the Proprioception that challenges the idealized autonomy of the father; the mother thus gains a measure of self-awareness. Then, with the retrograde action of S4, the mother more fully acknowledges her incomplete closure by acknowledging in concrete fashion her continuing reliance on the grandmother. In so doing, the mother's hidden dependence on the midwife is brought to an end. But while the dependence is no longer obscured, does it not still exist? How, then, can we claim that the lemniscatory mother completes her development in this stage? The dialectical logic that governs dimensional development tells us that even though simple independence indeed is not achieved in the climactic stage, neither does simple dependence merely continue.

We can in fact say that the lemniscatory vessel becomes hermetically sealed in S4, with the sub-lemniscate being truly contained by it. But this happens by virtue of a dialectic whereby genuine closure entails *dis*closure, or opening. The true containment of the sub-lemniscate involves lifting the repression of it that had been imposed with the matrix reduction of S2. In S4, the LB contains the SLB by acknowledging the latter's containment of *it*. We may conclude, then, that the containment realized in S4 is *mutual:* it is precisely when the one-dimensional mother thankfully pays homage to the zero-dimensional midwife's selfless containment of it that the mother wholly contains the grandmother! So if the mother gains genuine autonomy in S4, it is not of the phallic kind; true, she is no longer covertly dependent on the grandmother, but neither is she merely independent. The relation is best described as one of *inter*dependence.

Nevertheless, if we portray the containment of these dimensional vessels as mutual, we must be mindful of the *asymmetry* that is involved: whereas the grandmother's act of containment *is* utterly selfless, the mother's containment of the grandmother is carried out in the interest of completing the mother's own development, her individuation.

Can we clarify further the sense in which grandmotherly containment is "selfless"?

It seems the term "selfless" must be taken quite literally. For we have related zero-dimensional selflessness to absolute nothingness, to Non-Being. Therefore, rather than saying that the sub-lemniscatory grandmother is "selfless" in that she "generously gives of herself," it appears we must say that, in fact, she *has no* self. Then, if the grandmother is not a "giving presence" but an *absence,* her containment of the mother evidently must be viewed in negative terms. Possessing no substantive dimensional body of her own with which to supportively enclose the mother's body, thus unable to contain the mother in any positive fashion, the midwife can only "give the mother encouragement." That is, the midwife is limited to mediating the *mother's* Appropriation of 1d/0d and 0d/1d enantiomorphs in the interest of motherly self-containment. Since the 1d/0d enantiomorph associated with the midwife cannot actually *belong* to the midwife but must belong to the mother, the midwife can only "hold it in trust for her," as it were. Thus, when we speak of "mutual containment" in describing the LB S4 relation between the zero-dimensional grandmother and the one-dimensional mother, it must be understood that the grandmother, for her part, engages in a *negative* kind of containment: she contains the mother only in the sense of mediating the mother's *self*-containment. The elusiveness of this notion cannot be helped since what we are attempting to talk about here is the enigma of Non-Being that confounds language. In such an effort, the use of metaphor is indispensable.

Does LB S4 close the circle of Topogeny? Well, it closes the lemniscatory winding of the dimensional *spiral,* which means that there is also an opening into a new winding. With the Proprioceptive completion of the round of 2 × 2 matrix generation, logarithmic expansion carries us into the inaugural phase of Appropriation of the 3 × 3 Moebius round. The compass has now shifted so that action is once again nascently clockwise. In stage one of the two-dimensional Moebius winding, the midwifely capacity of the zero-dimensional sub-lemniscate has expanded. Through the new enantiomorph, 2d/0d, the SLB offers supportive containment (in its negative way) for the pregnant 0d/2d Moebius mother. The mother responds to this encouragement in S2 with her first manifest act of self-Appropriation. The matrix reduction that takes place signifies a decisive step in the MB's self-birthing. A quasi-closure (and opening) is achieved; 2d/0d and 0d/2d enantiomorphs merge, and, in so doing, are "annihilated," with the SLB being introjected by the MB, so that selfless midwifely support becomes motherly self-support. As a result, the ontical posture is assumed in a first projection of object-in-space on behalf of emergent two-dimensional subjectivity. However, whereas the one-dimensional lemniscate required but a single projection to complete its quasi-closure, the more complex Moebius pattern of development calls for a second such projection.

In the Moebius winding of the dimensional spiral, the lemniscate has achieved the status of midwife and so can offer assistance for the further development of the MB. The LB does this through a feature that it does not possess in its own cycle of Topogeny. It now has an enantiomorph, 2d/1d, which is paired with and supportively mirrors the still pregnant MB's "fractional" enantiomorph, 1d/2d. Responding to the lemniscatory midwife's offer of encouragement, the Moebius mother engages in her second act of self-Appropriation, thereby entering S3, the conclusive substage in her forward-directed self-nativity. A second quasi-closure thus occurs; 2d/1d and 1d/2d enantiomorphs are "annihilated" and the LB is introjected by the MB, as indicated by the further reduction of the matrix. This is the second and final Moebius projection of object-in-space on behalf of the two-dimensional subject.

Next comes the call to Moebius Being. Entering stage four, the "gears" are put into reverse and the Moebius is reoriented. In this first Proprioception, the divisive ontical posture of S3 is challenged by an initial self-awareness of Moebial (w)holeness. Then, moving concretely backward in S5, two-dimensional Proprioception is carried further. The Moebius mother now acknowledges the selfless giving of the lemniscatory midwife; in the process, the asymmetrically mutual containment of MB and LB is fully effected. Going still further back in S6, the contribution of the sub-lemniscatory midwife is recognized by the Moebius mother so that *these* two spheres of dimensional action become harmoniously interrelated. The MB's "reverse nativity" is thus completed. The 3 × 3 matrix reduction has now been wholly counteracted and the Moebius mother has genuinely given birth to herself. Her individuation has fully been accomplished.

Following the Proprioceptive closure of the 3 × 3 Moebius circulation, the dimensional spiral dilates once again, the compass reverts to clockwise, and we find ourselves in the opening phase of Appropriation of the 4 × 4 epoch of Topogeny. Beyond her incipient stage of pregnancy (S1), the Kleinian mother develops through three phases of self-Appropriation. Three quasi-closures occur, three quasi-introjections of lower-dimensional bodies into the three-dimensional Kleinian body: Entering S2, the 3d/0d sub-lemniscatory midwife and 0d/3d Kleinian mother of S1 fuse "annihilatively," with introjection of the zero-dimensional SLB; in S3, the 3d/1d lemniscatory midwife and 1d/3d Kleinian mother of S2 occlusively merge, introjecting the one-dimensional LB; finally, in S4, the 3d/2d Moebial midwife and 2d/3d Kleinian mother of S3 fuse in this way, with the two-dimensional MB now being absorbed into the Kleinian body. Three projections of object-in-space on behalf of the three-dimensional subject thus take place in the comparatively complex quasi-individuation of Kleinian dimensionality.

Subsequently, the call to Being is heard once more. The reversal of gears that this occasions brings us to stage five of the Kleinian winding. It is precisely here that we may situate the work that was done in the final section of chapter 2, to wit, the Proprioceptive reading of this Kleinian text. The incomplete character of this work should now be clear. If Kleinian individuation is to be realized in full, it is not enough for the Kleinian mother to gain knowledge of herself in this still-too-abstract way. To hermetically seal the three-dimensional vessel, to complete the mother's "reverse self-nativity," *three additional acts of Proprioception are required,* three further, more concrete acknowledgments of motherly self-Appropriation and the midwifely Appropriation that attends it (these additional Proprioceptions will be spelled out in due course). Thus we would advance from S5 through S8. With all this achieved, the four spheres of dimensional action would contain one another harmoniously.[2]

In general, the foregoing analysis brings to light the substages of dimensional development that serve to distinguish one topodimensional body from another. While the several windings of the Topogenetic spiral do overlap, each circulation of Being runs its own course, with distinct circulations being marked by differences in the number of substages that each requires to carry out its Appropriations and Proprioceptions, along with differences in the sizes of the dimensional action matrices corresponding to the stages. These differences reflect, in turn, differing degrees of dialecticity, of ontological complexity, of the capacity for individuation. What we see from this account is the array of dimensional lifeworlds that underlie the three-dimensional lifeworld more familiar to us. But I must acknowledge once more the highly abstract and skeletal nature of my presentation thus far. At this point, I ask you to bear with me as I reaffirm my intention of "putting some flesh on these bones" before I am finished.

2. THE CYCLONIC NATURE OF ONTO-DIMENSIONAL PROCESS

A major implication of this work is that ontological dimensionality must be understood in *processual* terms, must be grasped as *action*. Let us now attempt to further specify the nature of this action.

As should be clear by now, dimensional action is decidedly different from the classical form of action that takes place within a dimensional framework that is itself presumed to be *in*active. The latter kind of action is continuous; it consists of point-to-point mechanical motions through the continuum. But discontinuity is not simply absent from the classical scene. Rather, it makes its "presence" felt in the *negation* of continuity. And this is just what we have in the prime formulation of the ontical stance, viz. object-in-space-before-subject. Continuity lies on the side of object-in-space and discontinuity on the side of subject. A central problem for ontical thinking is that, while object and subject are categorially divided from one another, somehow they must interact. This is of course the *mind-body problem* that has haunted traditional philosophy from its inception. We may readily identify the dualism here with the juxtaposition of contradictories that Merleau-Ponty found in Sartre, the "intuition of being and negintuition of nothingness" for which Merleau-Ponty sought to substitute "a *dialectic*" (1968, 89). Stated in terms of action, ontical thinking juxtaposes continuous movement with its negation, thereby creating an implicit contradiction.

Earlier, I intimated that the non-classical action of dimensional Being is "holistic" or "integral," in that it possesses an aspect of indivisibility that precludes it from being simply contained in the infinitely divisible continuum. Of course, this is not to say that the action of Being is *just* discontinuous, discontinuous *as opposed to* continuous. Were we to say that, in effect we would merely be reducing Being to the "nothingness" of ontical subjectivity. Instead, onto-dimensional action must be thought as the *dialectical blending* (not contradictory juxtaposition) of continuity and discontinuity. In fact, such thinking has already been facilitated by topological observations made above. We have learned that, while the ordinary cyclical action of a structure like the torus embodies the non-dialectical splitting of opposites (opposing circulations of the torus are simply orthogonal), our topological bodies of paradox provide the necessary dialectical twist for a thoroughly blended action—action that, at the ontological level, does entail object and subject, wholeness and "hole-ness," continuity and discontinuity. In what follows, the dimensional dialectic will be clarified further by demonstrating that, when we look more closely at the cyclical action in question, its specific form proves to be *cyclonic.* Whereas simple cyclical action involves movement that merely is positive, i.e., movement from one delineable position to another, the action of a cyclone—or, more generally, of a vortex—blends "positive" and "negative," since it is a circulation through a center that is empty.

In ancient Greek cosmology, the vortex was employed as a fundamental model of natural process. Philosophers such as Anaximander, Anaximenes, Empedocles, Anaxagoras, and Democritus used the principle of vortical motion to explain the creation of the world. According to Anaximander, it is by the action of a vortex that all finite boundaries arise from the boundless archaic flux of the *apeiron.* In his interpretive translation of Anaximander, Heidegger (1946/1984) goes so far as to associate the *apeiron* with Being itself (see also my volume on *apeiron;* Rosen 2004). It is true that vortices are sometimes said to possess the topological structure of a *torus,* since the torus can plausibly be described as a ringlike circulation about an empty center. But the fact is that the torus does not provide us with the best model of natural cyclonic action. The "positive" and "negative" regions of the torus—its peripheral surface area or volume and its central hole—are strictly set apart from one another, for the hole in the torus is located *outside* the body of this simply closed structure. Generally speaking, this is not the way vortices actually manifest themselves in nature. In reality, the positive and negative portions of a whirlwind or whirlpool are blended dialectically in the fashion of the Klein bottle, whose hole is *internalized,* arising as it does from the uroboric act of self-containment.

In a book entitled *Sensitive Chaos,* Theodor Schwenk (1965) offers a detailed study of the manifold forms a vortex can take, and makes the claim that the manifestation of the vortex "is a fundamental process—an archetypal form-gesture in all organic creation, human and animal, where, in the wrinkling, folding, invaginating processes of gastrulation, organs for the development of consciousness are prepared. Forms arising out of this archetypal creative movement can be found everywhere in nature" (41).

Schwenk's primary example of vortex generation is the forming and cresting of a water wave:

> If we could watch the process in slow motion we would see how a wave first rises above the general level of the water, how then the crest rushes on ahead of the surge, folds over and begins to curl under . . . forming hollow spaces in which air is imprisoned in the water. . . . This presents us with a new formative principle: the wave folding over and finally curling under to form a circling vortex. (1965, 37)

Figure 5.1 is my adaptation of Schwenk's illustration of vortex generation. At the outset, we have a pool of water that is entirely flat and calm (fig. 5.1(a)); there is no circulation of liquid in this initial state of affairs, and no airy, curved, hollow space from which rotating water is excluded. Let us say that initially, the media of water and air are simply and completely partitioned from each other; they are juxtaposed without being opposed, so that, in effect, they are not dialectically engaged.

In the next stage of vortex generation, the water has begun to churn, resulting in the production of a wave train (fig. 5.1(b)). We now clearly do have the beginnings of a dialectical interchange. The two media have begun to enter into each other, the circulating water swelling up into the air, the air pressing down into the water to create a hollow trough in it; in the process of thus overlapping one another, the dialectical opposition of "positive" (i.e., aqueous) and "negative" (gaseous) domains is brought into play.

Figure 5.1(c) shows the waves now starting to crest. In so doing, they begin to fold over upon themselves. With their in-curving crests becoming elongated, what had been semicircular waveforms (fig. 5.1(b)) are now more fully circular. Commensurate with this increase in the positive

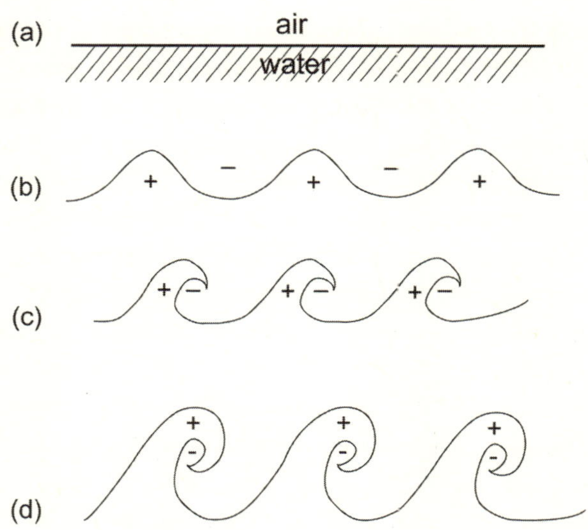

Figure 5.1. Stages in the formation of a vortex (adapted from Schwenk 1965, 37)

feature of the developing aqueous vortex, there is an increase in the negative: the semi-hollows of figure 5.1(b) have begun to deepen, to form themselves into full-fledged holes. In quantitative terms, the passage from stage b to c brings an increase in the angular velocity of the water, accompanied by a proportional decrease in its angular inertia and spin radius. The conservation of angular momentum that thus attends the folding in of the cresting wave is reminiscent of a skater tucking in her arms as she spins: her velocity increases in proportion to the decrease in the radius of her rotation.

Figure 5.1(d) depicts the completion of vortex generation. Here both the swirling aqueous "plenum" and the gaseous "void" have been brought to full expression. And this full-fledged differentiation of "wholeness" and "hole-ness" is at once an *integration*. For the water does not merely spin *around* its empty, airy center in the fashion of a torus; it spins uroborically *into* it, is "sucked" into the vacuous hub in a whirling, "screw-like spiral" (Schwenk 1965, 45). Thus, even though water and air are well differentiated in the mature vortex, there is no simple line of demarcation that partitions these media into totally separate regimes. Rather, in the self-intersecting, Klein bottle-like flow of the vortex, the "different media . . . flow together . . . the one [being] taken into the other so that the hollow space, like a vessel, is filled with [the medium] of a different quality" (40). Schwenk states, in general, that whenever different surfaces thus come into contact, "like the surface between water and air, these surfaces will become waved, overlap and finally curl round" (37). A little later, he observes:

> This important phenomenon—the curling in of folds or layers to create a separate organ with a *life of its own* within the whole organism of the water—does actually occur in the forming of organic structures. . . . Like vortices, organs have their own life: they are distinct forms within the organism as a whole *and yet in constant flowing interplay with it.* (41; emphasis added)

Continuing in the same vein, Schwenk asserts that a vortex is "a form which has separated itself off from the general flow of the water; a self-contained region in the mass of the water, enclosed within itself and yet bound up with the whole" (44). And again, a vortex is "a separate entity within a streaming whole, just as an organ in an organism is an individual entity, yet closely integrated with the whole through the flow of vital fluids. An organ is orientated in relation to the whole organism and also to the surrounding cosmos; yet it has its own rhythms and forms inner surfaces of its own" (47).

In the foregoing statements, Schwenk clearly is emphasizing that vortex generation—seen as an archetypal process that bespeaks the generation of all organic forms—entails both differentiation and integration. The vortex is differentiated and integrated within itself, and differentiated from/integrated with its surrounding environment. It is also clear that such differentiation and integration do not arise separately in development, one following the other in linear sequence, but are inextricably interwoven aspects of the same underlying process.

Our chief interest, of course, is with *dimensional* vortex generation, with *ontological* cyclogenesis. Therefore, were we to limit ourselves to dealing with the vortex as an *object in space,* we would not be addressing what primarily concerns us—regardless of whether we were describing vortex generation in the strictly quantitative terms of inertia, velocity, and momentum, or speaking of it more qualitatively as involving the relationship between the presence (+) and absence (–) of a circulating form (water, a living organ, etc.). The ontical production of vortices within a static spatial framework is a far cry from the vortical action of dimensional Being itself. Again, onto-dimensional action is "*pre*spatial" (Heidegger 1962/1972, 14; emphasis added); it is precisely the action that first opens up the framework within which objects appear for the scrutiny of the allegedly detached subject. (The ontological opening up of the framework is inconceivable from the ontical standpoint that *presupposes* the framework.) This means that the action of Being is neither itself objectifiable, nor is it an act of objectification carried out by a reflecting subject. Rather, it is the *prereflective differentiating/integrating* of object and subject—or, of the field of objectification (the space-time continuum or plenum) and the subjectivity that, in effect, constitutes a *hole* in the field. In accordance with our analogy, just as physical vortex generation entails a progressive increase in the differentiation/integration of the circulation of water and the airy hollow at the center of the circle, *dimensional* vortex generation involves an increasing differentiation/integration of object and subject that establishes prereflectively the possibility for reflective operations. To confirm that differentiation and integration are indeed wholly synchronous in this vortical process, let us consider what turns out to be a precise topological model of vortex generation: *expansion on a Moebius surface.*

A fundamental distinction between the Moebius structure and the cylindrical ring is that the former involves *torsion,* whereas the latter does not. As we know, a Moebius surface is constructed by endowing a narrow strip of paper with a 180° twist. Moreover, *expansion* upon the Moebius strip provides us with a model of the *generation* of torsion. For there is no apparent torsion at a *local cross-section* of the Moebius; here the Moebius is indistinguishable from its untwisted cylindrical counterpart. It is only when we expand our view of the Moebius and begin to consider its longitudinal extension that the twist becomes evident. Why does Moebius expansion furnish a model of *vortex* generation? It is simply because the creation of the vortex itself essentially entails the production of torsion. Figure 5.1 depicts several stages in this process, beginning with an initial absence of torque, when the media of water and air are not yet dialectically engaged (fig. 5.1(a)), and culminating with the realization of a full-fledged "twister," as vortices are called when they take the form of tornadoes. Thus, if we regard the respective local sides of the Moebius as symbolic of water and air, Moebius expansion to one-sidedness discloses the topological transformation that underlies the forming and cresting of a wave. In this dialectical process, sides undergo both differentiation and integration, these two functions being inseparably blended rather than linearly combined. We could in fact use the language for vortex generation provided by Schwenk to describe the results of expansion upon the Moebius: opposing surfaces are differentiated even as they "flow together" (1965, 40), as they "overlap and . . . curl round" (37).

However, analogies do have their limitations. In ontical experience, although water and air are not *dialectically* differentiated before the vortex forms, though there is neither churning water nor an airy void at its center, these media *are* differentiated: distinct elements are inertly juxtaposed (fig. 5.1(a)). Must the same not be said for the initial stage of *topological* "vortex generation"? At the local cross-section of the Moebius strip given prior to the expansion that brings one-sidedness, are not respective sides already fully differentiated from one another? And is this not clearly inconsistent with the state of affairs said to prevail at the outset of *ontological* development, where there is neither difference nor identity?

By now we are aware that the Ontogeny of the Moebius body can only take place in the *two*-dimensional context. In three-dimensional space, the Moebius strip is no more than an ontical structure, a well-differentiated object, an entity possessing sides that are simply distinguishable. In three dimensions, the merger of sides that occurs with Moebius expansion can rightly be taken as but a *global* effect, with the local differentiation of sides being strictly maintained from beginning to end. It is when we shift down into the two-dimensional "Flatland" environment and consider the interplay of Moebius *edges* (rather than sides) that the Moebius loses its onticality, its simple continuity; it is here that the Moebius possesses that special ontological hole, as does the Klein bottle in three-dimensional space. Therefore (continuing with the Moebius case), in order to express effectively the initial stage of ontological cyclogenesis (Cyclogenesis) wherein there is neither difference nor identity, we must perform a dimensional reduction.

As a first step, recall figure 3.6, the comparison of the Moebius strip and the cylindrical ring in terms of their edges. To repeat what I said in chapter 3, this figure illustrates the fact that we can view a cylindrical ring in such a way that we see only a single edge. This amounts to a perceptual reduction of the ring from a two-dimensional surface to a one-dimensional circle. It is clear from inspection of the Moebius that no such reduction is possible. The one-dimensional, strictly edgewise view obtainable over the full length of the cylindrical ring can only be realized in the Moebius case at a *cross-section* of the strip (which is consistent, of course, with the fact that the Moebius strip assumes the more limited character of the cylindrical ring at its local cross-section). I further noted in chapter 3 that, when viewing the Moebius in the edgewise fashion, we do not actually see the cross-sectional line itself but only the endpoint of this line that is nearest to us, illustrated in figure 3.6(b) by point P. However we position the Moebius, however we rotate it in three-dimensional space, at any given moment no more than a single point will be visible to us on the Moebius's edge at which extension in two dimensions will have vanished. It is at this singular point that the perspectival opposition of sides is, in effect, perceptually neutralized: with no outer surface visible here, neither can we speak of an inner surface on the opposing side. Employing this perceptual model for the present purpose of expressing Cyclogenesis, let us regard the singular point where sides of the Moebius cannot be differentiated as the point of departure for Moebius expansion. It is here that the topological vortex is but incipient, that opposing sides are not yet dialectically engaged—not because sides are juxtaposed, but because there *are no* sides.

As long as we maintain the edgewise view of the cylindrical ring given in figure 3.6(a), it is obvious that no degree of longitudinal expansion from any point on this edge will disclose the extension of the ring in two dimensions. In the contrasting Moebius case illustrated by figure 3.6(b), edgewise visual expansion from point P depicts the generation of higher-dimensional dialectical opposition from the undifferentiated lower-dimensional initiation point. And, as in vortex generation, the opposition of sides is at once their *integration*. In expanding from P, we do not just see a progressive differentiation of sides, one side becoming increasingly visible and the other being invisible. For, with continued expansion, the sides are twisted *inside out:* what was visible becomes invisible, and vice versa. The differentiation of sides is thus reversible; there is an interplay between them; an overlapping; a curling of them over upon each other; a mutual "insertion and intertwining" (Merleau-Ponty 1968, 138) that bespeaks the Moebius structure's aspect of *one*-sidedness. Hence, as Moebius expansion advances, we are not limited to the simple, nonreversible opposition of sides; rather, a vortical twisting over of sides upon each other occurs in which inside and outside exchange.[3]

Of course, the cyclogenetic process we observe through our perceptual model of Moebius expansion remains ontical, not ontological. We have been able to see that expansion from a singular point on the Moebius strip's edge brings into play a "higher dimension" of dialectical interaction, namely, that of the sides of the two-dimensional surface. But, in this appearance of planar extension from the unextended point, the Moebius remains an object-in-three-dimensional-space; the three-dimensional context is not itself affected by the event, nor does the observing subject lose its detached anonymity and enter into the process. In other words, the reflective exercise we do with the Moebius in order to symbolize dimensional vortex generation does not actually engage the *prereflective ontological dimension.* For the Cyclogenesis of the Moebius to be realized, a more complete dimensional reduction is required. It is not enough to switch from viewing the Moebius sidewise in three-dimensional space to viewing it edgewise in this context. The dimensional context itself must be reduced, if we are to bring Moebius Being to light (as we saw in chapter 3).

It is in the *two*-dimensional environment that Moebius Cyclogenesis would truly occur. We would have no Moebius *surface* in this "Flatland" milieu but a Moebius *line* (chapter 3), and the dialectic that would be generated would involve the perspectival *edges* of this line (not the sides of a surface). The generation of the Moebius vortex in the two-dimensional lifeworld clearly could not be objectified, since there would be a singularity or hole at the center of said topological vortex that would be no mere rupture in an object-in-space, or even in the spatial container per se. Rather, it would be the opening created by the uroboric *self*-containment of the two-dimensional Moebius, with this self-signification in fact serving to close the gap between object-in-space and subject. That is to say, the hole in the midst of the Moebius "object" would mark the concrete presence of the subject and at once would bring a *wholeness* of object and subject. To be sure, such (w)holeness would be *ontological,* unlike the interplay of water and air in the forming and cresting of an ocean wave described by Schwenk.

However, granting that the mature Moebius vortex would entail ontological (w)holeness, what of the initial stage of Cyclogenesis in this lower-dimensional lifeworld? The rule of neither/nor would apply here: neither whole nor hole, plenum nor void, object nor subject, positive nor negative. This negation of both the negative and the positive is what Tanabe named "absolute nothingness"; in cyclogenetic terms, we may call it absolute *holeness.* Its counterpart in our ontical model of Moebius expansion is the initiation point P where neither side of the surface is visible (fig. 3.6(b)). Viewing Cyclogenesis in this manner, we see that it certainly does not begin with well-delineated elements that are calmly juxtaposed (as are the air and water of fig. 5.1(a)), but, rather, with sheer negativity; at the outset of vortex generation there is only the calmness of the *hole.* Then, as the development of the two-dimensional wave unfolds, the dialectic is engaged, bringing the opposition and interaction of hole and whole. No longer do we have *absolute* holeness, for the hole in the vortex now exists only in relation to the whole. Ontologically, the relationship in question is that between subject and object (respectively). It is the prereflective differentiating and integrating of these terms that first opens the framework for the detached subject's ontical reflections upon the objects in space. When this happens, the reflective posture becomes dominant, the prereflective being overshadowed, relegated to the dim background of awareness, rendered preconscious. It is here that the categorially divisive framework of object-in-space-before-subject is manifested as always already given. This intermediary phase of Moebius Cyclogenesis corresponds to the first of the two main Moebius circulations: it is the *projective* circuit, the forward-oriented, "clockwise" cycle of Moebius action in which there is quasi-closure and opening. Projected in this phase is *idealized* wholeness and holeness; perfected objectivity and transcendent subjectivity. The uroboric Moebial interplay of whole and hole is thus obscured in favor of "cylindrical appearances."

The ontical counterpart of the forward-directed Moebius circuit is the stage of expansion upon the Moebius strip in which opposing sides have been brought out, and brought together, the passage from one side to the other having been completed. We know from the previous chapter that, in this phase, both sides have not been traversed in their *entirety;* in fact, only *half* of the one-sided surface has been covered. It is the *full* encompassment of enantiomorphically related sides that is required to complete the Moebius vortex in earnest, and, for this to happen, there must be the additional circulation, the backward-oriented, "counterclockwise" Moebius cycle that will bring *genuine* closure and opening.

Now, the rotary action depicted in the ontical vortex of figure 5.1 is moving only in a clockwise direction. This is surely not to say that all ontical vortices are so oriented, that objectively observable vortices cannot be twisted counterclockwise, or that there are no such vortices capable of turning in *both* directions. Indeed, in his richly illustrated book, Schwenk offers examples of vortices with all manners of twist. For our purposes, the point is that, whatever the direction of twist a vortex may possess, if the action is objectified, if it is ontical, then the direction of *ontological* twist is exclusively forward. This means that the ontological is deferring to the ontical, that Being is concealing herself in the interest of self-Appropriation.

Nevertheless, in many of Schwenk's "archetypal" formations, the ontological is symbolically adumbrated—and by spelling out the underlying topological structure of the vortex, the intimation becomes clearer. The double cyclicity of the ontical Moebius vortex echoes the process by which wild Being first conceals herself, then reveals herself. The forward circulation of the Moebius vortex can be taken as signifying the "premature closure" by which Being obscures her own nature: having gone but one time around the Moebius and moved to the other side in so doing, the contrary impression is conveyed of a circulation on the *same* side in which the circuit has simply and fully been closed. In this way, Moebius action would mistake itself for mere cylindrical action, the latter being the idealization of the former. Then, entering the second cycle of Moebius action, the reversal of direction can be taken to indicate that the mistake is now recognized.

But let me stress again that while ontical Moebius action can model or symbolize the ontological in an effective fashion, it cannot concretely embody the ontological. After all, the turnings to and fro upon a Moebius strip are continuous events in three-dimensional space that appear before a subject who is detached, exempted from these happenings, whereas the gearing entailed in the *ontological* would need to *include* the subject. Only when the subject is included can we speak meaningfully of projection and its withdrawal, of mistaken and accurate interpretations of Moebius action. Therefore, while it is true that the gearing of the action on the ontical Moebius structure changes in going from one cycle to another, because these directions of spin are merely objectified, they have no bearing on the *ontological* gearing, no implication for whether an action cycle is or is not oversimplified, idealized, or misconstrued.

We have seen that, for Moebius Cyclogenesis to be realized ontologically, dimensional reduction to the "Flatland" milieu is required. In this two-dimensional lifeworld, the subject is surely included, appearing as the unavoidable *hole* in the Moebius line. In the cyclonic interplay of whole and hole occurring in the Moebius environment, there would indeed be a forward, projective stage of integral action in which the vortex would appear "cylindrical," its central hole and its peripheral volume—subject and object-in-space—seeming to be fully formed and strictly set apart from one another. This stage of quasi-individuation would be followed by a gearshift to the backward cycle of ontological action. The transition would bring the Proprioceptive self-recognition required for the Moebius wave to reach its genuine climax. With the dimensional vortex flowing in reverse, the ontical projection of object-in-space-before-subject would be retracted and Moebius Being would emerge from eclipse.

Needless to say, the formation of the two-dimensional Moebius vortex would not be the only order of Cyclogenesis. The order of Topogeny that is actually of most immediate concern to us three-dimensional subjects is that of the *Kleinian* lifeworld (to the three-dimensional ontical observer, the Klein bottle is the only member of the bisection series with a necessary hole). A related Topogenetic process would occur in the one-dimensional milieu of the lemniscate. The three rounds of onto-dimensional vortex generation correspond precisely to the three windings of the dimensional spiral studied in the previous section. Table 5.1 specifies the *substages* of Topogeny for each winding. From the table, we can see how the evolving dimensional actions are interwoven in such a way that the substage structure of each winding is determined by the interaction of that winding with the other windings. The same substage patterning would have to obtain in the *vortical* forms of Topogenetic action. To see specifically how this works, we begin by considering a further limitation in the ontical analogy.

The formation of a whirlpool in space is pictured as a continuous process. Accordingly, the number of stages shown in figure 5.1 is arbitrary. Any number of intermediary stages could have been displayed, since what is being depicted is the development of a vortex in the *continuum*. Similarly, when the Moebius is approached ontically, as an object in space, the torsion-generating expansion is seen as a continuous operation. But we have learned that *onto-dimensional* action is not like that. Instead it possesses an aspect of discreteness or integrality that precludes its division into arbitrary numbers of separate substages. It might help to think of the passages between stages of Cyclogenesis as *phase transitions*. The transition from ice to water, for example, occurs *dis*continuously, as a sudden nonlinear shifting from one qualitative regime to another. Of course, in the case of *onto-dimensional* phasing, it is not only the transitions *between* phases that entail discontinuity, but also the phase action itself. Thus, the opening phase of any order of Cyclogenesis—far from being an event that could be located in a space-time continuum and subjected to analysis there—is a pre-spatiotemporal mode of utterly indivisible action in which the opposition of object and subject is but incipient. Then, as dimensional development ensues, rather than having the onset of a continuous twist-producing expansion, there is a "quantum leap" to a new mode of integral action, one that remains indivisible in itself, yet now projects the cyclonic division of whole and hole, of continuity and discontinuity.

Moebius Cyclogenesis correlates with the third winding of the dimensional spiral. Table 5.1(b) makes it clear that, beyond the initial phase in which the dimensional vortex is but nascent, there are two quantized (indivisible) phase actions directed clockwise or forward.[4] The first of these results from the interaction of the two-dimensional Moebius vortex with the zero-dimensional sub-lemniscatory wave. The Moebius "mother-wave" builds—i.e., a step is taken toward Moebius differentiation/integration of whole and hole, of object and subject—with the midwifely assistance of the zero-dimensional sub-lemniscate. The partially formed Moebius wave is generated via the quasi-fusion of midwifely SLB and motherly MB enantiomorphic phase actions. Like wave pulses that are "out of phase," the 2d/0d and 0d/2d enantiomorphs of S1 merge "destructively" in S2, canceling each other out, with the 1d/2d mother wave and its reciprocal 2d/1d midwife rising from the "annihilation." Then, in the final clockwise phase action of Moebius Cyclogenesis, the Moebius vortex is brought to quasi-completion by its fusion with the one-dimensional lemniscatory midwife, a merger wherein 1d/2d and 2d/1d enantiomorphs are "annihilated" to yield 2d/2d. Next there is a phase transition to the counterclockwise sub-waves of vortex generation (S4–S6), those in which the "destructive" fusions of the forward sub-waves are Proprioceptively *re*constructed in the process of completing Cyclogenesis in earnest. In much the same way, the phase actions that constitute the Kleinian and lemniscatory windings of the dimensional spiral can be expressed in cyclonic terms. In so doing, the dialectical nature of onto-dimensional action is clarified, for—as Anaximander hinted at the dawn of Western thought—the dynamics of wild Being (of the *apeiron*) entail a blending of whole and hole that is essentially cyclonic.

Does the relationship between vortices, on the one hand, and midwives, mothers, and nativity, on the other, still seem elusive? If so, let me underscore the connection between cyclogenesis and embryogenesis by reiterating Schwenk's claim that the manifestation of the vortex "is a fundamental process—an archetypal form-gesture in all organic creation, human and animal, where, in the wrinkling, folding, invaginating processes of gastrulation, organs for the development of consciousness are prepared. Forms arising out of this archetypal creative movement can be found everywhere in nature" (1965, 41). The claim that *I* would make is that cyclogenesis does not just govern the ontical realm of formative action but the ontological as well. The several orders of Being birth themselves via the Topogenetic twistings attendant to the formation of dimensional vortices.

In concluding this chapter, I offer a pair of images from Schwenk that will further elucidate the parallel between vortex generation and dimensional generation, and will shed additional light on the enantiomorphic feature of vortex formation.

In the course of his presentation, Schwenk depicts the generation of whole trains of vortices. Figure 5.2(a) shows a progressive strengthening of vortical form in advancing from one vortex in the train to another. In like manner, *onto-dimensional* vortices are better defined—i.e., there is greater dialecticity, greater differentiation and integration of subject and object, a higher level of individuation—as we expand to higher-dimensional circulations of the dimensional spiral. The growth of Ontogenetic potential is indicated in table 5.1 by the increasing number of developmental substages in the outer windings of the spiral.

Figure 5.2(a) can also be said to reveal how the clockwise action of a particular vortex in the train gives way to a counterclockwise action, and

a b

Figure 5.2. Train of vortices (a) and rhythmic array of vortices (b) (adapted from Schwenk 1965, 38 and 51)

how, from these two actions, there arises the next vortex in the train (note, however, that Schwenk apparently regards each action as a vortex unto itself). This is basically what happens in *dimensional* vortex generation. In the second winding, the formation of the one-dimensional lemniscatory vortex begins with the clockwise action cycle (S1–S2) and climaxes in counterclockwise action—or, to be more exact, in action that is *retrograde,* that "goes against the grain," and is thus both counterclockwise and clockwise (S3–S4). Then, with expansion of the dimensional spiral to the Moebius winding of Cyclogenesis, clockwise action resumes (S1–S3), shifts to counterclockwise action (S4–S6), and so on.

Now, clock sense entails *enantiomorphy:* clockwise and counterclockwise circulations are mirror images of each other. Figure 5.2 therefore provides a vortical model of enantiomorphic relations between dimensional cycles. Let us apply this model to the enantiomorphic interactions mirrored within those cycles.

In commenting on figure 5.2(b), Schwenk spells out the enantiomorphic aspect of vortex generation. Speaking of the "rhythmical arrangement of vortices" in a vortex train, he notes that "the vortices alternate in corresponding pairs, one slightly ahead spinning one way and the other, behind, spinning the other way" (1965, 51). We can view the intra-cyclic dimensional enantiomorphs of table 5.1 in the same fashion, as paired vortices spinning in opposite directions. And since Schwenk's allusion to the "rhythmical" character of vortex formation intimates the *musicality* of this creative process, perhaps, if we think of wave generation in *acoustical* terms, we could identify the member of the enantiomorphic pair that spins "slightly ahead" and clockwise as the midwifely "overtone" of the lower-dimensional wave pulse, and the member that spins "slightly behind" and counterclockwise as the motherly "undertone" of the higher-dimensional wave pulse. Our dimensional action matrices could then be taken as developing arrays of musical intervals. The *fundamental* tones of the arrays would correspond to the eigenvalues appearing on the matrices' principal diagonals, with series of overtones and undertones extending below and to the right of these fundamentals, respectively (the pattern is easier to discern in table 5.1(a) than 5.1(b)). In this reckoning, the "fundamental tones" would signify the relations of orders of dimensional Being to themselves, and the "overtones" and "undertones" would give the relationships *between* the different orders of dimensional Being, taken two at a time. Kleinian Being, for example, would correspond to the 3d/3d fundamental interval, and would relate to the 2d/2d Moebius fundamental via the enantiomorphic coupling of the 3d/2d Moebius "overtone" and the 2d/3d Kleinian "undertone." Generally speaking, in any pair of dimensional enantiomorphs, the fractional member constitutes an undervaluation of the higher dimension, one that lags behind its integral expression, and can be viewed accordingly as a "dimensional undertone" of it. The reciprocal member of the pair constitutes an overvaluation of the lower dimension, one pitched above its integral expression, and therefore a "dimensional overtone." By taking into account dimensional overtones, undertones, and the fundamental tones they link, all topodimensional relations can thus be rendered musically!

Harmonic theory indeed does allow that, to an overtone series, a mirror-inverted undertone counterpart may be added.[5] Interestingly, what was likely the first theory of musical enantiomorphy was that given by Pythagoras, centered on the idea of the "music of the spheres." The basic design of the so-called Pythagorean table is displayed in table 5.2.

The Pythagorean table is usually portrayed as an indefinitely expanding series of musical intervals. What is shown in the limited section of the table that I have selected for illustrative purposes is a set of relationships that essentially corresponds to our dimensional action matrix: there is a principal diagonal which contains a series of fundamental tones or

Table 5.2. Section of the Pythagorean table (adapted from Haase 1989, 94)

1/1	1/2	1/3	1/4
2/1	2/2	2/3	2/4
3/1	3/2	3/3	3/4
4/1	4/2	4/3	4/4

self-values, and these four principal intervals are coupled to each other two at a time by six enantiomorphically related pairs of overtone-undertone intervals. Perhaps, then, we could go so far as to speculate that the values provided in tables 3.2 and 5.1 are the onto-dimensional counterparts of the Pythagorean musical intervals. These topological relationships would then give us the "music of the dimensional spheres," and would show us how the "dimensional symphony" dynamically unfolds!

CHAPTER SIX

DISTILLING THE LOWER DIMENSIONS

1. INTRODUCTION

In chapter 2, three-dimensional flesh was realized concretely via Proprioception of the text. The ontological content of the writing was now signified in such a way that the text came to signify or contain *itself*. This reversal of gears counteracted the forward action by which Being is signified through linguistic abstractions that contain it as an object cast before a reflective subject. Being now drew back in upon itself. In this radical recursion, it reentered the prereflective bodily roots from which its reflection first originates. It was established in that earlier chapter that, for effective reentry, the self-signification of Being must be mediated by a topodimensional amplification in which the text no longer consists merely of intrinsically meaningless one-dimensional signifiers (strings of arbitrarily devised, conventionally agreed-upon graphic marks) that can only point outside themselves to disembodied meaning; the *three-dimensional Klein bottle* must do the signifying.

Located on the spiral of dimensional development, the work of chapter 2 corresponds to stage five of the Kleinian winding. In this phase of the ontological distillation, wild Being surely has been concretized, but in only an initial manner. What is now required is that it be embodied more fully in all its dimensions. The process will be carried out in two general steps. To begin, the abstract treatment of the lower dimensions given in previous chapters will be fleshed out over the next two chapters by giving the dimensions more palpable content. However, the concretization we seek must involve more than the *content* of an analysis that otherwise maintains the tripartite division among content, container, and uncontained (object, space, and subject). This limitation is surpassed in the second step of the embodiment process. Further acts of *self*-containment are called for here, additional Proprioceptions in which the lower-dimensional orders of Being signify themselves. The final chapter of the book addresses this task.

The work of Merleau-Ponty provides a clear-cut clue to the concretization of the lower dimensions. In his distinction between "first" and "second flesh" (see chapter 2), he associates the former with the "sensible world" (1968, 153), with the realm of the visible. When we enter the more abstract cognitive milieu, "It is as though the visibility that animates the sensible world were to emigrate, not outside of every body, but into another less heavy, more transparent body, as though it were to change flesh, abandoning the flesh of the [sensible] body for that of language" (153). Our prime concern in chapter 2 was with "second flesh," with the rarefied dimension of the *cogito* that governs this abstract text. What Merleau-Ponty's work suggests is that one of the three *lower*-dimensional realms we dealt with in chapters 3 through 5 corresponds to "first flesh," to the realm of the senses. Of course, to account for all the windings of the dimensional spiral, we require two other orders of the flesh. In this regard it is interesting that, when Merleau-Ponty sought to define the flesh, he noted that "we should need the old term 'element,' in the sense it was used to speak of water, air, earth, and fire. . . . The flesh is in this sense an 'element' of Being" (1968, 139). To be sure, in *The Visible and the Invisible,* Merleau-Ponty explicitly elaborated only *two* "elements of Being," the sensual and the linguistic or cognitive. Nevertheless, his distinctly alchemical allusion to the four elements of old resonates with our own requirement of *quaternary* flesh. If two of these lifeworlds are dialectically realized as dimensions of sub-objective or psycho-physical functioning (the sensing and thinking dimensions), we can expect that the other two may be concretized in a similar fashion. This is what I intend to demonstrate in the pages that follow. More generally, we are going to see that the four windings of the dimensional spiral gain palpable expression as distinctive patterns of sub-objective action whose different modes of functioning correlate with different degrees of consciousness, different regions of the body, and, ultimately, with different orders of the natural world. Let us begin by elaborating upon the cognitive order of the flesh.

2. THE FOURTH ORDER OF DIMENSIONAL FLESH: HUMAN COGNITION

In *The Ever-Present Origin* (1985), the cultural historian and philosopher Jean Gebser offers a detailed account of the dimensionality entailed in the development of four basic structures of human consciousness. In general,

> every mutation of consciousness that constituted a new structure of consciousness was accompanied by the appearance and effectuality of a new dimension. This clearly underscores the interdependence of consciousness and a space-time world; for each unfolding of consciousness there is a corresponding unfolding of dimensions. An increase in the one corresponds to an increase of the other; the emergence of consciousness and the dimensioning imply and govern each other. (117)

The first and most primordial form of consciousness is associated with the "zero-dimensional archaic" structure (43–45). Gebser relates archaic consciousness to a state of "deep sleep" (121). He describes this original condition as involving "the non-differentiation, indeed the non-distinguishibility [*sic*] of archaic man from world and universe—a non-awakeness by virtue of which he is still unquestionably part of the whole . . . an unconcerned accord, a consequent full identity between inner and outer . . . the perfect identity of man and universe" (44–45). We may wonder, however, whether "wholeness" and "perfect identity" are appropriate terms for this initial state of affairs. If the utterly undifferentiated situation prevailing at the outset of development is understood in the sense of zero-dimensionality established in chapter 3, the logic of *neither/nor* should obtain: the archaic individual should neither stand apart nor be simply immersed in a seamless whole—for there would be no such whole, any more than there would be separate parts. By the same token, rather than a state of "perfect identity" holding sway at the outset, there would be neither identity nor difference. Later, when we consider zero-dimensionality in greater detail, we will return to the enigma of "absolute nothingness" (Tanabe 1986). Note that Gebser's tendency to view human development as the differentiation of parts from a preexistent and ever-present "wholeness" implicitly preserves the *ontical separation* of part and whole by maintaining the former as evolving and the latter as not. From the standpoint of *Ontogeny,* we do not have a progressive separation of consciousness from a primal condition of wholeness that is itself exempted from developmental change. Rather than entailing the differentiation of the part from a preestablished whole, the self-birthing of wild Being involves the differentiation of both part and whole from that which is neither.

The next structure of consciousness identified by Gebser is the magic one. Of "magic man," Gebser says:

> He is distinguishable above all by his transition from a zero-dimensional structure of *identity* to one-dimensional *unity.* . . . The man of the magic structure has been released from his harmony or identity with the whole. With that a first process of consciousness began; it was still completely sleep-like: for the first time, not only was man *in* the world, but he began to face the world in its sleep-like outlines. . . . The more man released himself from the whole, becoming "conscious" of himself, the more he began to be an individual, a unity not yet able to recognize the world as a whole, but only the details . . . which reach his still sleep-like consciousness and in turn stand for the whole. Hence the magic world is also a world of *pars pro toto,* in which the part can and does stand for the whole. . . . In a sense one may say that in this structure consciousness was not *in man himself,* but still resting *in the world.* The gradual transfer of this consciousness . . . is something that man must master.
>
> Man replies to the forces streaming toward him with his own corresponding forces: he stands up to Nature. He tries to exorcise her, to guide her; he strives to be independent of her; then he begins to be conscious of his own *will.* (1985, 46)

"Man's" phallic quest to "stand up" to Mother Nature has thus begun and the human being is no longer simply "identified" with nature. Yet he does remain *united* with her, which means that the differentiation of person and world, subject and object, self and other, has not yet gone beyond the simplest mirroring. What is implied in the "pars pro toto" of magical life is that every part, every other, is a largely undifferentiated reflection of the whole or self. But parts and whole do appear now, indicating that a first step forward in fact has taken place, a first *objectification,* a first externalization of the relationship between self and other, however weak. The boundlessness of the archaic origin is obscured, giving way to the unitary self or subject, a subject that indeed appears bounded, closed within itself and closed off from what is other, though the boundary is exceedingly porous.

Gebser's third structure of consciousness is the two-dimensional *mythic* one. With its emergence, the boundary between subject and object becomes firmer, more sharply defined. If magical consciousness "was not in man himself, but still resting in the world" (1985, 46), in the mythic structure the human being is indeed now more clearly aware of an inner self, expressed as *psyche* or *soul.* Such self-awareness is a further step removed from the boundless archaic situation. Gebser sees the emergence of dreamlike mythic consciousness from the dreamless sleep of the magical structure as "also an emergence of the ego" (69). I would add that this growth of inner awareness just as much entails an enhanced objectification of the *outer* world. I am suggesting, in other words, that the passage to mythic consciousness is not simply a passage from "outer" to "inner" but to a *greater differentiation* of outer and inner (object and subject). Of course, in the mythic state, the *greatest* degree of differentiation of outer and inner—that associated with the three-dimensional structure—has not yet arisen.

Summarizing the first three structures of consciousness, Gebser says: "Just as the archaic structure was an expression of zero-dimensional identity and original wholeness, and the magic structure an expression of one-dimensional unity and man's merging with nature, so is the *mythical structure the expression of two-dimensional polarity*" (66). In focusing his attention on the latter, Gebser associates mythic consciousness with

> the circle, the age-old symbol for the soul. The individuated point of the [one-dimensional] magic structure is expanded into an encompassing ring on a two-dimensional surface. It encompasses, balances, and ties together all polarities, as the year, in the course of its perpetual polar cycle of summer and winter, turns back upon itself; as the course of the sun encloses midday

> and midnight, daylight and darkness; as the orbit of the planets, in their rising and setting, encompasses visible as well as invisible paths and returns unto itself. (1985, 66)

Yet, while mythic consciousness "encompasses, balances and ties together all polarities," the tension between poles is not simply resolved. No third element is introduced here to serve as a "synthesis of opposites" sublating the interplay of the two in a "higher-order oneness." Two-dimensional mythic awareness therefore should not be confused with three-dimensional mental consciousness. On the contrary, the mythic structure is an inherently ambivalent, "irrational" mode of experiencing; it flows to and fro, from one pole to its complement and back again, neither integrating the poles nor denying the validity of either. Gebser makes the observation that this burgeoning of the soul, with its cyclical, wavelike pattern, ushers into actuality the experience of *time.* Whereas magical consciousness is "timeless," time makes its first appearance in the mythic rhythm, and is characterized by a retrospectiveness, a continual "harking back to what was," a returning again and again to the "beginning," as the ocean waves rise, crest, and fall back down into their troughs, only to rise again. So as to distinguish cyclical/mythic time from the linear "progression into the future" that subsequently develops with the advent of mental/spatial consciousness, Gebser characterizes the former as "temporicity" (1985, 165–73).

Gebser's least primordial dimensionality is the three-dimensional mental structure, associated with "wakefulness" (121). This is the linear, discursive, highly reflective mode of the *cogito.* It is with the mental structure that the subject-object relationship assumes its categorially separative Cartesian expression. Gebser emphasizes that this three-dimensional structure is primarily defined by its *spatiality.* He observes that the manifestation of mental consciousness brings with it an opening up of the dimension of distance, of an objective world stretching before one's subjective perspective, extending to a vanishing point on a far-off horizon—i.e., the opening up of empty space.

What we see most essentially in the series of transformations by which higher-dimensional structures of consciousness emerge is a progressive enhancement of the differentiation of subject and object. However, according to Gebser, this "increasing expansion, extension, or growth of consciousness" (1985, 119) and concomitant concealment of lower-dimensional boundlessness is superseded with the appearance of a new structure, the "four-dimensional sphere." The contemporary manifestation of this "sphere" does not merely signify a new stage in the *expansion* of consciousness, one that divides subject and object even more completely than the three-dimensional mental structure has done, one that relegates the zero-dimensional origin to even greater obscurity. For the "four-dimensional sphere" that Gebser adumbrates is *transparent:* "with the sphere . . . the whole becomes transparent" (411). Instead of just indicating a new dimensioning, the sphere signals a new *direction* of dimensional action. If, with each previous dimensioning, an expansion of consciousness took place, now, for the first time, this inclination is reversed and consciousness *contracts,* so to speak.

Gebser associates the transparent sphere with what he calls "integral consciousness." The fourth dimension, he says, must not be "considered as a merely incremental, but as an integrating dimension. . . . The fourth dimension of the aperspectival world must serve consciousness as an integral function" (347). Thus, whereas previous dimensionings resulted in an increasing obfuscation of the less subject-object-differentiated lower-dimensional structures, the coming to presence of the integral dimension creates an openness of all dimensions to each other, a "diaphaneity" (6–7, 118, 147) in which all dimensions shine through.

Now, I suggest there is a close kinship between Gebser's "spherical dimension" and the dimension of *depth* described by Merleau-Ponty. To better understand the "four-dimensional sphere" and its relationship to depth, let us return to the work of Cézanne, considered by Merleau-Ponty as a prime example of the exploration of depth (see chapter 2). Gebser furnishes his own analysis of Cézanne.

According to Gebser, Cézanne revolutionized art by superseding classical perspective. With Cézanne, "painting is not simply unperspectival; it does not negate perspective, particularly the effect of depth that is attainable. Rather, it surpasses perspective, freeing itself of the dominance of merely sectorial linear perspective. Thus it is a-perspectival, and the former depth-effect restructures itself into transparency" (1985, 475). The "transparency" of Cézanne's work refers to the overlapping of multiple viewpoints. For example, in a Cézanne painting, a cup may appear to us as if we were viewing it both from the front and from above. It is this integration of ordinarily antagonistic perspectives that conveys a new sense of wholeness: "the aspired-for whole eludes perception as long as we see only one of its sides" (477).

Cézanne's transparent whole is decidedly dynamic. On this score, Gebser cites Thomas Herzog's characterization of Cézanne: "'He [Cézanne] wants the surface to become animated through the simultaneous frontal and top view, forcing upon the observer a kind of agitated vision as if he were moving in a space inhabited by the things depicted'" (473). In the same vein, Gebser notes Liliane Guerry's observation that, with Cézanne, "'the work of art accepts agitation . . . integrates itself with the universal rhythm . . . is an integrating component of the universe in motion'" (473). In sum, Gebser sees in Cézanne's painting a dynamic merging of perspectives in which opposing sides of figures become transparent to each other. Is this not in basic accord with Merleau-Ponty's reading of Cézanne's art as exemplifying the paradoxical experience of a "global 'locality'" in which "everything [is] in the same place at the same time" (1964, 180)? Another word from Merleau-Ponty confirms this judgment: Cézanne's space "radiates around planes that cannot be assigned to any [separate] place at all: 'a superimposing of transparent surfaces,' 'a flowing movement of planes of color which overlap, which advance and retreat'" (181).

The experience of "agitated transparency" is graphically illustrated by employing a well-known visual structure of Gestalt psychology employed by Merleau-Ponty himself (1962, 263): the *Necker cube* (fig. 6.1(b)). We first consider the mental/spatial or Cartesian principle of opposition, as expressed through visual perspective in figure 6.1(a).

If you were initially viewing a solid cube from the angle shown in the left-hand member of figure 6.1(a), you would obtain the point of view of the right-hand member by (1) moving 180° around the cube to the opposite side, and (2) moving above the cube, since the left-hand perspective gives the view from below. The faces of the symbolized solid that are visible from the right-hand perspective are precisely those which were concealed from the left-hand point of view, and vice versa. In our ordinary experience with perspective, it is of course impossible to view both the near side and the far side (or inside and outside) of an object simultaneously; all the faces of the cube cannot be apprehended in the same glance. Opposing faces are closed to each other.

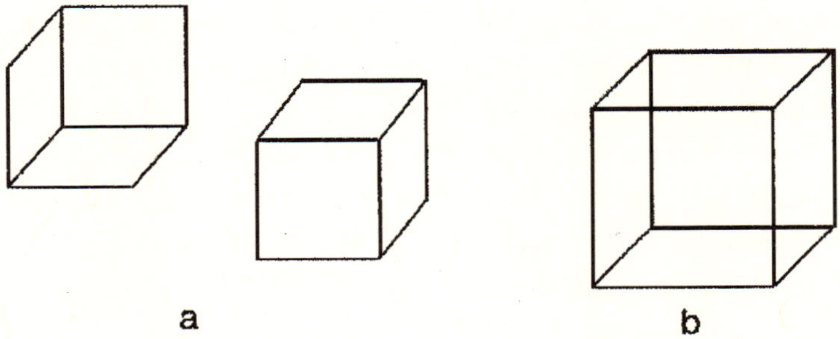

Figure 6.1. Opposing perspectives (a) and Necker cube (b)

The ordinary mode of perception is the Cartesian one. Here we perceive objects and events as extended in the world outside us, but have no immediate access to the inner, subjective ground of our perceptions: we cannot see our own act of seeing, touch our own touching. What figure 6.1(a) illustrates is that this underlying opposition between the subjective seat of perception "in here" and the objective realm "out there" is reflected in the external objects themselves, in the diametrical opposition we normally encounter between their concealed and exposed surfaces. Opposing sides of objects cannot be viewed at once.

Turning now to figure 6.1(b), you can see that both of the perspectives shown in 6.1(a) are encompassed in the body of the Necker cube. This creates visual ambiguity. You may be perceiving the cube from the point of view in which it seems to be hovering above your line of vision, when suddenly a spontaneous shift occurs and you see it as if it lay below.[1] Two disparate perspectives certainly are experienced in the course of gazing at the cube and this disparity reflects the continuing distinction between opposing sides. But the cube's reversing perspectives overlap one another in space, are internally related, completely interdependent (think of what would happen to one perspective if the other were erased).

To approximate more closely the kind of animated fusing of perspectives found in the work of Cézanne, let us attempt to go a step further in our perception of the cube. Ordinarily, our glance is limited to a mere oscillation from one perspective of the cube to the other. But we can actually break this visual habit and view both perspectives at once.

In figure 6.2, I've added some volume to the Necker cube, fleshed it out a bit. This modification should make it easier for you to see what I am talking about. When the cube's perspectives are integrated as I am

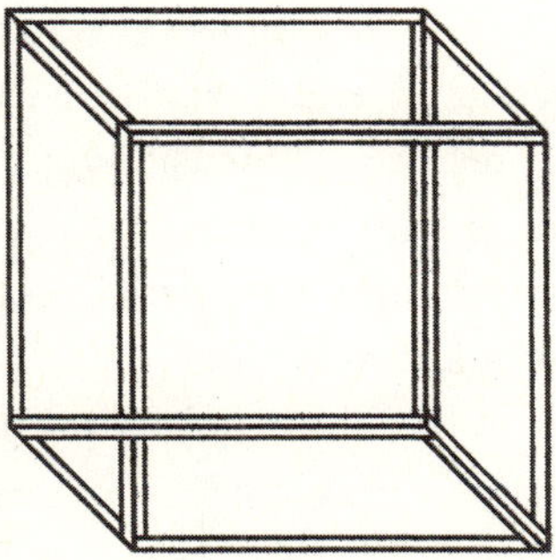

Figure 6.2. Necker cube with volume

suggesting, there is an uncanny sense of self-penetration; the cube appears to do the impossible, to go *through* itself. Here the division of sides is surmounted in the creation of an experiential structure whose opposing perspectives are simultaneously given.

Simultaneously? Well, that is not exactly the case. I am proposing that we can apprehend the cube in such a way that its differing viewpoints overlap in time as well as in space. But what we actually experience when this happens is not simultaneity in the ordinary sense of static juxtaposition. We do not encounter opposing perspectives with the same immediacy as figures appearing side by side in space, figures that coexist in an instant of time simply common to them (as do the words printed on this page, for example). And yet, there is indeed a temporal coincidence in the integrative way of viewing the cube, for perspectives are not related in simple succession (first one, then the other) any more than in spatial simultaneity. If opposing faces are not immediately co-present, neither do they disclose themselves merely *seriatim*, in the externally mediated fashion of linear sequence. Instead the relation is one of *internal* mediation, of the *mutual permeation* of opposites. Perspectives are grasped as flowing through each other in a manner that blends space and time so completely that they are no longer recognizable in their familiar, categorially dichotomized forms. You can see this most readily in viewing figure 6.2. When you pick up on the odd sense of self-penetration of this allegedly "impossible" figure, you experience its two modalities neither simply at once, nor one simply followed by the other, as in the ordinary, temporally broken manner of perception; rather, you apprehend the *dynamic merging* of perspectives. Just such a fusion is found in the paintings of Cézanne, and it gives concrete expression to Merleau-Ponty's depth dimension and Gebser's integrative dimensioning. (Have we not seen that Merleau-Pontean depth is embodied by the *Klein bottle?* The Necker cube may in fact be considered a perceptual prototype of the topological bodies. See the next chapter for more on this.)

Further confirmation of the kinship of Merleau-Ponty's and Gebser's approaches to dimension comes from considering the question of Cézanne's relationship to his own work. In Guerry's understanding of Cézanne, "'the boundaries between the space of the observer and the space of the work of art are eliminated. One identifies and harmonizes with the other'" (Gebser 1985, 479). Gebser goes on to quote Cézanne himself on this: "'I feel colored by all the nuances of the infinite. Henceforth I am one with my painting'" (479). Of this Gebser says, "We are dealing . . . with a supersession of time and space, object and subject, with a liberation from them [or, rather, from the dualistic splitting of them] and a conscious participation in the whole" (479). Thus, Gebser's integrative dimensioning is sub-objective, as is Merleau-Ponty's dimension of depth.

Now, what about the *spherical* character of the integrative dimension? Gebser credits Guerry with the "extremely important insight" that Cézanne's work is "based on a conception of spherical space" (472). The sphericality of Cézanne's painting is seen as resulting from the fact that he was tapping into a "fourth dimension," the dimension of time. Thus, says Gebser, Guerry "has demonstrated that Cézanne established a space which 'integrates the work of art into the breath of time'" (472). It is the "fourth dimension" that "curves space, rendering it spherical, and makes transparency possible rather than [classical] depth effects" (476). Gebser had made a similar point earlier in his book: "The simple sphere is merely three-dimensional; only the moving, transparent sphere is four-dimensional. And only the transparency guarantees the aperspectival perception" (346). In other words, the "four-dimensional sphere" is what accomplishes perspectival fusion, the integration of inside and outside, container and contained, subject and object. Gebser gives evidence that this sphere is at play in many areas of contemporary cultural expression, not just in art. He cites Werner Danckert's analysis of the music of Debussy, to name one example: "Debussy abandons the old 'box image. . . . Nearness and distance coalesce into an inseparable unity.

. . . [Debussy's] *tonality is spherical*'" (463, Danckert's italics). Gebser concludes that in "Debussy's music, which demonstrates an inceptual liberation from all spatial 'box images' [i.e., containers] . . . the transparent sphere interconnecting nearness and distance, has . . . become a reality" (463). Going on to architecture, illustrations are given in which the "'interior and exterior of a building . . . are presented simultaneously'" so that the "'boundary between inside and outside becomes negligible'" (468).

Another of Gebser's examples of the operation of the perspectivally integrated "four-dimensional sphere" comes from the field of philosophy and Heidegger's work with Parmenides' notion of Being as a "well-rounded sphere" (411). Gebser comments that "Perceiving the whole as a 'sphere of Being,' is definitely aperspectival. Perspectivally we can see only one side; the completeness of all aspects of Being is perceptible only aperspectivally. With the sphere, the whole becomes transparent" (411). What I find particularly evocative is the phrase "well-rounded sphere." Just such a term is used in alchemy to describe the hermetic vessel. C. G. Jung says of the vessel that "it must be completely round . . . (the spherical or circular house of glass)" (1968, 236). Elsewhere he speaks of the "house of the sphere" as the "*vas rotundum,* whose roundness represents the cosmos" (1970, 279), this "rotundity" being associated with the realization of wholeness; evidently, the "roundness" must be "simple and perfect" (1968, 88). What I brought out in my article on the vessel (Rosen 1995) is that this "roundness" is decidedly paradoxical in nature, as is plainly evident from alchemy itself, since the vessel is associated with the self-containing *uroboros.* I then went on to correlate the paradoxically "spherical" vessel with the Klein bottle. As for Gebser, it is clear to me that the transparency and aperspectivity of his integrative, "four-dimensional" sphere make it identifiable as a *Kleinian* kind of sphere. Bearing in mind our observation of the kinship between Merleau-Pontean depth and the Klein bottle, I am proposing that the idea of originary dimensional process is a common thread in the works of Merleau-Ponty and Gebser, and that their two distinctive versions of this process come together in our concept of topodimensional embodiment.

But I have established that the Klein bottle alone does not suffice to give full expression to onto-dimensional process since there are denser orders of dimensional flesh that can only be brought out through lower dimensions of topological corporality. Again, it is not just the Kleinian body of paradox that constitutes primal dimensionality, with the lower dimensions being merely juxtapositional, inherently non-dialectical. What we have seen is that each dimension comprises an order of dialecticity unto itself; each entails a primal dimension of depth corresponding to a unique topological corpus. Thus, the full integration of the dimensions cannot be provided solely by the Kleinian "transparent sphere"; the integrative action must be repeated more concretely at several dimensional levels. We know that ontological action of this sort is *Proprioceptive.* It is through the series of Proprioceptions described in chapter 5 that we are to go beyond the mental sphere of the Kleinian body to the fleshier life-worlds of the lower-dimensional bodies.

Of course, Gebser, for his part, does not limit himself to a single qualitative dimension but describes a total of *five.* There are the four dimensionalities involved in the *expansion* of consciousness, i.e., the series of transformations in which the archaic zero-dimensional structure is progressively obscured. And then there is the fifth structure of consciousness, the dynamically integrative "four-dimensional sphere." This is the dimensional action that reverses the process of expansion and brings to light all hitherto concealed dimensions, so that all become diaphanous to each other. However, if the "transparent sphere" is not just the next higher member in the sequence of expanding dimensions (the one that follows the three-dimensional structure), is it really appropriate to designate it as *four*-dimensional? Should not the *dimension number* of the sphere reflect the fact that, rather than simply introducing a separate new dimension in the expansion series, it effects a dimensional contraction or condensation in which old dimensions are integratively illuminated?

From the onto-topological standpoint, Kleinian dimensionality is not four-dimensional. Rather than involving a separate dimension arising beyond the three-dimensional mental structure, it *completes* that structure by concretizing it. Gebser emphasizes that the mental structure is essentially perspectival. He does not make clear that such perspectivity implies the *incompleteness* of three-dimensional space. In the experience of classical perspective, what we actually have is not the perception of three-dimensionality as such but the partial perception of the two-dimensional surfaces of objects. The incompleteness of this experience is glossed over by the idealization discussed earlier: the whole object is assumed to be "out there in space," though one cannot *see* all of its surfaces at any given moment. With the emergence of the aperspectival transparent sphere, the classical idealization of perspectival perception is surpassed via an integration of the opposing perspectives of objects—as well as a fusing of the objects (contents) with the containing space, and with the uncontained subject. In this coming to presence of Kleinian dimensionality, the three-dimensional order of flesh is made a reality, brought to concrete culmination (albeit in a preliminary way, since additional concretizations will be required).

To counteract the persistent tendency toward ontical idealization, let me stress once more that Kleinian realization is not just another step in the phallic quest for unity or purity. It is not about the "fatherly transcendence" of all divisions and attainment of perfect wholeness. Rather, it is concerned with embodying ontological *(w)holeness* in the act of motherly self-birthing carried out by wild Being. The Kleinian fulfillment of three-dimensionality is therefore at once an *emptying* of it, an opening of it that makes it transparent to itself (via Proprioception), and, in the same process, takes the first step in making transparent the hitherto concealed lower-dimensional sub-objective structures, thereby preliminarily bringing about the integrative "dimensional contraction" of which I have been speaking.

Now, we have associated Kleinian three-dimensionality with thinking and language, the domain designated by Merleau-Ponty as "second flesh." Merleau-Ponty's "first flesh," his "sensible world" (1968, 153), is a denser, more concrete sub-objective action sphere that requires a lower order of topodimensionality for its proper embodiment. We are going to see that the sensuous lifeworld is that of the one-dimensional lemniscate. We will also explore the sub-objective orders of flesh that correspond to the two other windings of the dimensional spiral: the two-dimensional Moebius vortex and the zero-dimensional sub-lemniscatory circulation. In this expansion to four orders of dimensional flesh, thinking is renumbered as "fourth flesh" and sensing as "second flesh." Do Gebser's lower-dimensional structures of consciousness—archaic, magic, and mythic—actually correspond to lower topodimensional bodies? I propose that while these structures do involve differences in dimensional density related to their differential attunement to lower-dimensional bodies, they themselves should not be identified with those bodies in a primary way. For I suggest that the structures described by Gebser are in fact all members of the same "dimensional family," viz. the three-dimensional Kleinian family associated with the human world of cognition. In developmental terms, we may regard the archaic, magic, and mythic structures as immature forms of three-dimensional mental functioning that precede the appearance of the rational form. As a matter of fact, all five of Gebser's structures of consciousness can be precisely located within the Kleinian winding of the dimensional spiral shown in table 5.1(b). If the Gebserian dimensionings that climax with the appearance of the Klein bottle-like "transparent sphere" indeed correlate with stages of Kleinian Ontogeny, then the progression from archaic to mental-rational consciousness should correspond to the S1→S4 matrix reduction, and the realization of "four-dimensional integral consciousness" should correspond to the Proprioceptive gear reversal bringing us to S5 of the Kleinian winding. In this winding of the spiral, the lower-dimensional bodies serve chiefly in the capacity of selfless midwives functioning to facilitate the individuation of the motherly *cogito*. But we have found that the lower-dimensional bodies undergo their own individuation processes in the denser milieus of the sub-Kleinian windings (the exception is the sub-lemniscatory body, which undergoes no Ontogeny whatever; see chapters 3 and 5). Moreover—as table 5.1(a) brings out, and as we shall further discuss in the next chapter—the lower-dimensional windings actually overlap the Kleinian winding. Gebser does not address himself to the workings of the lower dimensions in their own right.

We may better appreciate the dimensional limitations of Gebser's structures by considering again what he says of the most primitive structure of consciousness. It involves "the non-differentiation, indeed the non-distinguishibility [*sic*] of archaic man from world and universe—a non-awakeness by virtue of which he is still unquestionably part of the whole . . . an unconcerned accord, a consequent full identity between inner and outer . . . the perfect identity of man and universe" (1985, 44–45). The focus of the foregoing passage is upon "man," not on the "universe" within which this primal human being is undifferentiatedly enmeshed. Although Gebser intimates that the archaic human is asleep in the womb of the nonhuman, nonmental world, is utterly embedded in it and thus is inseparable from it, Gebser does not go so far as to engage with that surrounding world per se; he does not directly address the question of *its* dimensionality. Upon considering said question, we see at once that it must be the noncognitive environment of the archaic cognitive structure that is zero-dimensional, rather than the archaic structure itself. For archaic consciousness *is* a mode of cognition, albeit a highly deficient one; as such, it must be regarded as the embryonic form of *three*-dimensionality associated with stage one of Kleinian Ontogeny. Stated in terms of maternal self-nativity, archaic consciousness defines the S1 condition of the Kleinian mother (0d/3d), "negatively contained" as she is by the selfless sub-lemniscatory midwife (3d/0d) who pervades this primal milieu (see discussion of negative containment in previous chapter). Similar conclusions appear warranted for the other immature forms of mentality: magical consciousness is not itself one-dimensional but is the S2 form of Kleinian three-dimensionality (1d/3d) contained in the one-dimensional environment of the lemniscate (3d/1d); the mythic structure is not itself two-dimensional but is a still-immature or "fractional" form of three-dimensionality (2d/3d) contained in the two-dimensional Moebius matrix (3d/2d). This reinterpretation of Gebser appears consistent with his frequently *negative* way of describing the dimensions of consciousness in question: They are to varying degrees undifferentiated; are unconscious or sleeplike (1985, 121), spaceless and/or timeless (117), without perspectivity (117), prerational and/or irrational (146), pre-causal and/or non-causal (146), egoless (149), and so on. In other words, Gebser's archaic, magic, and mythic structures of consciousness, rather than constituting separate dimensionalities unto themselves, are undeveloped forms of cognitive dimensionality, of fourth flesh.

3. THE THIRD ORDER OF DIMENSIONAL FLESH: ANIMAL EMOTION

The task at hand is to gain a more concrete grasp of third flesh, of the two-dimensional Moebius order of wild Being. We know that this noncognitive dimension plays the role of selfless midwife to the Kleinian *cogito* in the three-dimensional winding of the dimensional spiral. Now we want to understand better the nature of this lower dimension in its own right. We are seeking a more tangible insight into Moebius dimensionality per se.

What sort of dialectical activity is carried out within the Moebius lifeworld? In chapter 3 we saw that, whereas the perspectival opposition of planar surfaces is a basic feature of the three-dimensional dialectic, in the two-dimensional environment the dialectic should entail a form of perspective involving *opposing edges of the line.* It is this lower-dimensional action that is concealed from us three-dimensional *cogito*s; since the edges of the line are simultaneously apprehensible to us (unlike the sides of the plane), they appear to lack dialectical tension; to us, the edges of the line are not opposed but merely *juxta*posed. Evidently, this is not the case in the denser environment of two-dimensional space. But what would it actually mean to speak of a "perspectival line" enacting a dialectic of opposing edges in two-dimensional space? To better comprehend the two-dimensional situation, we must go beyond the merely abstract geometric characterization of it. At bottom, the dialectic of the line in two-dimensional space is not some objective process observed by a detached *cogito;* it is a *sub*-objective order of the flesh that functions noncognitively. If the *thinking* subject presides in three-dimensional space, what is the subjective quality, the functional modus operandi characteristic of two-dimensional space? In seeking an answer to this question, the work of Gebser provides us with some clues. Although Gebser spoke of the mythic structure of consciousness as two-dimensional, I have just suggested that we regard the mythic function as an immature form of *three*-dimensionality, of cognition, that is embedded in a "two-dimensional environment." Despite Gebser's greater concern with mythic cognition than with its noncognitive milieu, his investigation does shed light on the functional nature of said milieu.

Gebser variously associates mythic consciousness with the realm of the soul, the dream state, the element of water, the astral and lunar worlds (stars and moon), movement and kinesis, and the organ of the heart (see 1985, 61–73 and 165–73). I propose that, in functional terms, all of these qualities are related to *feeling* or *emotion.* The relationship between emotion and the heart is a commonplace that I will not elaborate on here, since my immediate purpose is only to note that there *is* a relationship. For now, we can also take for granted the link between emotion and dreams; indeed, it is this that forms the basis of classical psychoanalysis. Most of the other cognates of the mythic given above are touched on by Jung in *Mysterium Coniunctionis* (1970), in a section where he examines the symbolism of the moon in alchemy.

First we have the tripartite link between the moon, the soul, and water: "The relation of the moon to the soul, much stressed in antiquity, also occurs in alchemy, though with a different nuance. Usually it is said that from the moon comes the dew, but the moon is also the *aqua mirifica* . . . [the] 'Mercurial water' and 'fount of the mother'" (1970, 132). Continuing in a similar vein, Jung associates Luna with "moisture" (140) and with the "'waters under the firmament'" (143). He also intimates a connection between the moon and the stars: "The moon appears to be in a disadvantageous position compared with the sun. The sun is a concentrated luminary: 'The day is lit by a single sun.' The moon, on the other hand—'as if less powerful'—needs the help of the stars" (143). Thus the moon is linked to the *astral* domain. In the esoteric literature, much is made of the "astral plane" and the "astral body." And, typically, astrality is related to the moon (e.g., Steiner 1972) and to the element of water (e.g., Metzner 1971, 89).

Now, a primary characteristic of the astral body is kinesis or motion; the literature is filled with references to "astral travel" and "astral projection." And, indeed, this is an essential feature of the *soul,* which Jung generally takes to be synonymous with the archetype of the *anima.* The *anima* is an aspect of the unconscious often symbolized by the snake, which bespeaks the "active, animal principle" and is related to "emotionality and the possession of a soul" (Jung 1967, 257); Jung links "animation of the body and materialization of the soul" (257). So we have here the correlation of the soul or *anima* with animation or motion, and with *e*-motion. Regarding the latter, Jung says elsewhere that "since the soul animates the body . . . she tends to favour the body and everything bodily, sensuous, and emotional" (1970, 472). Also on the subject of emotions, Jung says, "The appetites . . . pertain to the sphere of the moon: they are anger (*ira*) and desire (*libido*). . . . The passions are designated by animals because we have these things in common with them" (1970, 143–44). As for Gebser's own assessment of the role of emotion, most often he groups it with "instinct" and "drive," placing it under the heading of "basic attitude and agency of energy" (1985, 144). In so doing, he relates emotion to the *magical* structure of consciousness, not the mythic. However, in an apparent contradiction, Gebser says, "With the mythical structure an upward relocation of emphasis on the organs considered numinous becomes evident: a shift from the vital to the *emotive* sphere symbolized and designated by the diaphragm and heart" (271;

emphasis added). This seeming contradiction may actually be no more than an artifact of translation, and an English word other than "emotion" may provide a more accurate rendering of the "basic attitude and agency of energy" of the magical. In any case, the evidence seems to favor the hypothesis that the subjective character of the two-dimensional environment of mythic consciousness is essentially emotional. Accordingly I propose that, whereas three-dimensional space is cognitive, two-dimensional space is emotional.

Jung's association of the soul or *anima* with the "animal principle" and his remark that the "passions" are what we have in common with animals point significantly to phylogenetic implications which, for the most part, are neglected by Gebser, whose focus is primarily on the evolution of *human* consciousness. In this regard, the old idea of the "animal soul" is noteworthy. Philosopher Antony Flew defines this as an "analogue in animals of the human soul or mind" (1979, 14), one that suggests a principle of animation in the world existing independently of the human sphere of action. A more general expression of the "animal soul" is the "world soul" or *anima mundi,* a concept "founded on the view that the [nonhuman] world is productive of life and animation, and can therefore be regarded as itself animate" (377). In fact, Jung makes frequent reference to the *anima mundi,* demonstrating that its liberation is a central goal of alchemy (1967, 307). I venture to suggest that the imprisonment and subsequent freeing of the *anima mundi* portrayed in the alchemical literature metaphorically expresses the phylogenetic aspect of the developmental drama described in previous chapters. (The present section deals chiefly with the *animal* aspect of the *anima mundi.* In succeeding sections, other aspects will be explored.)

The alchemical theme is encapsulated in the Grimms' fairy tale "The Spirit in the Bottle," characterized by Jung as a story that "contains the quintessence and deepest meaning of the Hermetic mystery" (1967, 193). A powerful spirit is trapped in the earth beneath an oak tree, enclosed within "a well-sealed glass bottle" (193). Hearing the spirit cry, "Let me out!" a passing youth opens the bottle, whereupon the spirit rushes forth, identifies itself as mighty Mercurius, and threatens to strangle its liberator. But the boy tricks the spirit back into the bottle, and the tamed Mercurius then promises that, if freed again, it will serve the boy in a beneficial way.

Jung relates the glass bottle of the fairy tale to the hermetic vessel of alchemy, the container in which the transformation of the *prima materia* was supposed to take place: "The bottle is an artificial human product and thus signifies the intellectual purposefulness and artificiality of the procedure, whose obvious aim is to isolate the spirit from the surrounding medium. As the *vas Hermeticum* of alchemy, it was 'hermetically' sealed" (1967, 197). In Jung's interpretation, the Mercurial spirit represented to the alchemist the initially unconscious, wildly irrational power of embodied nature, of the *anima mundi* (214). In turn, the "bottling up" of Mercurius signified the necessity of gaining cognitive control over nature. Jung took the fashioning of the hermetic vessel to symbolize the process of individuation that culminates in the *unio mentalis* (1970, 465), a state of intellectual maturity through which human development reaches a climax. This is the challenge symbolically faced by the boy in the Grimms' fairy tale, and the challenge the alchemist faced. Prior to "properly sealing the bottle," Mercurius could always escape—a regression to the primal past that would overwhelm the alchemist. In the meanwhile, the alchemist had to keep this nonhuman "spirit" imprisoned as best he could.

In Topogenetic terms, the initial imprisonment of the "animal soul" corresponds to the matrix reduction enacted in the fourth stage of Kleinian development (KB S4), that wherein Moebius dimensionality is obscured by containment within the Kleinian vessel (see table 5.1(a)). The animal emotionality that suffuses mythic consciousness in the third stage of Appropriation is thus "bottled up" by the intellect in S4. But we know that, in passing from S3 to S4, the fusion of 3d/2d and 2d/3d enantiomorphs is but a *quasi*-fusion. The Kleinian vessel is not yet sealed *hermetically,* and, as a consequence, there is indeed a danger of regression to S3. It is not until the Kleinian mother's S5 turn from self-Appropriation to Proprioception that the vessel is sealed more convincingly. The containment in question is paradoxical, of course; it is a closure that is at once an *opening*. KB S5 is the stage in which the *cogito*'s phallic fantasy of simple autonomy is exploded. And the act of Proprioception that illuminates the Kleinian mother brings a healthy recognition of the midwifely assistance she has received from the Moebius *anima,* along with the understanding that the tie to the midwife or grandmother has never really been severed. The S5 containment of the *anima* therefore also lifts the repression of it. However, the merely *cognitive* acknowledgment of the link to the noncognitive *anima* does not in itself go far enough toward fully liberating the Mercurial *anima*. A second sealing of the Kleinian vessel is thus required, one that invites a more concrete liberation of the Moebius body.

In this connection, consider Jung's observation that "the alchemists rightly regarded 'mental union in the overcoming of the body' [i.e., the *unio mentalis*] as only the first stage of conjunction or individuation. . . . In general, the alchemists strove for a *total* union of opposites" (1970, 474–75). This meant that, once mental integration was attained (KB S5), there would need to be additional "distillations," alchemical processes entailing a reunion of the now concretely realized mind with the more concrete orders of embodiment that had been repressed. Jung further notes that "the second stage of conjunction, the re-uniting of the *unio mentalis* with the body, . . . consists in making a reality of the man who has acquired some [abstract] knowledge of his paradoxical wholeness" (1970, 476). The hermetic closure occurring in KB S5 is "made a reality" in the sixth stage of Kleinian Ontogeny. The Proprioception called for in KB S6 is *emotional*. It entails a backward movement through the history of Being that goes further than the first moment of the *cogito*'s quasi-maturation (in KB S4).

Gebser sets the emergence of the mental structure at the middle of the first millennium BCE, which is roughly the time set by Heidegger (1964/1977) as the beginning of Being's concealment. For the Proprioceptive culmination of cognitive development realized in KB S5, we must switch gears and go back to this beginning. The "first must come last," as Heidegger prescribed in his "eschatology of Being" (1946/1984, 18). To bring the thought of Being to proper closure, we must reprise its opening moment (see chapter 4). But this will not suffice for the uroboric closure that is called for in KB S6. Here a still earlier beginning must be Proprioceived. "Proceeding in reverse," we need to apprehend the first *mythic* moment (KB S3). By Proprioceptively reprising this heartfelt dawning of the soul, we give concrete thanks for the two-dimensional *anima*'s midwifely ministrations to the not-yet-mature three-dimensional *cogito*.

Only with this second, emotional Proprioception would Moebial Mercurius truly be sealed into the Klein bottle so as to preclude the *cogito*'s regression, its loss of rational consciousness. And only with this KB S6 movement backward would Mercurius truly *leave* the bottle (it had been more or less "bottled up" since KB S3), now taking its leave as a contributor to (w)holeness and inter-dimensional harmony. For the departure of the embodied *anima* from mental containment would entail a full-fledged *conjunction* of body and mind. Now, rather than the body containing the mind—as in the KB S3 mythic situation wherein human cognition is still weak and animal emotion is the prevailing force—or the mind containing the body—as happens with the KB S4 hegemony of the mental structure—we have the *asymmetric mutual containment* (chapter 5) of mind and body; that is, of Kleinian cognitive Being and Moebial emotional Being. The first "notes" in the "music" of the dimensional spheres thus are struck. Through "dimensional mirror fifths," viz. 3d/2d and 2d/3d, humankind's 3d/3d cognitive dialectic and the denser, 2d/2d emotive dialectic of the animal lifeworld become attuned to one another. In harmonic resonance now are the *unio mentalis* and what we may call the *unio emotionalis,* the fulfillment of 2d/2d animal feeling realized in the Moebius winding of the dimensional spiral. Alchemically speaking, the "circle is being squared."

We need to be clearer about the nature of lower-dimensional emotionality. For there can be no doubt that emotion is experienced in KB S4, the stage of Ontogeny in which the *anima* is repressed and the *cogito* is ascendant. In this stage, I can feel love and hatred, frustration and anger, elation and depression. Yet whatever emotion I experience here is indeed governed by the "I," by "that *I think* that must be able to accompany all our experiences" (Merleau-Ponty 1968, 145). That is to say, feeling is ruled by the *cogito* in KB S4. The psycho-cultural historian Richard Lind (2001) helps us to see better how said rule was established, and what it means. Lind sums up his basic hypothesis as follows: "Between two historical periods, prior to about 500 BCE and the modern era beginning around 1700 CE, the location and quality of self-awareness was gradually, but over this extended period radically, altered by the *migration of subjectivity from heart to head*" (27). With the transfer of the seat of identity into the head has come the tendency "to objectify the body":

> The body is included in self-awareness but individuals rarely attribute subjective experience to bodily locations. . . . In the modern era, although emotions and sensations are still experienced

> in the body, this shared sense of awareness is notably limited by the sense that the location of the "I" of consciousness is restricted to the head. The unmediated experience of the body remains relatively unconscious. When the statement is made that "I" feel or "I" touch, for example, an abstraction is being expressed by a self-construct. Bodily experience is first translated into language and evaluated according to its compatibility with the self-construct. The self-construct then "owns" (one etymological root of the word self is *possession*) or disowns (e.g., represses, denies, or ignores) these bodily experiences. Acknowledged experience of the body is "possessed" by head-consciousness and bodily experience incompatible with the self-construct is dispossessed. (Lind 2001, 26)

Contemporary emotional life, constrained as it is by the continuing dominance of the KB S4 *cogito,* is far removed from the heart-centered mythic emotionality of KB S3, where the *anima* holds sway. In *The Origins and History of Consciousness,* the Jungian anthropologist Erich Neumann states that "primitive man experiences an 'animated' world, while modern man knows only an 'abstract' one. Pure existence in the unconscious, which primitive man shares with the animal, is indeed nonhuman and prehuman" (1954, 329). Neumann asserts that "[t]he dawn man lives his affects and emotions to the full" (331). The transpersonal philosopher Michael Washburn, taking his cue from Neumann, notes that these primordial emotions (today still experienced in human infancy) include a feeling of "exuberance, or overflowing well-being" (1988, 45), "waves of bliss" (47), and a sense of "delight" in a world that is enchanted with "an aura of the miraculous" (48). However, Neumann points out that primal emotionality also includes "instinctive reaction[s]" involving "flight or attack . . . rage, paralysis, etc." (1954, 332). He contends that human consciousness has evolved "from the primitive emotional man to the modern man" in a manner that "protects—or endeavors to protect—him from this access of primitive emotionality" (331). The protection is needed because "primitive man . . . like the child . . . was forced into total reaction by any and every content that emerged, and, overpowered by his emotionality and the underlying images, acted as a totality, but without freedom" (332). This "total reactivity of primitive man is no subject for romanticism" (332). Nevertheless, given "the present crisis of modern man, whose overaccentuation of the conscious, cortical side of himself has led to excessive repression and dissociation of the unconscious . . . it has become necessary for him to 'link back' with the medullary region" (331) of the brain, or, what amounts to the same thing, with the "heart-soul which animates the body" (237). I suggest that this "linking back" to primal feeling is what happens in the S6 Proprioception of Kleinian Topogeny. Neumann's alchemical rendering of the process is noteworthy:

> As in alchemy the initial hermaphroditic state of the *prima materia* is sublimated through successive transformations until it reaches the final, and once more hermaphroditic, state of the philosopher's stone, so the path of individuation leads through successive transformations to a higher synthesis of ego, consciousness and the unconscious. While in the beginning the ego germ lay in the embrace of the hermaphroditic uroboros, at the end the self proves to be the golden core of a sublimated uroboros, combining in itself masculine and feminine, conscious and unconscious elements, a unity in which the ego does not perish but experiences itself, in the self, as the uniting symbol. (1954, 414–15)

Now, Neumann places strong emphasis on the overriding importance of the animal in primordial human affairs. He notes that "the animal forms of the gods and ancestors originally symbolized and expressed man's oneness with nature" (338). In native bands of the Pacific Northwest, for example, the initiation rites crucial to the entire course of a person's development centered on

> induction into the spiritual world of the ancestral totem . . . [which] is very often an animal. . . . The totem is an ancestor, but more in the sense of a spiritual founder than a progenitor. Primarily he is a *numinosum,* a transpersonal, spiritual being. He is transpersonal because, although an animal, a plant, or whatever else, he is such not as an individual entity, not as a person, but as an idea, a species. (1954, 144–45)

Through totemic initiation, the individual acquires a "guardian spirit": "This spirit, who may be lodged in an animal or a thing, introduces into the life of the initiate who experiences him a whole sequence of ritual obligations and observances, and plays a decisive role among all shamans, priests, and prophetic figures in primitive societies" (145). Later, Neumann comments on how the expansion of human consciousness via "secondary personalization leads finally to the local deities becoming heroes and the totem animals domestic spirits. . . . In consequence, the human and personal sphere is enriched at the expense of the extrahuman and transpersonal" (337). Taking the history of ancient Egypt as an example, Neumann documents "the increasing humanization of the gods." He shows how, from the First Dynasty to the Third, the animal figures that predominated were gradually transmuted into human ones, so that, "from the Third Dynasty onwards the human form becomes the rule. The gods establish themselves in human form as lords of heaven, and the animals retire" (338).

In *Coming to Our Senses,* Morris Berman (1989) makes clear the extent to which modern industrial society has lost touch with its animal heritage. According to Berman, the Paleolithic human being saw the animal as an Other, but one with whom s/he was intimately identified. In fact, the earliest humans were so deeply immersed in the animal world that it might be plausible to say that, in a sense, they were more animal-like than human. "Animal life was everywhere, even in the skies," says Berman, and "animal movement, the animal body, was the model of human expression in hunter-gatherer society" (68). It is not surprising, then, that "the first art, which can be seen in the caves of Altamira and Lascaux, is about animal subjects rather than human ones" (66). Berman goes on to say of hunter-gatherer culture that "human life . . . had no special significance apart from the animal world" (69). Even in the next historical epoch, with the advent of agriculture and the domestication of animals, their profound influence on the human psyche was maintained: "Animal cults and symbolism continue into the Neolithic period, still strongly coloring human processes of cognitive and psychological development" (70).

However, the agricultural revolution also brought with it a momentous change: the animal realm was now divided into the "Wild" and the "Tame." Animals in the latter category, pressed into the service of human needs, were stripped of their Otherness, whereas those in the former category presently tended to be regarded as *merely* what is Other, as alien beings to be hated and feared. According to Berman, this splitting of the human and animal worlds constituted the basis for all subsequent dualisms attendant to human affairs. In the earlier hunter-gatherer society, "If there was no sharp divide between Wild and Tame, or Self and Other, there was also no such divide between sacred and profane, or heaven and earth. . . . Domestication changed all this. The fundamental categories that presented themselves were now two—Wild and Tame—and eventually all forms of thought, down to the present day, came to be based on this model" (71).

On Berman's account, the transition to modern industrial society brought with it a further significant transformation of the relationship between animals and human beings. Now, the very awareness of the animal as an Other is eclipsed: "[F]ear of the animal Other acquires a whole new dimension during the modern period; it becomes vague, unspecified, repressed, and all the more terrifying as a result" (73). Thus, in modern culture, "Nonhuman Otherness is not merely degraded . . . but absent" (85). Berman concludes that "[f]ear of organic life and the existence of the Tame/Wild distinction is so central and pervasive a feature of modern technological societies that it is, paradoxically, almost invisible. Like . . . the mind/body split [which derives from it], it is virtually everywhere, so it seems to be nowhere" (97). Zoos, pets, and Disney characters certainly abound in today's world, but all these give us are "humanized form[s] of ourselves, not a true Other" (91).

Significant for our purposes is that the animal Other prominent in Paleolithic times is an *emotional* Other: "Paleolithic men and women took their cues from body feelings and the movements of animals. This was a life governed by shifting moods rather than the demands of the ego" (Berman 1989, 69). The contemporary social phenomenon of psychotherapy makes it more than clear that *modern* emotional life tends to be "bottled up." When the "cork" is in place, emotions are tame, well socialized, controlled by cognition. Of course—prior to the genuine maturation of the *cogito* (in KB S6)—the "cork" does tend to "pop" from time to time, whereupon emotions rush out regressively and we are flooded by them, as the youth of the Grimm brothers' fairy tale experienced. Only with the hermetic sealing of the cognitive Kleinian container can the "Spirit Mercurius"—i.e., genuine animal emotion—emerge in a full-fledged and constructive fashion. In topodimensional terms, what emerges from imprisonment in the three-dimensional Klein bottle and enters into harmony with it is the two-dimensional Moebius body. While Berman says nothing on the matter of dimensionality, confirmation of the two-dimensional character of the animal world of emotion comes from two independent sources.

Comparing human consciousness to that of animals, the nineteenth-century philosopher Jules Lachelier concludes that animals "are provided with the same senses as we, but it is probable that these senses *move* them much more than they teach them and that these impressions themselves are entirely subordinated to their organic feelings" (1962, 155; emphasis added). Lachelier associates animal being or the "animal *self*" (172) with "desire" and "purpose" (168), with the "will to live," which he views as the "fundamental emotional state" (172). This emotional order of consciousness is seen as played out in "two-dimensional space or surface" (168). By contrast, human consciousness involves a higher order of reflection in which "we project outside of ourselves solid objects by adding to the two dimensions of visible space that which is only the imaged affirmation of existence—depth. . . . Three-dimensional space, individual reflection and reason—these are the elements of a . . . consciousness which we have . . . called intellectual" (169).

The other confirmation of the two-dimensionality of animal emotion comes from the Russian theosophical philosopher P. D. Ouspensky, whose views on space we considered in chapter 3. Ouspensky also saw the animal world as a world of emotion: "In reality the animal does not reason its actions, but lives by its emotions, subject to that emotion which happens to be strongest. . . . Its actions are directed not by thoughts but principally by emotional memory and motor perceptions. . . . Any perception of an animal, any recollected image, is bound up with some emotional sensation or emotional remembrance—there are no non-emotional, cold thoughts in the animal soul" (1970, 79–80). Ouspensky then asks, "How does the world appear to the animal?" (89). His answer:

> The world appears to it as a series of complicated moving surfaces. The animal lives *in a world of two dimensions*. Its uni-

> verse has for it the properties and appearance of *a surface.* And upon this surface transpire an enormous number of different movements of a most fantastic character.
>
> Why should the world appear to the animal as a surface?
>
> First of all, because it appears as a surface to *us.*
>
> But we *know* that the world is not a surface, and the animal cannot know it. It accepts everything just as it appears. It is powerless to correct the testimony of its eyes—or it cannot do so to the extent that we do.
>
> We are able to measure in three mutually independent directions. . . . The animal can measure simultaneously in two directions only—it can never measure in three directions at once. This is due to the fact that, not possessing concepts, it is unable to retain in the mind the idea of the first two directions, for measuring the third. (1970, 89)

In the analysis of Ouspensky given earlier, we found that his view of dimensionality was not without its contradictions. From the passage just cited, it would seem that animals are capable of perceiving surfaces, of experiencing "simultaneously in two directions" (89). And yet, at an earlier point in his text (53–54), Ouspensky claims that the two-dimensional being could *not* perceive surfaces, that its perception would be limited strictly to the one-dimensional line. We concluded in chapter 3 that Ouspensky's confusion came from adhering to the classical idealization of dimensionality, and, as a consequence, neglecting the dialectical character of actually lived experience. In attempting to describe the dimensional nature of ordinary human and animal perception in idealized terms, Ouspensky glossed over the phenomenological facts of perspective. Governed as it is by the dialectic of opposing sides of planes, perspectival human experience is in fact neither simply two-dimensional—as it would be if strictly limited to but a single side of the plane—nor fully three-dimensional—as it would be if opposing sides could be apprehended in simple simultaneity. Were humans able to perceive opposing surfaces as simply simultaneous, opposition actually would be reduced to mere juxtaposition. This cannot happen in our perceptual experience. The irreducibility of dialectical opposition speaks, in turn, to the fact that human perspectivity, while being greater than two-dimensional, is but *fractionally* three-dimensional. What Ouspensky could not see is that perspectival *animal* experience, also being dialectical, is governed by an interplay of *opposing edges of the line* that makes it fractionally *two*-dimensional.

Naturally, this does not take into account the full course of dimensional *development,* with its *pre*perspectival and *trans*perspectival stages. The Topogeny of the *anima* is mapped in the Moebius winding of the dimensional spiral given in table 5.1. Generally speaking, animal experience begins with the preperspectivity of S1 and S2, where immature emotional functioning is nurtured in the midwifely matrices of still denser orders of functioning (to be explored in the following sections). We might say that, geometrically, the *anima*'s initial experience of the line is strictly limited to a single edge. But what this must mean is that edges are *undifferentiated,* since, dialectically, a given edge could be experienced distinctly only *in relation to* an edge that would oppose it. By S3 of the Moebius winding, the fractionally two-dimensional interplay of opposing edges of the perspectival line would ensue: while these edges could not be encompassed in simple simultaneity, there would be a *sense* of the "far edge," similar to the way we humans possess a sense of the other side of surfaces during the perspectival phase of our development (S4 of the Kleinian winding). Evidently this would mean that, just as human perspectivity can encompass more than a single surface of the solid cube but fewer than all six surfaces, the lower-dimensional perspectivity of the animal would be able to encompass more than a single side of a square but fewer than all four sides; thus, contra Ouspensky, animals *would* be able to detect angles in lines, albeit not as completely as can human beings. Then, in the S4→S6 transperspectival stages of animal development, opposing edges of the line would now be integrated—not in simple simultaneity but in paradoxical interpenetration, and this would yield the fully two-dimensional one-edged structure known as the *Moebius.* Only with the subsequent transition into the human epoch of dimensional generation (the Kleinian winding of the spiral) would both edges of the line (and all four sides of the square) become simultaneously apprehensible in the juxtapositional manner.

Of course, by itself, the foregoing analysis is but a geometric abstraction. To appreciate it in more concrete terms, we must keep in mind what has been said about the *qualitative* nature of dimension. Again, the dialectic of the line in two-dimensional space is indicative of an entirely different order of sub-objective functioning than that of the plane in three-dimensional space. Whereas the latter points to the human realm of cognition, the former bespeaks the denser, emotive transactions of the ontologically less differentiated animal sphere. In this regard it is important to recognize that, while the emotional dialectic of animal sub-objectivity indeed entails less reflective intricacy than humankind's cognitive dialectic, this does not mean that the animal should be judged to be "more primitive" than the human being in the sense of being an imperfect, less mature version of what a human is. The relatively weak subject-object differentiation of the nonhuman realm must not be confused with the lack of differentiation characteristic of the mythic, magic, and archaic structures of human functioning. The latter does necessarily point to deficiency, to immature three-dimensionality; the former, on the other hand, can signify the mature and efficient expression of lower-dimensionality. We can say that the maturity of a given dimensional action essentially depends upon whether it has realized its own potential. To be sure, within the animal lifeworld there are immature phases of emotionality involving little differentiation. But, by S6 of the Moebius winding, animal emotion has reached its fruition. And this two-dimensional dialectic is fully as mature as the three-dimensional Kleinian dialectic of S8, even though the *anima* is not as differentiated or integrated as the *cogito*. The following generalization thus seems warranted: whereas lack of differentiation *within* a dimensional sphere signifies immaturity, the lack of differentiation of one sphere relative to another does not.

The generalized gloss of animal emotional development thus far given is clearly no substitute for a detailed and concrete account. But we are not yet prepared for such a rendering. Recall the approach to understanding inter-dimensional relationships previously followed. In chapter 3, we examined the basic associations among topodimensional bodies by working with a matrix (table 3.2) in which dimensional connections are "averaged over," i.e., abstracted from, the actual facts of dimensional development. Then, in chapter 5, by "setting the matrix in motion" via table 5.1, we proceeded to fill in the details of how the several dimensional bodies evolve in relation to one another. In our present attempt to flesh out those bodies, we have so far considered the functional and phylogenetic meanings of the Kleinian and Moebius bodies, and have taken a partial look at how they develop. For the remainder of this chapter, I will focus primarily on the phylo-functional identification of the other two non-ontical bodies. Then, in the next chapter, I will return to table 5.1 with the idea of systematically distilling all of its abstractions into a concrete reckoning of phylo-Ontogeny. By thus setting the matrices in motion in a newly coagulated way, we shall gain a more tangible grasp of how the several developmental windings of the dimensional spiral turn in synchrony, of how "Ezekiel's wheels" mesh their gears! (See epigraph of preamble to part II.)

4. THE SECOND ORDER OF DIMENSIONAL FLESH: VEGETATIVE SENSUALITY

Let us now inquire further into the *second* order of the flesh. This lemniscatory lifeworld would feature the dialectic of the point in one-dimensional space. We have seen that the three-dimensional dialectic entails the perspectival interplay of opposing sides of the plane and that the two-dimensional dialectic involves opposing edges of the line. In like manner, the one-dimensional dialectic would involve opposing "facets" of the point. As a rough approximation, picture a point dividing a line into two segments. A "Linelander," a one-dimensional being confined to said line, could not apprehend both "sides" of the point at once but would be limited to the perspective given on the side in which it was located. To be sure, we detached, three-dimensional observers of Lineland see a very different picture. Our "bird's eye view" encompasses all pointal perspectives in a single glance. To us, the one-dimensional line appears as a closed continuum, one whose point-elements are not internally opposed but merely juxtaposed. We are thus unable to see the real dialectical tension that presumably exists between the pointal perspectives of a lower-dimensional space that is not simply closed but that possesses an aspect of openness. This leaves us to wonder what it would actually mean to speak of a "perspectival point" enacting a dialectic of opposing "facets."

To gain a better insight into the one-dimensional dialectic, let us go beyond a merely abstract geometric characterization of it. The dialectic of the point in one-dimensional space constitutes the second order of the flesh, of sub-objective self-containment. If the fourth flesh is the cognitive order of functioning and the third flesh is the emotive order, what of the second flesh? What is the subjective quality, the functional modus operandi that presides in one-dimensional space? A preliminary answer has already been given. What we are calling "second flesh" Merleau-Ponty named "first flesh": it is the "sensible world" (1968, 153). However, *one-dimensional* sensibility is hardly the same as the sense perception that is governed by reflective, three-dimensional consciousness in the perspectival stage of human development. In seeking a fuller comprehension of the one-dimensional world, the work of Gebser again provides us with some clues. Although Gebser spoke of the magical structure of consciousness as one-dimensional, I have proposed that we regard it as an immature form of *three*-dimensionality—of cognition—that is embedded in a one-dimensional environment. Despite Gebser's greater concern with magical cognition than with its noncognitive milieu, his investigation does shed light on the functional nature of said milieu.

(As we follow Gebser's lead, it may help to remember that lower-dimensional flesh possesses less dialectical tension than its higher-dimensional counterpart; there is less opposition among contained object, uncontained subject, and the mediating spatial container. The lesser differentiation of the one-dimensional dialectic is reflected in what Gebser says about magical consciousness. For Gebser, magical awareness is completely absorbed in the point-like details of its one-dimensional environment. And yet, according to the principle of *pars pro toto* that governs magical participation in the world, every point of experience is inseparably united with every other point and with the whole. Thus, "magic man" inhabits a "point-like unitary world" [1985, 48], a "point-related unity in which each and every thing intertwines and is interchangeable" [48]. Through its complete immersion in the discrete and concrete details of its world [particular things, events, and actions], magical awareness fuses, is undifferentiatedly entangled with, the whole world.)

The magical lifeworld is indeed the "sensible world," but this does not refer to the sublimated, constricted form of sensibility controlled by the wakeful *cogito*. For magical experience is such that reflective consciousness is *asleep* (Gebser 1985, 121). Yet Gebser can still speak of the "sharper senses" that prevail in the one-dimensional action sphere, of the "heightened, natural, sensory apperception of magic man—superior to our own" (55). Evidently, the form of sensibility prevailing in the one-dimensional milieu entails the vital force of life, of sexuality and sensuality.

The sense experience of the one-dimensional domain is driven by impulse and instinct (Gebser 1985, 46, 60, 67) rather than by a higher reflectiveness, and these biological directives, in turn, are of course related to sexuality (208, 230). Time and again, Gebser associates the one-dimensional world with *vitality,* with life-energy or life-force. Whereas water is the operative element in the two-dimensional mythic soul, "blood and semen . . . [are the] pre-eminent vital forces" (270) of magical one-dimensionality. Gebser describes an interesting manifestation of vital life-energy: the *aura.* Commenting on prehistoric cave drawings from Australia, he notes the "clearly drawn radiations" emanating from the heads of the figures. To Gebser, "it seems obvious that they do not represent head ornaments . . . or the attributes of a sun-god" (55); rather, they signify the externalization of "man's inner potency" (64). Through the aura (shown "around the head and even the entire body" of magical figures), there is "a seamless transition to the flux of things and nature with which he [i.e., the magic person] is merged" (64). Gebser goes on to say that "aura drawings clearly show that early man's visual perception took in at once more and less than ours. He perceived in any event an emanation—the energy in motion. . . . [Early] man sees the energy that emanates from objects" (202).[2]

For Gebser, an important correlate of a structure of consciousness is its "organ emphasis." We have seen that mythic consciousness is associated with the heart. Gebser would relate the magic structure to the viscera or intestines: "viscera represent the intertwined unity which we have ascertained as characteristic of the magic structure of consciousness" (145). No organ is specified for the archaic sphere. Conspicuous by their absence from Gebser's account are the *genitalia.* Given the strong link Gebser establishes between magic and sexuality, one would think that the region of the body accentuated in magical functioning might be the groin more than the viscera or guts. And, indeed, there is an element of ambiguity in Gebser that opens the possibility of relating the magic structure to the genital organs rather than the guts. In discussing the latter, Gebser says that they "are evidence of unity *or identity*" (145; italics mine). But we have seen that, elsewhere in his text, Gebser brings out the fundamental *distinction* between unity and identity, defining "magic man" in terms of the former: "he is distinguishable above all by his transition from a zero-dimensional [archaic] structure of *identity* to one-dimensional *unity*" (46). Then could it not be that the guts would be the organs emphasized in the archaic identity structure, leaving room for us to view the genital organs as those of one-dimensional unitary magic? Perhaps I am making too much of the phrase "unity *or identity*." But it does seem that an obvious relationship exists between the magic structure and the genitals, that the sexual organs are conspicuously absent from Gebser's account, and that Gebser fails to provide an organ correlate for the archaic structure. Accordingly, I propose that we amend Gebser's analysis. Let us link the one-dimensional domain to the reproductive organs and the zero-dimensional realm to the guts. We will touch on the latter relationship again when we consider zero-dimensionality.

Now, while it is true that Gebser focused primarily on the evolution of *human* consciousness, his description of the magic structure connotes a broader phylogenetic meaning. Gebser characterizes the magic realm as essentially *vegetative* in nature: "the realm of vegetative energy" (49), that of the "vegetative intertwining of all living things" (49). In the same vein, Gebser speaks of "the late drawings, paintings, and frescoes of magic man, in which man's merger with nature is portrayed so graphically that the entire picture is nothing but a plant-like amalgam" (52). Elsewhere Gebser alludes to magic man's "merger with the primeval forest" (51). These clues support the conclusion that the one-dimensional sphere is the sphere of simple vegetative life.

A few words on etymology may be instructive here. As previously noted, the *anima mundi* is a general concept "founded on the view that the [nonhuman] world is productive of life and animation, and can therefore be regarded as itself animate" (Flew 1979, 377). The word "animal" is of course linked with the idea of "animation," but so is the word "vegetable." In its adjectival form, "animal" derives from the Latin *animalis,* "living, animate,"[3] and "vegetable" is "from LL. *vegetabilis,* animating, hence, full of life, from L. *vegetare,* to enliven, quicken."[4] Given the order of evolutionary emergence, the *vegeta* would be the older, more primal aspect of the *anima mundi,* that which is associated with reproduction, sexuality, and vital life-energy.

Above, in seeking to elucidate animal emotion, we turned to Morris Berman's *Coming to Our Senses* (1989). Recall Berman's basic distinction between the wild and the tame. The domestication of animals by human beings corresponded to the human denial of true animality. What we were not able to tame, we eventually repressed. Does Berman's analysis extend to the vegetable realm? His primary focus is on *animal* life. Thus, he does no more than make passing reference to the parallel between tame and wild animals, on the one hand, and edible and inedible plants, on the other. Nevertheless, it appears reasonable to conjecture that if the early human being was intimately attuned to the two-dimensional realm of animal emotion, at an even earlier point in the development of the three-dimensional *cogito* there was an attunement to the one-dimensional "vegetative-vital" order of nature, the sphere of primordial sensuality, life-energy, impulse and instinct. It would seem to follow that, if the wild emotionality of the *anima* is repressed in current human affairs, the primal sensuousness of the *vegeta* is even more thoroughly overshadowed. Indeed, assuming the correctness of Gebser's conclusion that the vegetative domain entails the complete merger and intertwinement of life, it is clear that the present-day self-absorption of the isolated individual is far removed from such selfless interrelatedness.

Let us now look for confirmation of the *one-dimensionality* of the vegetative sphere. We saw above that Lachelier (1962) regarded animal consciousness as essentially emotional and as played out in two-dimensional space. While he was not as explicit about vegetable consciousness, the implication is clear that, indeed, it would be one-dimensional. For Lachelier set forth a dimensional hierarchy of being, from line to plane to three-dimensional space (166–70); he described the hierarchy of nature, from mineral to vegetable to animal to human (154–55); and he explicitly associated humans with three-dimensionality and animals with two-dimensionality. It therefore follows that vegetables must be one-dimensional. But did Lachelier relate the vegetative sphere primarily to sensation and sensuality, as we have done? He did not.

First of all, while Lachelier identifies the two-dimensional animal world as one suffused with emotion, he appears to combine or conflate emotion with sensation. He says, for example, "the *self* is at once the will to live and the fundamental emotional state. . . . Such is . . . our sensuous *self* or the animal *self* in us" (172). Similarly, Lachelier associates sensation with "desire, or purpose," and the latter, in turn, with "two-dimensional space or surface" (168). As for the one-dimensional sphere, Lachelier claims that "the vegetable has no outer senses and nothing external can exist for it. There is, therefore, room in its consciousness only for the obscure feelings which doubtless express in it the slow evolution of nutritive and reproductive tendencies" (155). So Lachelier evidently would conclude that while the feelings and sensations of the two-dimensional animal world are well developed, in the one-dimensional vegetative realm, they are faint and incipient.

In the foregoing account, Lachelier seems to presuppose a dualism of inner and outer reality. Because the "vegetable has no *outer* senses and nothing *external* can exist for it," the assumption appears to be that it is limited to an insular, *inner* world where there is "room in its consciousness only for . . . obscure feelings." Apparently, then, concrete experiences depend entirely on exposure to outer reality: without external senses, there can be little or no sensation or feeling. In contrast to this view, what I am proposing is that the very division between inner and outer reality only first arises in earnest with the emergence of two-dimensional being, and that, in this domain, sensation is subordinated to feeling. To be sure, we can say with Lachelier that the "vegetable has no outer senses," but this hardly means that it simply does not sense. Rather, vegetative consciousness would be *less* limited than animal consciousness in this regard. The world of the *vegeta* is governed by the relation of *pars pro toto* that Gebser ascribed to magical consciousness, the relation of total intertwinement that precedes the articulation of the inner-outer split. It is this split that calls forth emotion, for emotion requires a *psyche,* an individualized soul over against which there is cast an exterior world; when the division occurs, mature emotion takes precedence over sensation, just as mature cognition rules both sensation and emotion in the three-dimensional domain. But *prior* to the division, there is *sheer sensuality,* an all-encompassing vital participation in a world that, far from being external to consciousness, is fused with it.

The completely sensuous, vegetative character of the one-dimensional sphere is intimated by Ouspensky. We have already heard Ouspensky's conclusion that the animal realm is governed by emotion and is two-dimensional. With respect to the lower-dimensional domain, he says, "For the being possessing sensations only, the world is one-dimensional" (1970, 74). Although Ouspensky does not go into detail on the *vegetative* nature of the one-dimensional sphere, he does note the following: "Those influences which proved to be beneficial for a given species during the vegetative life, with the transition to the more active and complex animal life begin to be sensed as *pleasant,* the detrimental influences as unpleasant" (77–78). The implication is that while the two-dimensional sphere of animality is ruled by emotion (the experience of pleasure and displeasure being fundamental), the lower-dimensional realm of the vegetable is pre-emotional; indeed, it is purely sensuous.

Vegetative sensuality is fundamentally concerned with life as such, and this life principle is one of *reproduction.* I have examined the issue elsewhere:

> The evolution of life from the inorganic depends on the emergence of a classification or description of events that is distinct from the events themselves. In other words, life depends on the separation of a genotype from phenotypic being. Living systems are defined as systems that can reproduce themselves, and self-reproducing units must possess self-descriptions from which the reconstruction can be carried out. (Rosen 1994, 37)

Now, the appearance of a genotypic self-description, one in which the macrocosm is now "mirrored in microcosm" (37), essentially requires a *hyperdimensional reflection.* That is, in reproducing itself, the system must turn back upon and reenter itself via a movement through an added dimension. An indication of this is found in biologist Clifford Grobstein's (1973) concept of "neogenesis," the emergence of novel properties at higher levels of organization. Grobstein's case in point is the process of folding in the protein molecule. The emergent property is that of the enzyme, a function that is vital to life. According to biologist Howard Pattee, "Without enzymes there would be no DNA replication, no transcription to RNA, no coding, and no synthesis" (1973, 99)—that is, no life. Grobstein demonstrates that the enzyme function appears, in effect, as a hyperdimensional connecting of local areas that are widely separated in lower-dimensional space.

In Grobstein's illustration (fig. 6.3), we see how a linear chain of amino acids (basic constituents of the protein) is folded into a three-dimensional configuration called a "native protein." The significant fact about the folded structure is that:

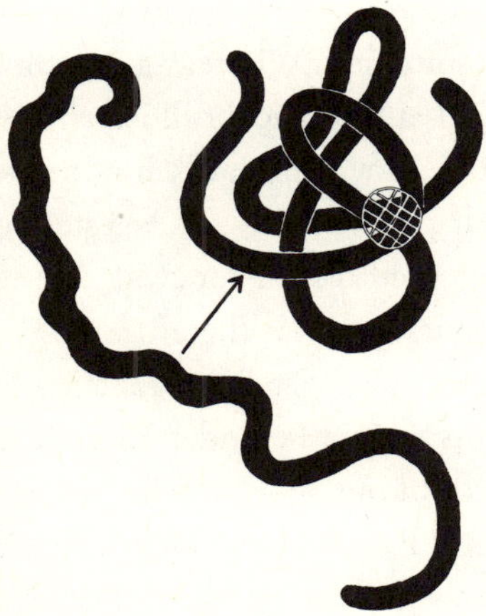

Figure 6.3. Conversion of linear chain of amino acids into three-dimensionally configured native protein (reprinted from *Developmental Biology Supplement*, vol. 2, C. B. Anfinsen, "Self-Assembly of Macromolecular Structures," pp. 1–20, © 1968, with permission from Elsevier)

> In this "native configuration" a critical "new" property appears—an enzymatically active site [the cross-hatched region of fig. 6.3] has been formed which was not present previously. Note that the site consists of several local areas of the linear order which are widely separated in . . . the unfolded . . . configuration, but are brought together in the so-called native one. . . . The linear distribution of amino acids does not by itself produce enzymatic activity—suitable folding is necessary for the activity to *arise*. (Grobstein 1973, 40–41)

Grobstein's demonstration that the emergence of the life function requires "suitable folding" points to the reflexive, self-referential character of the process. The extra dimension is introduced not merely by an act of extension in a new direction perpendicular to the old, but by a turning over upon itself of the old in which "widely separated" elements are "brought together."

The idea that life essentially entails self-reference is the crux of the well-known principle of *autopoiesis* articulated by Varela, Maturana, and Uribe (1974). In the words of systems theorist Erich Jantsch (1980), "Autopoiesis refers to the characteristic of living systems to continuously renew themselves and to regulate this process in such a way that the integrity of their structure is maintained. Whereas a machine is geared to the output of a specific product, a biological cell is primarily concerned with renewing itself" (7). In the same vein, Jantsch notes that "*autopoiesis* . . . is from the Greek for 'self-production.' . . . A system is autopoietic when its function is primarily geared to self-renewal. . . . An autopoietic system refers in the first line to itself and is therefore called *self-referential*" (33). More recently, the anthropologist Peter Harries-Jones (1996) observed that autopoietic systems cannot be understood in terms of the old dyadic logic of mechanistic thinking, but require a "triadic," self-referential logic; Harries-Jones cites Paul Ryan's "kleinforms" as illustrative of such a logic. In chapter 2, we encountered Ryan's (1993) threefold logic of contained, containing, and uncontained in his most *basic* kleinform, his blueprint for the Klein bottle (fig. 2.4). What I would add, however, is that if the self-reference we are dealing with does not involve human or animal life but is limited to relatively simple, vegetative life, a self-referential system of lower dimension than the three-dimensional Klein bottle would be required, namely, that of the one-dimensional lemniscate. So the new dimension associated with the emergence of life would be embodied in the lemniscate, our first and simplest self-reflective body of paradox (the *zero*-dimensional situation involves no self-reflection, no differentiation of subject and object whatsoever). It is at this level that, while a microcosmic mirroring of the macrocosm has been achieved, microcosmic beings have no individuality in themselves. Here, "each and every genotype is a description of the same phenotypic truth; the microcosmic reflection upon macrocosm is identical from one microcosm to another. With individuality only serving to mirror the universal, there is no individuality per se" (Rosen 1994, 40).

However, doesn't the work of Grobstein contradict the above conclusion about the emergence of life? Whereas I am proposing that this event be associated with the transition to one-dimensionality, doesn't Grobstein relate it to *three*-dimensionality, in that the linear chain of amino acids is folded into a three-dimensional configuration? The apparent contradiction is resolved when we remind ourselves of a distinction that bears repeating at this juncture: the *ontological difference*. Grobstein's dimensional observations are strictly *ontical* in nature. The production of the "three-dimensional" native protein implies no transformation of sub-objective dimensionality as such but is limited to the transformation of *particular objectified beings,* "one-dimensional" amino acids, appearing within the purportedly changeless three-dimensional framework of the human *cogito.* In contrast, when we relate the emergence of life to the movement from zero-dimensional spacetimelessness to one-dimensionality, we are referring to an *ontological* happening, to the creation of a self-referential order of dimensional Being. Such a process is invisible when the ontical posture is assumed. In adopting that stance, the "inner life of the line" is completely obscured.

5. THE FIRST ORDER OF DIMENSIONAL FLESH: MINERAL INTUITION

First flesh corresponds to zero-dimensionality. We have seen that each of the higher-dimensional orders finds its flesh as a topodimensional body of paradox. Each thus entails a dialectic of object, subject, and space; of contained, uncontained, and containing. Can the same be said of the zero-dimensional body? The answer must be no.

Brought to light in chapter 3 is the clear-cut difference between zero-dimensional and higher-dimensional structures attesting to the non-dialecticity of the former. We have considered the dialectic in terms of the interplay of planes in three-dimensional space, of lines in two-dimensional space, and of points in one-dimensional space. The relevant generalization suggested by these relations is that the $n - 1$-dimensional bounding element of n-dimensional space itself serves as space in relation to a lower-dimensional bounding element (see chapter 3). What we encounter in the zero-dimensional domain is the absence of bounding elements and, thus, the absence of space. The zero-dimensional point that bounds one-dimensional space does not translate as a space that possesses its own, lower-dimensional bounding elements because there is no dimensional element below zero. Thus, if the dialectic essentially entails the interplay of opposing $n - 1$-dimensional bounding elements in n-dimensional space, the zero-dimensional world would not be dialectical, since there would be no bounding elements and no containing space. Indeed, as I have already intimated, the zero-dimensional situation is characterized by a complete lack of differentiation of the containing, contained, and uncontained.

The undifferentiated character of zero-dimensional reality does not make it "primitive." As I proposed above, the term "primitive"—which connotes immaturity, deficiency—should apply only to the initial stages of development in a dimensional regime in which the potential for further development does exist. In the zero-dimensional realm, there is no potential for development; no growth can take place, and none is required. The possibility of growth, of developmental progress (indeed, of time itself), only first arises with the emergence of higher dimensionalities. It is therefore not appropriate to apply developmental criteria to the zero-dimensional realm.

For the higher dimensionalities, development does not just involve achieving fulfillment, realizing completion or wholeness. That is because the transpermeative wholeness that results from the "sealing of the hermetic vessel" is paradoxical; it is a wholeness that is just as much a partiveness, i.e., it is a *(w)holeness* (contra idealism, we do not have *simple* wholeness here, wholeness *as opposed to* partitiveness). This attests to the ontological nature of higher-dimensional flesh. Motherly Being is what is realized with the individuation of the flesh, not merely a "fatherly self." In the case of zero-dimensional flesh, the situation is still different. While we surely cannot speak of an idealized quest for the self, neither are we dealing with the realization of paradoxical Being. There is no attainment of (w)holeness in the sphere of "absolute nothingness" (Tanabe 1986) since what we have here at all times is *neither* whole *nor* part. When Gebser equated zero-dimensionality with "original wholeness" (1985, 66), he apparently failed to grasp its logic of neither/nor.

Let us also note that zero-dimensionality is the only order of the flesh that is non-symbolic. For we know that, like three-dimensional cognitive flesh, both one-dimensional and two-dimensional orders of the flesh entail subject-object differentiation, though of a lesser degree than that of the *cogito*. It is this differentiation that opens the gap between an external world and one who reflects upon said world. The subject who thus reflects, being a step removed from the now-objectified world, is not simply present in the world but knows it through *re*-presentation. Animal and vegetable subjectivity, then, could be said to be *symbolic,* to have "language," though of an order that is radically less reflective than, and different from, the human. Perhaps it is especially difficult to think of the *vegeta* as having language, yet we have found that one-dimensional flesh, in effect, does re-present the world—not cognitively, as in human thinking, but biologically, by *reproducing* the world, mirroring the macrocosm in a microcosmic description of it. Undoubtedly, the differentiation of microcosm and macrocosm is exceedingly weak in the *vegeta,* yet a modicum of differentiation does exist. It seems, then, that only zero-dimensional flesh, having no subject-object differentiation whatsoever, is without language, is non-symbolic.

Now, while we cannot articulate zero-dimensionality in categorical terms of any kind, I suggest that there is indeed a mode of human functioning that gives us some access to it. Consider the case of the one-dimensional realm of the senses. In the previous section I indicated that the perspectival sensory experience of the human adult is but a pale reflection of the unbridled sensuality of the *vegeta.* Yet we humans obviously do experience the world through our senses; we are not just thinking creatures but sensing creatures as well, even though our senses bear the stamp of thinking in the ontical stage of development and are attenuated accordingly. The same idea applies to the two-dimensional realm of feeling. The emotion experienced by the perspectival *cogito* may only weakly reflect the pure feeling of the *anima,* but experience emotion we do. Then should it not similarly be true that, although the three-dimensional *cogito* cannot describe zero-dimensionality in definitive symbolic terms, there should be a form of human awareness that corresponds to zero-dimensionality, though the perspectival *cogito* would actuate it in a highly attenuated fashion that bears the stamp of symbolic functioning? What is this mode of awareness? We may wish to turn once again to Gebser for some clues. From our earlier examination of his work, we were able to see that his mythic structure is related to emotion and his magical sphere to sensuality. But Gebser provides us with scarcely a hint about the functional correlate of *archaic* consciousness. Let us turn, then, instead to C. G. Jung.

In *Psychological Types* (1971), Jung maintained that the functioning of the human psyche entails four basic forms of activity. These functions cannot be reduced to each other and are invariant relative to the specific content of experience, which changes from moment to moment. The four functions of which Jung spoke are *thinking, feeling, sensing,* and *intuition.* Jung viewed thinking and feeling as diametrically opposed to each other and as "rational." That is, both are *representational* functions;

they involve a reflection on, and an (e)valuation of, externally experienced reality from the inner perspective of the subject. In the case of thinking, the subjective base is abstract and disembodied (e.g., discursive reasoning, logical deduction, mathematical calculation), while in the case of feeling it is more concrete and embodied (I value what lies outside me in terms of the likes and dislikes of my body).

Beyond these "rational" operations lie the "irrational" functions of sensing and intuition, also seen to make up a pair of opposites. Irrational activity can be characterized as more "presentational" than representational; that is, it entails a more immediate (less reflective) reaching out to, grasping, and taking in of that which is other in the field of experience. In sensory perception, we experience the world discretely, dividing the objects we encounter into units that are bounded from each other. On the other hand, intuiting the world means apprehending it as an undivided whole, as in cases of "hunches" or "visions," where we are seized by a nebulous impression about the general course of events but are unable to account for the source of the presentiment. Note that, while the irrational functions are attenuated in adulthood, being constricted by the ascendant influence of rational thinking, according to Jung these functions originate in an "infantile and primitive psychology" that serves as the "matrix out of which thinking and feeling develop as rational functions" (Jung 1971, 454). Evidently, intuition would be the *most* primal form of experience, for while sensation involves "perception via conscious sensory functions" (538), intuition entails "perception by way of the unconscious" (518), through "dreams and fantasies" (539), through a mode of operating in which one surrenders oneself "wholly to the lure of possibilities" (519).

I propose that the four basic functions described by Jung can meaningfully be correlated with the four basic lifeworld dimensions, our four orders of the flesh. The connection is already evident for the three higher orders of dimensional flesh, since we have established that fourth flesh is indeed the domain of thinking, third flesh the emotive sphere, and second flesh the realm of the senses. It appears reasonable to hypothesize, then, that the final, most primordial order of the flesh, zero-dimensional first flesh, is tied to Jung's most primordial function, *intuition*. It is true that Gebser, for his part, says relatively little about the functioning of zero-dimensionality, describing it largely in negative terms (it has no perspectivity, no "organ emphasis," no "sign," etc.), or simply leaving blanks in the zero-dimensional rows of the tables he uses to characterize the various attributes of the dimensions. However, Gebser does leave us with the cryptic intimation that zero-dimensional consciousness expresses itself through "presentiment" (1985, 142), and this corresponds to what Jung says about the way in which intuition is manifested. Moreover, Gebser associates zero-dimensionality with "deep sleep," which is consonant with Jung's association of intuition with the deepest stratum of the unconscious.

But the intuitive function as described by Jung must be distinguished from the classical intuition I have spoken of in previous chapters. The latter is a sublimation of the former. Consider Descartes' definition of intuition:

> *By intuition* I mean, not the wavering assurance of the senses, or the deceitful judgment of a misconstructing imagination, but a conception, formed by unclouded mental attention, so easy and distinct as to leave no room for doubt in regard to the thing we are understanding. It comes to the same thing if we say: It is an indubitable conception formed by an unclouded mind; one that originates solely from the light of reason, and is more certain even than deduction, because it is simpler. . . . Thus, anybody can see by mental intuition that he himself exists, that he thinks, that a triangle is bounded by just three lines, and a globe by a single surface, and so on. (1628/1954, 155)

This penetrating mode of insight—which was carried forward by Kant in his a priori certitude about space and time, by neo-Kantian intuitionists such as Brouwer and Weyl, and by Husserl in his rationalist approach to phenomenology—reflects the ascendancy of thinking over the intuitive function. In this intuition dominated by the head, little remains of the *body;* in this crystalline sphere of "pure consciousness," the unconscious that Jung deemed so crucial to concrete intuitive functioning is largely repressed.

Now, in the previous section, I proposed that Gebser's account of the one-dimensional and zero-dimensional realms be amended with regard to the organ emphasis with which they are respectively associated. Rather than relating one-dimensionality to the guts and not specifying an organ correlate for zero-dimensionality, I argued for the plausibility of correlating one-dimensionality with the genital organs or groin, and zero-dimensionality with the guts. Is the gut not associated with concrete intuition? Do we not commonly speak of intuitive experience as occurring at the "gut level"? Apparently, this connection is borne out by esoteric systems such as Sufism, where individuals who derive their energy primarily from the gut are identified as intuitive types (whereas the brain and heart are associated with thinking and feeling types, respectively). However, the esoteric literature is filled with ambiguity on this matter, both within and among the various systems set forth. Therefore, any conclusions we reach must be taken as provisional.

What may we say about the *phylogenetic* implications of first flesh? If three-dimensional flesh is human, two-dimensional flesh is animal, and one-dimensional flesh is vegetable, I propose that zero-dimensional flesh corresponds to the *mineral* realm. Of course, the ordinary *human* experience of the mineral kingdom is a far cry from the zero-dimensional sphere itself. From the standpoint of the perspectival *cogito,* the world of minerals consists merely of discrete inanimate matter: crystals, metals, rocks, the stuff of inorganic chemistry and physics. Such an objectifying, deadening representation of the mineral domain hardly does justice to minerality as such. To apprehend the zero-dimensional *minera* in a deeper way, we turn once more to the concrete intuitions of *alchemy,* that ancient source of chemistry and physics.

In alchemical terms, the mineral kingdom gains ultimate symbolic expression in the *lapis philosophorum* or *philosopher's stone.* This mysterious substance certainly does not consist of matter that is merely inert, objectively delineable. According to Jung, the lapis was "the *filius macrocosmi* as opposed to the 'son of man,' who was the *filius microcosmi.* This image of the 'Son of the Great World' tells us from what source it was derived: it came not from the conscious mind of the individual man, but from those border regions of the psyche that open out into the mysteries of cosmic matter" (1967, 96). Later in the same volume, Jung says that "alchemical projections . . . point back to something primeval, to the apparently hopelessly static, eternal sway of matter. . . . They show us, as the redemptive goal of our active, desirous life, a symbol of the inorganic—the stone—something that does not live but merely exists or 'becomes'" (238). Correlated with the stone is the unconscious psyche, which "is refractory like matter, mysterious and elusive, and obeys laws which are . . . nonhuman" (239).

In *Psychology and Alchemy* (1968), Jung notes:

> The Christian projection acts upon the unknown in man. . . . The pagan projection, on the other hand, goes beyond man and acts upon the unknown in the material world, the unknown substance which, like the chosen man, is somehow filled with God. And just as, in Christianity, the Godhead conceals itself in the man of low degree, so in the "philosophy" [i.e., in alchemy] it hides in the uncomely stone. In the Christian projection the *descensus spiritus sancti* stops at the *living body* of the Chosen One, who is at once very man and very God, whereas in alchemy the descent goes right down into the darkness of inanimate matter. (304)

Although the alchemist's projection may make it appear that matter requires the descent of spirit in order to be animated, Jung proceeds to relate the "inanimate matter" of the apparently "uncomely stone" to the *anima mundi* (304), which we know to be a fundamental principle of animation in its own right. What this suggests is that the *anima mundi* is not only associated with the animal and vegetable domains of nature, as established in previous sections, but also, and most primordially, with the "inorganic" mineral realm. It would seem then that *all* of nature's dimensions entail dynamic activity. However, this conclusion is blunted by the logic of absolute negativity that governs the zero-dimensional sphere. While the *minera* surely is not inert, neither can we really describe it as active. For, when it comes to such terms of opposition, in the sphere of absolute nothingness the rule is *neither/nor.*

What is the *topological* correlate of first flesh? If the fourth flesh constituting the human *cogito* is embodied by the three-dimensional Klein bottle; if the flesh of the *anima* is given by the two-dimensional Moebius body; and if the vegetative lifeworld is expressed by the one-dimensional lemniscate, our topological bisection series tells us that the corpus of the *minera* must correspond to the zero-dimensional *sub-lemniscate*. Of course, when limited to the three-dimensional observational framework of the *cogito,* the sub-lemniscatory body (fig. 3.3) will appear as do the lemniscate and the Moebius strip within this framework: as a two-dimensional object embedded in three-dimensional space. I have established that, in their *own* spheres of operation, the Moebius structure and lemniscate actually are not merely planar objects (surfaces) contained in three-dimensional space. The former, when fully developed, is the *sub*-objective order of Being associated with self-containing two-dimensional space, and the latter is the sub-objectivity that involves self-containing one-dimensional space. What of the sub-lemniscatory body? In its native sphere, it clearly would not be a planar object. But would the sub-lemniscate constitute a sub-objectivity? Would it entail a self-containing zero-dimensional space? To both questions the answer of course must be no, for we have learned that the zero-dimensional body is non-dialectical, is utterly undifferentiated with respect to object, space, and subject. The true sub-lemniscate, then, is neither an object, nor a subject, nor is it a space.

The sub-lemniscate is indeed a far cry from the planar model shown in figure 3.3. We know that three-dimensional cognitive subjectivity, having become differentiated from its object-world, subsequently reenters that world via the Proprioceptive movement through the hole in its "special object," the Klein bottle. Although the *anima* and *vegeta* entail less subject-object differentiation than does the *cogito,* they undergo analogous processes of dialectical development; in both cases, it seems there would have to be a lower-dimensional, nonhuman counterpart of the "special" or "holy object," i.e., of the topological container with its necessary hole.[5] In the case of the *minera* it must be different. Because the mineral realm involves no differentiation of subject and object, there could be no "special object" for a mineral subjectivity to reenter through the object's hole, and thus no individuation. Perhaps we might say that, in the absence of both subject and object, there is "only the hole," that the *minera* is the "holiest" of dimensional bodies, is "all hole." This would accord with the notion of "absolute holeness" previously put forward as the cyclogenic counterpart of Tanabe's "absolute nothingness": at the outset of dimensional vortex generation, there is only the vortex's hole. With this in mind, I wonder whether it is merely an etymological coincidence that the Latin word *minera* is associated with a *mine,* defined in the dictionary as "a large excavation made in the earth"—i.e., as a large *hole!* (We shall see in the coming chapter that while the *minera* enacts no Proprioception of its own, its climactic derepression does require additional Proprioceptions on the parts of the other topodimensional bodies.)

CHAPTER SEVEN

CO-EVOLVING LIFEWORLDS FLESHED OUT

Having carried out an identification of the functional and phylogenetic attributes of the four main dimensions of the flesh, and having obtained a preliminary and partial idea of their patterns of interrelationship and development, we shall now undertake a detailed, comprehensive study of how these orders of embodied Being evolve in relation to one another. What this means is that the topodimensional matrix of table 3.2 that was abstractly "set in motion" via table 5.1 will now be fleshed out phylofunctionally. Each entry of table 5.1 will be condensed into a specific mode of sub-objective (psycho-physical) functioning, each with its place in the drama of inter-dimensional development, each as a "note" in the unfolding "symphony" of the dimensional spheres.

Before proceeding, I must acknowledge the limitations of my analysis. Given the purpose and scope of this book, in no way could I even have begun to do justice either to the science of phylogeny or to the vast bodies of psychological and phenomenological research on the functions of intuition, sensation, emotion, and cognition and their respective patterns of development. For that, an additional volume would be needed, at the very least! Also, while I hope to make an adequate case for all of the dimensional specifications I am about to offer, there is no denying that conjecture plays some role. Be that as it may, the ultimate intention is to transmute the contents of these specifications into *self-containing acts of Proprioception.* In the process, all hypotheses are to be tested in the crucible of lived experience.

1. PHYLO-FUNCTIONAL DISTILLATION OF THE TOPODIMENSIONAL SPIRAL

Table 7.1 maps the concrete intricacies of Topogeny, distilling in phylo-functional terms the several windings of the dimensional spiral originally given in table 5.1. Notice that the new table is formatted after the design of table 5.1(a) rather than 5.1(b). Whereas the latter parses the circulations of the dimensional spiral, teases them apart for easier identification, the former more effectively brings out the nonlinear interwovenness of the windings. It is the intertwinement of the Topogenetic circulations that we must now come to better appreciate.

Where shall we enter table 7.1? Let us begin *where we are now.* It is early in the twenty-first century and human culture is still primarily governed by the perspectival *cogito,* by the "I think." For the most part, the three-dimensional Kleinian mother, operating through her "fatherly child," continues to be engaged in the "clockwise," forward-oriented, ontical activity of self-Appropriation. Proprioceptive acknowledgments of the gifts of wild Being have not yet commenced in society at large. In table 7.1, this situation is reflected in the matrix corresponding to the fourth stage in Kleinian Topogeny (KB S4). Here, with all lower-dimensional, nonhuman matrices having been reduced, "man," the "rational animal," holds sway.

The dominant functions of KB S4 are given in the single cell that is filled. Chief among them is rational thinking. It is this "mental-rational structure of consciousness," as Gebser calls it, which makes up the core of the quasi-mature *cogito.* The other three modes—mental emotion, visual perception, and mental intuition—correspond respectively to the three noncognitive functions: feeling, sensing, and intuition. By KB S4, these functions have been appropriated by the *cogito* and serve on its behalf; thus they have taken on a sublimated form that is over-colored by cognition. To bring out the meaning of the integral fourfold structure of KB S4 and confirm its operation on the contemporary scene, let us turn to the research of Trigant Burrow. (Be assured that all the terms and features of table 7.1 will be explained as we work with it in this chapter.)

In chapter 2, I mentioned Burrow's call for human beings to gain proprioceptive awareness of the organismic basis of thinking and language, these being the activities he deemed responsible for the fragmentation of human society. Burrow's term for the "I think" operating behind the scenes to foster division is the *"I"-persona.* Through ceaseless acts of objectification, this anonymous thinking subject governs every facet of human experience and behavior in today's world. Importantly, the functioning of the "I"-persona has a distinct anatomical locus. It is centered in what Burrow called the "cerebro-ocular" region (*Science and Man's Behavior,* 1953, 526), that is, in the cerebral cortex of the brain and in the organ of vision associated with it. Elaborating on Burrow, I proposed elsewhere (Rosen 1999) that the objectifications carried out by the "I"-persona or *cogito* are facilitated by the well-established neurophysiological habit of *binocular convergence:* our eyes continually operate in concert to bring particular objects into focus as well-bounded figures that stand out from their backgrounds, stand apart from one another, and stand before our detached gaze.[1]

Burrow was not alone in observing an intimate link between abstract cognitive activity and the visual system. Speaking of the "sight-intellect" equation, communications theorist Walter Ong contends that "we would be incapacitated for dealing with knowledge and intellection . . . without conceiving of intelligence through models applying initially to vision" (1977, 134). Moreover, "The drive to . . . consider intellection and understanding by analogy with vision . . . corresponds to the drive to objectify knowledge, to make it into something which is clearly thing-like" (140). Ong's conclusions are confirmed by phenomenologist Drew Leder, who stressed the fact that "vision lends itself to . . . analogies with intellect" (1990, 117). According to Leder,

> The "mind's" knowledge of a stable, copresent, external world is . . . largely derived from the eye. Thus, in a model such as Descartes's, which emphasizes the epistemological subject [the *cogito*] and regards truth as involved with definiteness and permanence, visual experience will be attended to above all others. It emerges as "the noblest and most comprehensive of the senses" as Descartes writes at the beginning of *Dioptrics,* his work devoted to vision. (117)

Table 7.1. Phylo-functional distillation of spiral of dimensional development

Appropriation					Proprioception
S4					S5
(KB)					(KB)
				3d/3d **(Quasi)Mature Cogito** Rational Thinking Mental Emotion Visual Perception (II) Mental Intuition (III)	
S3					S6 (KB)
(MB, KB)			2d/2d **(Quasi)Mature Anima** Love Hearing Emot. Intuition (II)	2d/3d Mythic Thinking Visual Perception (I) Mental Intuition (II)	S4 (MB)
			3d/2d [Mental Emotion]		
S2		1d/1d **(Quasi)Mature Vegeta** Smell (Taste) Sensuous Intuition	1d/2d Anger Emot. Intuition (I)	1d/3d Magical Thinking Mental Intuition (I)	S7 (KB)
		2d/1d [Hearing]			S5 (MB)
(LB, MB, KB)		3d/1d [Visual Perception]			S3 (LB)
	0d **Minera** Null Intuition	0d/1d Touch	0d/2d Fear	0d/3d Archaic Thinking	
S1	1d/0d [Sensuous Intuition]				S8 (KB)
(LB, MB, KB)	2d/0d [Emotional Intuition]				S6 (MB) S4 (LB)
	3d/0d [Mental Intuition]				

Now, Burrow's work makes it clear that visual perception is not the only concomitant of cognitive rationality. The objectifying projections of the "I"-persona also entail an *emotional* involvement. Burrow terms this "affect," and defines it as "a reaction that is subjectively biased through the artificial linkage of feeling and symbol" (1953, 525). Elsewhere, Burrow (1937) describes affect as an "amalgamation of feeling and mental image." Affect results when "the organism's total pattern of motivation is overtly superseded by the symbolic or partitive pattern of behavior" (411). Partitive processes are those "mediated by the organism's part-functions restricted to the cephalic segment [or cerebro-ocular zone]. . . . The artificial supremacy of man's partitive or affecto-symbolic mode of behavior over the primary motivation of the organism as a whole constitutes the physiological basis of both individual and social neurosis" (Burrow 1953, 530–31). So, from the "total feeling" of the organism (Burrow 1937, 411) that had prevailed early in human development, a constricted form of emotionality comes to the fore, one that is linked to the symbolic or mental operations of the "I"-persona (accordingly, this abstract affect is termed "mental emotion" in table 7.1). The result, as Burrow sees it, is an order of neurosis that far surpasses the aberrant behavior of particular individuals:

> There is futile conflict and dissension at each step up the ladder of complexity in social interaction. There is prejudice and hostile competition within the close family group, between one family and another, between groups of differing social, educational and economic status. There is bigotry and conflict between peoples of differing religious convictions, political belief, racial extraction and national boundaries. At the most massive level of human interaction is the deadly and implacable conflict between vast aggregates of peoples with divergent ideological concepts and forms of government—between totalitarianism and liberalism, communism and democracy. (1953, 27–28)

These dire social consequences all are rooted in the "partitive" emotional functioning that has taken hold of humankind.

The mental emotion arising in S4 of the Kleinian winding is "partitive" in more than a neurophysiological sense. An *ontological* segmentation occurs in making the transition to S4. The objectifications enacted by the "I"-persona in this stage of Ontogeny depend in the first instance on the *splitting of object from subject* that has taken place. The subject that does the objectifying now constitutes an isolated center of identity from whose narrow sphere of interests the interests of the other are excluded. With this partitioning of self from other, only the interests of the self are of real concern and the other is exploited to serve those interests. Thus Burrow, in speaking of the evolution of "man's feeling, man's motivation," asserts that "[p]art-interest or self-advantage—advantage accruing to the 'I'-persona—came to alter and obscure man's total outlook or attention. Attention became partitive, digressive. It became self-attention. . . . This mode of attention . . . tends to divert interest in the object to interest in oneself—in one's self-image" (1953, 204). Given the truncation of identity that reduces it to an exclusionary self, feelings now revolve around relations with *externalized* others. My feelings about you will be "positive" only if our relationship reinforces *my* self-image; otherwise, I may react to you with irritability, be arrogantly dismissive, or downright hostile. Typically, there seems to be the need for me to prove myself to you so as to earn your affirmation of me, or, failing that, to show how "right" I am compared with you. These are some common examples of what Burrow meant by *affect,* of what I mean by *mental emotion.* I should add, however, that Burrow tended to regard the "partitive affecto-symbolic reaction" as a merely adventitious, pathological deviation from "man's organism as a whole" (1953, 203), from the "solidarity of the species" (71). In contrast, the coming to prominence of mental emotion is here viewed as a natural concomitant of Being's self-Appropriation. To be sure, the projections of the "I"-persona will need to be retracted in the Proprioceptive stages of Ontogeny, and this will bring to an end the blind rule of the "I." But the initial occurrence of projection cannot simply be deemed a mistake, an errant deviation from an original condition of "wholeness," "purity," and "perfection" (cf. the biblical "Fall"). As a matter of fact, I do not think Burrow would have denied that at least a part of the underlying motivation for his valuable questioning of the "I"-persona *derived* from the "I"-persona!

Does Burrow's research also shed light on the fourth mode of functioning of KB S4: *mental intuition*? Recall Descartes' rationalist definition of intuition: it is "a conception . . . so easy and distinct as to leave no room for doubt in regard to the thing we are understanding. . . . It is an indubitable conception formed by an unclouded mind . . . and is more certain even than deduction, because it is simpler. . . . Thus, anybody can see by mental intuition that he himself exists, that he thinks, that a triangle is bounded by just three lines, and a globe by a single surface, and so on" (Descartes 1628/1954, 155). The sequencing of the last sentence is revealing. Does Descartes not place the mental intuition of one's own existence and one's capacity to think before the particular intuitions of triangle and globe because one's sense of intuitive certainty, of the absolute "rightness" of any particular apprehension one may have, is rooted in the prior, never-to-be-questioned certainty of the "I think"? As Heidegger puts it in his analysis of Descartes, the *cogito* engages in an indubitable act of "self-binding and self-grounding" (1962/1977, 271) wherein its positing "has the peculiarity of first positing that about which it makes an assertion. . . . What it posits in this case is the '*I*'" (279). Although Burrow, for his part, does not deal explicitly with intuition as far as I know, he has much to say about the sense of absolute "rightness" that accompanies the partitive functioning of the "cerebro-ocular" and "affecto-symbolic" systems. According to Burrow, the "absolute, authoritarian mood . . . the sense of right and wrong that is unimpeachable and that is not to be questioned" (1953, 38), derives from the "I"-persona. Similarly, when Descartes was struck by the intuitively self-evident "rightness" of his abstract propositions, the basis for his convictions was the taken-for-granted certitude of the "I."

Note that, in Burrow's analysis, the operation of an unshakable sense of "rightness" cannot be separated from the partitive functioning of the affecto-symbolic and cerebro-ocular systems. All systems act together as interlocking facets of a complex that serves the interests of the "I"-persona. We may say, then, following Burrow, that in the fourth stage of Kleinian Ontogeny, mental emotion, visual perception, and mental intuition function cohesively to support the cognitive activities of the rational thinking subject. But, to repeat, whereas Burrow viewed the workings of the "I" complex as an adventitious deviation from the "solidarity of the species," we presently regard the fourfold action that gains ascendancy in KB S4 as critical to the *individuation* of the human species.

Having situated ourselves in stage four of the Kleinian winding, we are now prepared to explore concretely the evolutional process by which we arrived. But before examining the specific meanings of all the functions given in table 7.1, let us consider again the general pattern of nonlinear interwovenness of the evolutional gyres. We see from the table that S1 is the opening stage of all phylo-functional orders of Topogeny. Here the four windings of the flesh are overlapped. (Bear in mind that, for the zero-dimensional *minera,* in effect we have a *null* winding, since the *minera* undergoes no development.) The prevailing state of affairs in S1 is that of timelessness. Higher-dimensional orders of temporality are but nascently present within the wholly nontemporal milieu of the zero-dimensional *minera.* In their common lack of differentiation with respect to time, the dimensional windings of S1 are undifferentiated from each other. Then, in S2, dimensional divergence commences as the *vegeta, anima,* and *cogito* take their first steps away from the timelessness of the *minera.* Proceeding to MB S3, the stronger temporality of the *anima* arises, after which the *cogito*'s still higher degree of temporal differentiation is manifested in KB S4.

What happens with the subsequent shifting of gears from self-Appropriation to Proprioception? The switch from "clockwise" to "counterclockwise" action brings a *convergence* of dimensional gyres. After the initial Kleinian Proprioception of KB S5, Kleinian and Moebial orders of temporality come into accord. (To fully appreciate the interplay of the S6 Kleinian body and S4 Moebial body, the table 7.1 transition from KB S5 to S6 must be read as carrying forward the mature, integral form of the KB. In S6, the 3d/3d KB does not merely regress to its S3 fractional form, 2d/3d; instead, it *Proprioceives* that form.) The Kleinian-Moebial liaison is followed by the entrance of the lemniscatory body into the harmonic round (KB S7–MB S5–LB S3). In the finale (KB S8–MB S6–LB S4), all four temporal windings are turning together. Therefore, by S8, human and nonhuman development both are, and are not, temporally differentiated from each other, since, now, rather than time or timelessness holding sway, temporal and nontemporal "wheels" have "meshed gears."

To acknowledge the obvious, the preliminary overview I am offering of the stagewise sequential development of several orders of temporality is given solely through the auspices of the *cogito*'s sense of time. While the perspective of the KB S4 *cogito* does have its validity, it is by no means *universally* valid. In the less time-differentiated realms of the *minera, vegeta,* and *anima* themselves, my developmental account naturally would have little meaning. Of course, the ultimate aim is to achieve a *meshing* of cognitive and noncognitive spheres. But for that to happen, we must go beyond a strictly cognitive description of the meshing. In the next chapter, I will attempt to provide a better indication of what this might mean.

Now, the initial entanglement of evolutional windings applies just as much to space as it does to time. Stage-one minerality is utterly spaceless—something that table 7.1, as an ontical, KB S4 form of representation, obviously fails to reflect. In the table, each phylo-functional order is set off from all others by being enclosed in its own well-bounded cell. How different from this is the actual situation that prevails in S1. Despite the impression of simple boundedness conveyed by the compartmentalized layout of the table, topodimensional boundaries are in fact nascent, seedlike, unformed. As a consequence, no spatial extension can exist. There can be no "here" and "there," any more than there can be "earlier" and "later." Dimensional windings are as spatially undifferentiated as they are temporally undifferentiated in this opening stage.

Subsequently, a process of spatial differentiation ensues that is so closely entwined with temporal differentiation that the divergence of dimensional gyres is best described as *spatiotemporal.* The same can surely be said for the stages of dimensional convergence. The four circulations that come into synchrony by S8 express differing degrees of *spatiotemporality.* This abstract sketch of dimensional co-evolution must of course be filled in.

Let us resume our investigation of the specific modes of functioning indicated in table 7.1. The contents of the table reflect the proposition that each of the four main functional modalities—thinking, feeling, sensing, and intuition—consists of four sub-functions. The 16 sub-functions are summarized in table 7.2. Some of the sub-functions given here have already been discussed. In working with Gebser, we studied the four forms of thinking. With the help of Burrow, we examined the four abstract modes of operation dominant in KB S4 (row 1 of the table 7.2 matrix). What remains to be considered are the nine more concrete sub-functions.

Let me begin by conceding another shortcoming of table 7.1. Owing to the table's linearity, it seems that only four functions are operative in the KB S4 matrix. Because the four 4 × 4 Ontogenetic matrices are shown

Table 7.2. Matrix of ontological functions

THINKING	FEELING	SENSING	INTUITION
Rational	Mental Emotion/Affect	Vision	Mental Intuition
Mythic	Affiliative Passion/Love	Hearing	Emotional Intuition
Magical	Anger or Rage	Smell (Taste)	Sensuous Intuition
Archaic	Fear or Foreboding	Touch (Taste)	Null Intuition

as simply separated from one another, the impression is given that, in advancing upward to the fourth matrix, functions appearing in the first three matrices have merely been left behind. We must not read the progression of matrices in this linear fashion. Just as we have seen that different windings of Ontogeny overlap in the same stage, we must now recognize that, in a similar manner, the stages themselves overlap. This actually follows from the discontinuous character of the stage-to-stage movement. Yes, the divisibility of the linear space-time continuum is *generated* in the course of dimensional development, but by a form of action that is itself indivisible. Thus, rather than involving a continuous passage between juxtaposed points that are externally related to one another, Ontogeny entails a "quantum leaping" to new modes of integral action (see chapter 5), with each novel mode related to its predecessor in an *internal* way.

The overlapping entailed in discontinuous phase transitions is well modeled by the visual structure employed in the previous chapter: the Necker cube (fig. 6.1(b)). In taking the leap[2] from one perspective of the cube to the other, the first perspective is certainly eclipsed, but, because of the intimate interdependence of perspectives, the occlusion is incomplete. As Merleau-Ponty put it, "the hidden face of the cube radiates forth somewhere as well as does the face I have under my eyes, and coexists with it" (1968, 140). For Merleau-Ponty, the "coexistence" of the hidden face with the visible face means an intertwinement so thorough that the "hidden" face could not *merely* be hidden; it, too, must fully be present, albeit in a less explicit, background fashion. This is the sense in which antecedent matrices and their functions are not simply left behind with the passage from one stage of Appropriation to another (we will see that the stages of Proprioception are modeled by the *perspectival integration* of the cube, as discussed in chapter 6). Instead of being totally muted, they continue to "echo in the background." Or we may use Sartre's visual metaphor and say that the stage-to-stage matrix reduction given in table 7.1 signifies a stratification of functioning in which "translucent layers of consciousness" are created.

Sartre's idea of the "translucency of consciousness" (1943/1975, 309) constitutes a rejection of the Freudian dualism of unconscious and conscious wherein the former is opaque to the latter. For Sartre, consciousness is not categorically compartmentalized in this way. He would have us think instead that, while there are areas of our awareness that are not fully transparent to us, some "light" always gets through, as if through a translucent screen. Similarly, in advancing by "quantum leaps" to subsequent stages of clockwise Ontogeny, antecedent matrices are overshadowed but by no means entirely eclipsed; some "light" from each one of them manages to get through. Functioning becomes gradated in such a way that a differential penumbra is formed, with the more primordial sub-functions being shaded more darkly, yet never completely obscured. It is no accident that the stage-to-stage matrix reduction of table 7.1 needs to be read in this Sartrean/Necker-cube-like fashion. The cube is a prototype of the topological bodies that—when expressed ontologically—make up the principal windings of the Ontogenetic spiral. So the Necker-cube reading of the table is a *topological* reading. Just as the quantum leap to a whole new perspective of the cube also implicitly maintains the initial perspective, the transition from one side of the Moebius strip or Klein bottle to the other paradoxically keeps us on the same side. Applying this to relations among the 16 sub-functions of the table (not counting the *repetitions* of functions, which will be discussed below), we may say that, while the transition to KB S4 brings four functions to dominance—rational thinking, mental emotion, visual perception, and mental intuition—all other functions continue to operate as well, albeit in a penumbral fashion in which their "light" is attenuated in direct proportion to their primordiality.

By KB S4, four sets of four functions each have become operative. Thus, beyond the four types of thinking already identified, there are four forms of feeling, of sensing, and of intuition. I am proposing, then, that in the course of Ontogeny each of the four orders of dimensional flesh is differentiated in a fourfold way. Interestingly, just this pattern of development is found in alchemy. Jung summed up the whole alchemical opus as an "unfolding of totality into four parts four times" (1959, 259), and, elsewhere, he associated this multiply quaternary structure with the "4 × 4 faces in the vision of Ezekiel," noting that, of the 16 faces, "one quarter was human and three quarters animal" (1967, 280). So the link between phylo-functional circulations and Ezekiel's wheels actually might entail a great deal more than a vague metaphor! However, while I am sorely tempted to pursue these tantalizing esoteric leads, I will save such an expedition for another forum. Our immediate priority is a detailed understanding of the exact manner in which the 4 × 4 pattern evolves. To prepare ourselves for this, let us become acquainted with the three quaternities not yet considered.

2. THE NONCOGNITIVE QUATERNITIES

2.1. *Fourfold Emotional Flesh*

Our analysis of Ontogeny presently requires us to specify four forms of evolving emotionality. With the assistance of Burrow, we have established *mental emotion* or "affect" as the most abstract of these, and the latest to appear (it is only first manifested in KB S4; however, it is *latent* in KB S3, as indicated by its enclosure in square brackets, a feature of table 7.1 to be clarified below). What are the other three forms of feeling?

Among Charles Darwin's classic works is *The Expression of Emotions in Man and Animal* (1872/1965). This study enumerates the variety of ways that emotions are evidenced in the gestures and motoric expressions of humans and other organisms. In Darwin's account, three emotions in particular appear to stand out from the rest: fear, anger, and love. It is true, however, that Darwin did not explicitly underscore the primacy of these emotions. Yet, consider the implications of the comment by Schultz and Schultz (1992) that, for Darwin, "self-preservation and sexual gratification were the only two instincts in human physiology" (420). The behavioral counterpart of the drive toward self-preservation is "flight or fight." That is, when the survival of an organism is threatened, it either will move instinctively to avoid the danger or it will strive to counteract the threat by aggressive means. According to the Darwinian medical researcher G. W. Crile, the respective *emotional* correlates of these two actions are *fear* and *anger* (or rage). Crile succinctly summarizes all three basic emotions as follows: "A phylogenetic fight is anger; a phylogenetic flight is fear; a phylogenetic copulation is sexual love" (1915, 76). Note that while Crile did amply acknowledge the operation of many other emotions, he stressed the relative importance of fear, anger, and love by referring to other emotions as "lesser" (1913, 291).

Support for the central role of fear, anger, and love is implicit in the writings of Sigmund Freud. Darwin's influence on Freud is well established (Sulloway 1979), and Freud himself readily acknowledged it (see Schultz and Schultz 1992, 420). Like Darwin, the founder of psychoanalysis regarded self-preservation and sexual satisfaction as the prime directives of biology. These instinctive forces are played out in Freud's theory of development.

For Freud (as well as for several existentialist and metaphysical writers), the genesis of emotion ultimately can be traced back to *angst,* to the primordial fear of nonbeing. The first stirring of emotion comes with the newborn organism's first inkling of itself as separate from the matrix of nature in which, hitherto, it had been totally embedded. In *Emotion* (1960), psychologist James Hillman echoes Freud's view that this anticipation of a "rupture in the natural order of things, an alienation of the Ego . . . from the primordial Id" (162), entails a sense of foreboding at the prospect of nonbeing. According to Hillman, the Freudian view was foreshadowed in the writings of Kierkegaard, who took dread as fundamental, associating it with humankind's first glimmer of the possibility of being "cast out of Eden." Hillman also sees a similarity with Heidegger, whose "concept of *Angst* . . . [is] linked with a concept of separation, 'thrown-ness' (*Geworfenheit*), and with a concept of not-being ('Dread reveals Nothing')" (163).

Hillman elaborates further on the primacy of fear. He cites the work of Lacroze as one of the clearest examples of "a theory of emotion based upon a root concept of *Angst*" (163). Lacroze believed that "all emotion, normal or pathological, in man as well as in animal, in the primitive as in civilized man, in the child as in the adult, has its source in a specific anxiety which represents, consequently, the primary affective fact" (Hillman 1960, 163). Hillman comments that Lacroze's "elevation of anxiety to such a metaphysical position reminds one of the role of fear in Buddhist thought [wherein] . . . 'Fear is innate, not only in man, but in everything which exists'" (163). "The conclusion to all this," says Hillman, is that, "In spite of completely different centuries and nationalities, and even more, completely different systematic approaches . . . there is such a striking parallel that we take the model of thinking to be similar in all, *in one respect—emotions can ultimately be traced to a root concept of not-being*" (163). (See also Gebser's characterization of the emotional tone of the archaic zero-dimensional milieu as one of "foreboding" [1985, 142] and Jung's association of alchemy's opening stage of individuation—the "nigredo"—with the emotion of "fear" [1970, 229].)

Resuming the familiar account of development provided by Freud, in the second half of the first year of life, the primordial experience of fear gives way to anger or rage, manifested behaviorally in oral activities such as biting the mother's nipples. According to Freud, the extent of these "sadistic impulses" is "far greater in the second phase [commencing around the age of two], which we describe as the sadistic-anal one . . . satisfaction is then sought in aggression and in the excretory function" (Freud, cited in Schultz and Schultz 1992, 446). Finally, around the age of three or four, the negativism of the anal period takes a positive turn and the tendencies toward anger and aggression are sublimated into an erotic attachment to the mother that serves as the prototype of romantic love.

An additional line of support for the triadic theory of primary emotion comes from an unexpected source. Darwin stressed the continuity of the species and the innate character of organismic action. This tendency to account for all behavior in biological terms was countered by the advent of behaviorism. The movement was founded by John B. Watson (1913). As a proponent of strict empiricism, Watson minimized the influence of heredity and placed strong emphasis on the acquisition of responses through contingent interactions with the external environment. It is no surprise, then, that Watson should seek to explain all emotional reactions as the results of conditioning. However, his research on the behavior of infants led him to conclude that, in fact, there are three kinds of emotion that are *not* learned: fear, rage, and love. When an infant loses physical support, when it is falling or being dropped, or when it suddenly encounters a loud noise, it reacts with fear. For Watson, of course, this term did not primarily refer to an inwardly experienced subjective state but to an externally observable behavior pattern: a sudden spasmodic movement, a puckering and trembling of the lips, an outburst of tears, etc. When an infant's bodily movements are restricted, it responds with anger or rage: its body stiffens; it moves its hands and arms in a staccato, slashing manner; it pumps its feet and legs up and down and holds its breath. And when the baby is touched gently, it responds automatically in a pacified manner and may coo or smile, responses that Watson termed "love." Each reaction is unconditioned. Each occurs irrespective of the environmental contingencies to which the infant has been exposed. Watson believed that the complex and subtle patterns of human emotion are all conditioned from the three basic emotions of fear, rage, and love.

Having made the case for the special status of fear, anger, and love among the vast array of possible emotions, let us see how this relates to the Topogenetic requirement of a *fourth* basic form of emotionality, viz. *mental emotion.* According to table 7.1, the first three modes of feeling constitute stages in the Ontogeny of the two-dimensional *anima.* Although a new, "post-Oedipal" form of emotion does subsequently arise in KB S4, this is not a stage in the development of the *anima* per se. Rather, mental emotion is the form of feeling that emerges out of the interaction of the *anima* with the *cogito.* More specifically, this abstract emotional sub-function is what the 3d/2d animal midwife selflessly offers for the self-Appropriation of the 2d/3d human mother. Until the mother *accepts* the midwife's offer via the enantiomorphic fusion of KB S4, mental emotion remains latent, as indicated by its KB S3 appearance in square brackets. The phylo-functional transactions of dimensional midwives and mothers are elaborated on in section 3. Suffice it to say for now that animal emotion as such is trinary, not quaternary. Naturally, the nonhuman origin of fear, anger, and love does not preclude human beings from experiencing these emotions. The overlapped, nonlinear relation among dimensional gyres makes it clear that human and animal lifeworlds hardly evolve along simply separate tracks.

At this point I must acknowledge that the theory of primary emotion we have been considering is certainly not the only alternative. From the time of ancient Greece to the present day, a great many theories of basic emotion have been offered. No doubt the potential for semantic confusion attending the proliferation of these ideas is considerable, and methodological problems abound. Furthermore, not only has there been a welter of competing views regarding the numbers and types of basic emotions, but the very notion that certain emotions are "basic" or "essential" has been challenged by postmodern analysts (see Ratner 1989). At the end of this subsection, I will address briefly the onto-phenomenological alternative to both the "essentialist" (i.e., classico-modernist) view of emotion and that of postmodernity. But I trust the reader will understand why, at this late juncture in the book, I have no intention of attempting to sort out all the ambiguities (semantic and otherwise) that pervade the field of emotional research. Instead I will take the liberty of pursuing the fear-anger-love approach—expanded to *four* basic modes of feeling for the human context, and radically reinterpreted along topodimensional lines.

To better understand the way emotion develops, let us examine more closely the earliest emotional state (stage one of table 7.1). We have noted a convergence of opinion on the primordiality of fear. However, in view of the absolute ontological negativity ("absolute nothingness") that prevails in the opening stage of Ontogeny, it is clear that primal emotionality cannot just be affirmed categorically as fear, despite the "positive" format of table 7.1. Though the table unequivocally indicates "fear" for S1, the negative logic appropriate to this stage suggests that, rather than *positing* primal fear as a distinct, well-differentiated emotion, or flatly denying said emotion, we must say "neither/nor": the emotionality of S1 is not *other* than fear, yet neither is it fear in any sharply differentiated sense of this term. Note that, despite the above-mentioned consensus on the originality of fear, there are researchers who dissent from this outlook, stressing the *undifferentiated* character of early emotion. But these investigators go on to posit their own distinct alternatives to fear. Psychologist Katherine Bridges (1931), for example, puts forward the root emotion of "excitement." Other theorists have characterized primal emotionality as "the startle pattern" (Landis and Hunt), the "unconditioned affective response" (Harlow and Stagner), and "spasm" (Wallon) (see Hillman 1960, 154). What I am suggesting in contrast is that, while we indeed cannot say that primal emotionality is distinctly that of fear, neither can we say that it is anything other than fear.

But exactly what does this mean? Is this "triply negative" bare declaration all that we can manage in seeking to describe the nature of stage-one emotionality? To begin, let us say that fear is "implied" in S1. I use this term in the special sense brought out by Eugene Gendlin in his essay "Thinking Beyond Patterns" (1991b). Gendlin offers the example of a poet searching for just the right words to effectively express the next line of a poem. The poet appeals to a place in her body that Gendlin calls a "blank" or a "slot," signified by ". . . . ." (61). In the . . . . ., the next line of the poem is *implied:* "This . . . . . demands and implies a new phrase that has not yet come" (61). "Yes, the next line is *implied,*" says Gendlin, "although it does not exist and never has" (63). By way of explaining further, Gendlin distinguishes the conventional meaning of "implicit" from his own:

> What we have once thought explicitly can become implicit when we stop thinking about it. Much previous thought is also implicit; it has been built into our situations and our lives although we have never thought it explicitly ourselves. But in our instance (of a poet), what *was* implicit can be new to the world. Let it stand that something quite new can have been implicit. (1991b, 81)

Gendlin's concept of *implying* entails an unconventional way of thinking about time.[3] For an event occurring at time T_2 truly to be new, yet also to have truly been implied at T_1, time must work *retroactively:* "Implying can imply something new of which we *then* say that it *was* implied. This is not wrong; rather this temporal relation works backward into time. . . . The . . . . . did not contain the line. Now, from the arrived line backwards, the blank *was* the implying of that line. But it is not an error. . . . It is not a confusion of memory" (81).

It seems that Gendlin's concept of "implying" is well suited to the primal negativity of stage one. By stating that fear is "implied" in this nascent stage, we surely do not say that this emotion existed in any well-differentiated, positive sense: "The . . . . . did not contain the line." And yet it is clear that we cannot just say that fear did *not* exist in S1, since, from the retroactive vantage point, the differentiated emotion of fear that emerges in S2 *was implied* in S1 ("from the arrived line backwards, the blank *was* the implying of that line").

At one point Gendlin associates the notion of "implying" with being "pregnant" (75). Such a term accords well with the present approach, given that, in this book, we are essentially dealing with *birthing processes.* We may say, then, that original fear is *embryonic.* Understood by the logic of neither/nor, an "embryo" is not an Aristotelian potentiality. It is not a preformed, miniature version of the mature being, one that is "in there" already and just waiting to come out. Nor is the embryo a *mere* negativity, a state of simple and total formlessness from which any arbitrary form could be created, the creation having to be *ex nihilo.* Rather, an embryo—embryonic *fear,* in the case we are considering—is an *implication.*

But what of the *other* primary emotions in the opening stage of development? Are anger and love mere negativities in stage one? Are they simply nonexistent, which would mean that, in subsequent stages, they would have to arise *ex nihilo,* like rabbits being pulled from a hat? If fear is undifferentiated in S1, if it is an emotion possessing no distinctive characteristics that would set it apart from other emotions, it seems we can say that it is undifferentiated with respect to *anger and love,* with these latter emotions also existing embryonically in this primal stage. Let us conclude, then, that all emotions are initially merged in a kind of "emotional synaesthesia" signifying an absence of clear-cut boundaries among them. It is only in subsequent stages of development that, retroactively, each primary emotion becomes uniquely associated with its own developmental stage. To fully grasp this process, an *ontological* reading is required.

Whereas Gendlin is concerned with the timing of particular events (e.g., the line of poetry), the present concern is with the "timing" of time itself. For Ontogeny addresses the question of *Being*'s evolution, and Being is no particular event transpiring *in* space and time but is itself inherently spatiotemporal. In our concretization of the several orders of dimensional Being, we have found them to be organismic processes with certain phylo-functional properties. The primary emotions we have been dealing with are ontological functions of this kind; as such, they are *spatiotemporal* functions. What, then, is the ontological counterpart of Gendlin's ontical analysis, where an event can be both truly "new to the world," meaning that it did not occur at an earlier time, and also, retroactively, can truly have occurred at that earlier time? Whereas Gendlin would say that, from the retroactive standpoint, a particular event truly did occur in the past, ontologically we are saying that a certain temporal structure itself existed previously, that the dimension of fear constituted the opening stage in an onto-emotional sequence consisting of several clearly delineable stages (as shown in table 7.1). And whereas Gendlin would say that, from the non-retroactive standpoint, the particular event truly did not occur earlier, the ontological equivalent is that there was no differentiation of earlier and later stages of emotional temporality; that there were no clear-cut distinctions among fear, anger, and love; i.e., that emotional temporality itself existed but nascently (owing to the linear format of table 7.1, the non-retroactive aspect is not displayed and must be interpolated by the reader). In the timeless, spaceless womb of the origin, the several ontological emotions (associated with several degrees of spatiotemporality) are inchoately fused. Then, with the "quantum leap" out of the embryonic stage to S2, time becomes differentiated retroactively such that the emotion of fear becomes associated with what *had* taken place in S1, and the emotion of anger is related to the present situation, viz. that of S2. The process of temporal differentiation advances further in S3, where ontological love emerges as more distinctly present than anger had been in S2, vis-à-vis the now more distinct posteriority of anger and fear. Finally, in S4, aided by the emotional midwife's selfless sacrifice on behalf of the cognitive mother, mental emotion (which had been latent in S3) gains presence as maximally distinguished from the "emotional past."

We may view all this alternatively in terms of the *stratification of consciousness* that transpires in progressing through the stages of Appropriation. Remember that, in passing from one matrix to another, the antecedent matrix is eclipsed. But rather than being totally occluded, the earlier matrix persists as a penumbra, a preconscious layer whose "light" filters through to consciousness as if through a translucent screen. Thus, when the new emotion of anger gains sway in S2, fear is differentiated in such a manner that, while being overshadowed by anger ("relegated to the past"), it continues to operate "penumbrally." Fear is not all-pervasive now, as it was (retroactively but truly!) in its primal S1 form. In being sublimated as a distinct emotion, its range of operations becomes limited. Next, in passing to MB-KB S3, another layer of consciousness is formed. The emotion of love now comes to the fore, anger is partially eclipsed, and fear is overshadowed still further. Finally, in KB S4, the penumbral gradient is extended in such a way that love, anger, and fear are all backgrounded in favor of abstract mental emotion.

Note that only with the transition to KB S4 does all emotion take on an exclusively *human* coloration. This results from completing the process of phylo-dimensional divergence discussed above. At the outset, the evolutional windings are thoroughly interwoven. The undifferentiated fear of S1 pervades in a phase of Ontogeny when all higher-dimensional gyres are enmeshed in the zero-dimensional mineral sphere. The rage of S2 is an experience attendant to the embedment of both animal and human lifeworlds in the sensuous matrix of vegetative life. In S3, ascendant love is essentially an animal emotion, and the mythic human being who experiences it is steeped in the animal lifeworld (recall Berman's observation that, in hunter-gatherer society, "animal life was everywhere. . . . Animal movement, the animal body, was the model of human expression"; 1989, 68). Only with the KB S4 disentanglement of the human being from the rest of nature are fear, anger, and love eclipsed by abstract mental emotion, whereupon the older emotions assume the subdued and attenuated form familiar to the perspectival human adult.

In closing this subsection, let me address the issue of essentialism that I broached above. From a postmodern perspective, any theory that reduces emotional experience to a finite set of invariant operations is "essentialist." This kind of reduction is typical of the classico-modernist approach, and it has the effect of oversimplifying human experience in a way that strips it of its diversity and nuanced richness. At the hands of classico-modernism, emotion loses its vitality, its unpredictable transmutability. Then is the present work, which specifies several basic emotional functions, not "essentialist"?

In earlier chapters, I attempted to make clear how the approach adopted in this book differs from modernism and postmodernity alike. From the standpoint of the former, the tendency is indeed to view emotionality as fixed, determined in advance by inexorable forces. The theories of Freud and Watson are good examples of such an absolutist outlook. Each, in its own way, subscribes to the notion that, when it comes to emotional expression, "biology is destiny." The postmodern reaction is to view emotions as arbitrary constructs, adventitious productions of a language-bound society. Emotional experience thus can be *anything*—a judgment effectively tantamount to saying that it is nothing.

In the onto-phenomenological alternative I am proposing, emotions are neither arbitrary constructs, nor are they limited to a rigid set of biological imperatives. Rather than being immutably preestablished or proliferating with capricious abandon, emotions *grow organically.* To be sure, there *are* certain fundamental feeling modalities. In fact, emotions function *ontologically.* Yet these ontological actions are certainly not fixed Platonic forms. Far from constituting objectively predetermined expressions of eternally changeless Being, they are *sub-objective spatiotemporal processes, sub-dimensions of emotional Being that play their parts in the drama of maternal evolution by which Being gives birth to itself.* Importantly, the basic set of emotional sub-functions is not closed. But doesn't table 7.1 give us four emotions only? If the gyration of Being were simply circular, the emotional set could be limited in this manner. But we have seen that, in fact, it is *spiralic.* Ontogenetic process must therefore possess a certain openness that cannot be exhausted.

When all four dimensional windings have been brought into harmony; when all 4 × 4 of "Ezekiel's wheels" are spinning in synchrony; when the alchemical circle has fully been squared; when the second nativities of the several dimensional mothers have been completed in earnest and they have truly given birth to themselves (in KB S8–MB S6–LB S4)—they will now be able to serve as selfless midwives for the self-birthing of a *higher-dimensional* mother, in a new round of Ontogeny. The dimensional spiral will be expanded then in logarithmic fashion, with the dialectical matrix now growing from 4^2 to 5^2. In this wider turning of the spiral, a new and more dialectically intricate, *four*-dimensional order of the flesh will come into play beyond the Kleinian—a "meta-Kleinian" onto-topological action pattern laid out in 5 × 5 matrices. Evidently, *five* forms of emotion will operate in this novel lifeworld. In fact, each of the four functional families that are known to us should consist of five sub-functions, and there should be an entirely new family beyond thinking, feeling, sensing, and intuition. Indeed, there should be a *new order of nature* beyond the *minera, vegeta, anima,* and human *cogito.* In concluding this chapter, I shall have a little more to say about the opening up of these vast new possibilities for maternal evolution. Then, in the next chapter, we will obtain a sense of what *cannot* merely be said in an abstract fashion, of what must be *Proprioceptively lived,* if the way is truly to be paved for the next turn on the spiral.

2.2. Fourfold Sensuous Flesh

In the foregoing subsection, four orders of emotional flesh were identified: fear, anger, love, and mental emotion. This is consistent with the general proposition that human Ontogeny is characterized by a 4 × 4 pattern in which each of four main functions—intuition, sensation, feeling, and thinking—is associated with four sub-functions. What now appears called for is the identification of four sub-functions for the sensuous aspect of individuation. But are there only *four* modes of human sensation? Does conventional wisdom not speak of *five:* touch, taste, smell, hearing, and vision?

According to Walter Ong,

> man's senses can be arranged in a scale as follows:
> Touch—taste—smell—hearing—sight
> Movement in this direction → is:
> toward greater distance from object physically;
> toward greater abstraction . . .
> toward objectivity, nonsubjectivity;
> toward idealism, divorced from actual existence.
> Movement in the opposite direction ← is:
> toward propinquity of sense organ to source of stimulus;
> toward concreteness;
> toward matter, potency, indistinctness (sight and hearing are not easily confused with one another; smell and taste are, taste and touch are . . .);
> toward subjectivity . . .
> toward actual extrasubjective (as well as subjective) existence ("real as this stone I clutch").
> (1977, 136–37)

Ong's primary criterion for distinguishing among the several senses is the distance they span. Sight clearly operates over a greater range than sound (I see the airplane, though it may be too far away for me to hear the drone of its engines). Sound waves obviously traverse a greater distance than the short-range molecular interactions involved in smell (I can hear the sound of the cars passing on the road beyond my window far more easily than I can smell their exhaust fumes). The distance over which sensory interaction takes place is reduced to zero only in the case of touch. As Ackerman puts it in identifying the uniqueness of this sense, "no other part of us makes [direct] contact with something not us but the skin" (1990, 68).

What about taste? When Ong speaks of senses that are "confused with one another," he says, "smell and taste are, taste and touch are" (1977, 136). Many sense experiences we associate with taste involve short-range molecular interactions that engage the *olfactory* nerve; i.e., they are actually *smells*. Other gustatory sensations require immediate contact with the taste buds and thus may be regarded as specialized forms of *touch*. Taste is evidently the one sense modality that does not possess a range of operation uniquely its own. It seems, then, that if we adhere to the criterion of operational scale in seeking to draw unambiguous distinctions among the senses, *four* basic senses are identified: sight, hearing, smell (which includes experiences of taste), and touch (which includes certain gustatory sensations).

Of course, there is a vast difference between sensation as experienced by the perspectival human being of KB S4 and the primal sensuousness of S1. In the opening stage of Ontogeny, the sense of touch is but a *nascent implication* (as we said of primal fear). What this means is that primordial touch does not exist in any well-differentiated, positive sense of the term but only in a "negative," embryonic sense. The other senses, too, are undifferentiated in S1, creating a state of synaesthesia, an inchoate fusion of the senses. Yet, from the *retroactive* vantage point of subsequent stages, we can truly say that touch *was* preponderant in S1, and can associate each of the other three senses with their own stages of ascendancy. (Again we are challenged by the radically nonlinear logic of ontological development!) Note, however, that while each of the four emotions is uniquely associated with a single stage of Ontogeny, according to table 7.1 there is one sense modality that appears in two different stages (not counting S2, the stage in which it is latent): visual perception. What does this mean?

In the previous subsection I intimated that in KB S4, mental emotion emerges from latency when this functional gift proffered by the emotional midwife in KB S3 is accepted by the cognitive mother; the "acceptance" is enacted via enantiomorphic fusion of the 3d/2d midwife and 2d/3d mother. In MB-KB S3, two modes of *sensation* are distilled from the fusions of the sensuous midwife with emotional and cognitive mothers. The fusion of the 2d/1d *vegeta* and 1d/2d *anima* consummated in MB S3 brings hearing out of S2 latency; and the fusion of the 3d/1d *vegeta* and 1d/3d *cogito* accomplished in KB S3 manifests visual perception. Whereas the *anima*'s self-Appropriation is completed in S3, the *cogito* requires one more stage to achieve its quasi-maturation. In passing to KB S4, the *cogito* not only receives the gift of mental emotion from the now-individuated *anima,* but she takes with her what she previously obtained from the *vegeta,* namely, the gift of sight. In this transition, the visual faculty is sublimated. The diffuse visual imagery of the S3 mythic *cogito* becomes the more sharply differentiated, more abstract visual perception of the S4 rational *cogito.* Borrowing a term from chemistry for our "alchemical" purposes, we may regard these two forms of vision (denoted I and II in table 7.1) as "isotopes" of one another. From *isos,* meaning "equal" or "alike," and *topos,* meaning "place," isotopes are atoms that possess different masses but derive from the same element. The common origin of isotopes makes them more akin to each other than are atoms of different elements. In a similar fashion, the modes of visual functioning associated with KB S3 and KB S4, while being distinct, have more in common than do "non-isotopic" modalities. In the next subsection we shall see that a total of seven functional isotopes come into play in the course of development.

Let me attempt to describe the Ontogeny of the sensuous dimension in a little more detail. Consider the most immediate, concrete, and existentially grounded of the senses, viz. the sense of touch. Ong refers to it as "generic." He is not speaking here of an *abstract* sort of generality. On the contrary, "the generic is . . . corporeal, allied to material indeterminacy" (1977, 139). The concretely nonspecific, undifferentiated character of tactility is borne out in the ontical functioning of the KB S4 adult: touch is the only sense modality whose receptors are not restricted to specialized zones of the body. Whereas I see only with my eyes, hear only with my ears, and smell only with my nose, every part of my body is capable of touch. And yet, though touch is "the first used and the most . . . basic of the senses" (141), as development progresses, the more abstract orders of sensation gain supremacy and tactility becomes attenuated. But if tactile receptors operate over the entire surface of the body even in adulthood, in what way does tactility become narrowed?

While every part of the body possesses the potential for touch, in the perspectival adult this potential can be actualized only locally, one part of the body at a time. It seems it is different at the *outset* of development. In its undifferentiated way, the newly conceived organism evidently contacts its world with the *whole* of its body. In fact, the organism is inseparable from that world, since, *in utero,* the boundary between self and other does not yet exist. We know it should also be true that boundaries among the *several* sense modalities are synaesthetically blurred in the earliest stage of development. All this is generally supported by Michael Washburn's (1988) theory of early development, although Washburn was speaking of both the prenatal organism and the neonate, whereas the global tactility in question may actually only be present in the former. As Washburn puts it, nascent awareness "is both 'unaimed' and 'fully dilated.' Like a mirror, the consciousness of the newborn [*sic?*] unselectively registers everything that comes within its range" (42). According to Washburn, the newborn "senses no difference between self and other, inner and outer . . . makes no distinctions or exclusions," so that infantile experience "is altogether without limits or boundaries" (42). It is therefore "unlikely that any of the sensations or feelings that pass through [neonatal] awareness is distinguished from any other" (44). Washburn conveys an inkling of the pan-sensuous, whole-body tactility of incipient life by portraying it as a state of "dynamic saturation . . . a liquid-like plenum or, as Neumann calls it, a pleroma, a condition of dynamic fullness" (43).

With the phase transition to S2, the olfactory sense gains ascendancy and the primal sense of touch becomes overshadowed and constricted in its functioning. No longer is undifferentiated tactility all-pervasive. Touch now plays a secondary role, with the more sublimated sense of smell taking precedence. What are the *phylogenetic* implications of this?

The account of sensory development we are considering cannot refer exclusively to the *human* sphere, since we have found that, in S2, it is the *nonhuman vegeta* that holds sway. Evolutional windings are interwoven in such a way that the human being of S2 is actually more plant-like than human. Our phylo-dimensional analysis suggests that we take this *literally,* despite the S4 *cogito*'s tendency to peremptorily divide the several orders of nature. Needless to say, plants do not have noses! In fact, they have no specific sense organs at all. Yet plant biologist Thomas

Boller (1995) recently demonstrated that, at an unspecialized cellular level, plants are capable of short-range chemical interactions with their environment that are the equivalents of olfactory activities in animals and humans. Presumably, human development would occur in such a way that, very early on, it would "recapitulate" the kind of generic cellular "chemoperception" (Boller 1995, 189) of which plants are capable. Only later would this primal mode of olfaction be superseded by the more differentiated kind associated with the specialized organ of smell.

In MB-KB S3, the process of sensory sublimation is carried further, and both tactility and olfaction are eclipsed and attenuated in favor of hearing and imaginal sight (visual perception I). Since dimensional divergence is not yet complete at this stage, the auditory and visual experience of the S3 human being should have a dreamlike quality that is more animal than human. Finally, by the fourth stage of Appropriation, the more sublimated "isotope" of visual perception has achieved dominance and all the other senses have become relegated to the penumbra, their degrees of eclipse being directly proportional to their primordiality. Thus, by KB S4, the "cerebro-ocular" region (Burrow 1953, 526) is in command and the human being, having cut himself off from the rest of nature, is now primarily guided by "that central vision that joins the scattered visions . . . that *I think* that must be able to accompany all our experiences" (Merleau-Ponty 1968, 145).

2.3. Fourfold Intuitive Flesh

The zero-dimensional *minera* is the only order of the flesh that does not develop. We know, of course, that it *contributes* to higher-dimensional development in an indispensable way. It does this through the gift of intuition, as we shall see in more detail below. But this does not mean that the *minera* itself engages in intuitive functioning. Though we can reasonably claim that the *cogito* itself thinks, the *anima* itself feels, and the *vegeta* itself senses, we cannot speak similarly of the *minera,* since it possesses no self, no subjectivity that could function epistemically in this fashion. Indeed, the *minera* is the one order of Appropriation that is not an order of Being. In the language of Heidegger, this non-ontological or meontological modality supports the ontological by "a delivering over" to Being and time of "what is their own" (1962/1972, 19). In our topo-dimensional elaboration upon Heidegger, the meontological *minera,* through its gifts of intuition, gives midwifely assistance to *three orders* of spatiotemporal Being. Considering the fact that, while the *minera* is the source of intuition, it does not itself function intuitively, I designate its primary functional mode as "null intuition" (see tables 7.1 and 7.2). This term is compatible with the thoroughly negative character of the zero-dimensional sphere.

Now, of the four basic functions to which Jung devoted so much attention, he had the least to say about intuition. This is not surprising, given that intuition is the most recondite of the functions, entailing as it does "perception by way of the unconscious" (Jung 1971, 518). Of course, Descartes' highly sublimated form of intuition was hardly clouded by immersion in the unconscious, but was a model of lucidity. Our term for this kind of intuition is *mental intuition.* If it is true, as I have claimed, that Ontogeny is governed by a 4 × 4 pattern of individuation, a total of four intuitive sub-functions must be identified. In addition to the null intuition of the *minera* and the mental intuition of the *cogito,* two other varieties must thus exist. What are they?

In *Awakening Intuition* (1979), the transpersonal psychologist Frances Vaughan proposes the following:

> The broad range of intuitive human experiences falls into four distinct levels of awareness: physical, emotional, mental and spiritual. Although any given experience may have elements of more than one level, experiences are usually easy to categorize according to the level at which they are consciously perceived. For example, mystical experiences are intuitive experiences at the spiritual level, and as such they do not depend on sensory, emotional or mental cues for their validity. Intuition at the physical level is associated with bodily sensations, at the emotional level with feelings, and at the mental level with images and ideas. (66)

The entries for intuition appearing in table 7.1 only partially reflect Vaughan's classificatory scheme. The divergence is due to differing interpretations of the body, and of spirituality. Vaughan defines "spiritual intuition" as "a holistic perception of reality [that] transcends rational, dualistic ways of knowing and gives the individual a direct transpersonal experience of the underlying oneness of life" (78). For Vaughan, such transcendence entails a clean separation from the body. This "pure intuition" is "devoid of feeling . . . is not associated with the body" (77). By contrast, "physical intuition" involves "a strong body response. . . . [It is] the kind of jungle awareness which enables primitive people to sense danger when there are no sensory cues of its presence" (66). To illustrate "physical intuition" among "people living in an urban environment" (66), Vaughan speaks of paying attention to bodily symptoms such as tension, headaches, or stomachaches, whereby "you may find that you are indeed in a situation which is unhealthy and which is creating undue stress on the organism" (68).

Reading between the lines of Vaughan's exposition, what came through to me was her apparent grounding in the patriarchal tradition that equates the material body (matter, mater, mother) with *maya,* with illusion, personal hang-ups, a dualistic state of impurity—i.e., with the imperfections of the finite. Spirit, on the other hand, is seen as transpersonal purity and perfect truth, the unitive awareness of the infinite that has left the body behind. Vaughan thus concludes her discussion of the four types of intuition with the assertion that "[a]ctivating spiritual intuition means focusing on the transpersonal rather than the personal realms of intuition. . . . Forms of intuition [that are] focused on sensation, feeling, and thinking become obstacles to pure awareness" (80).

When Vaughan ties bodily experience to what is "personal," in effect she limits it to the sphere of the finite particular, to that which is objectified. I would certainly agree with Vaughan that spirituality or wholeness could not be found in such ontical embodiment. But I have maintained that neither is genuine wholeness evident in the "purity" of a disembodied spirit. Vaughan's interpretation presupposes the set of interrelated polarities so long at play in our culture: spirit and matter, infinite and finite, father and mother, good and evil, perfection and imperfection, unity and diversity, fullness and emptiness, wholeness and hole-ness, etc. Within each pair of opposites, the first term is one-sidedly idealized, extolled as pure positivity, while the second term is one-sidedly negated. What we have seen is that these opposites actually *imply one another.* They are the twin aspects of dialectical Being. Thus the search for "pure wholeness" has never failed to put us in the *hole,* a fact to which history bears eloquent testimony. The "purity of spirit" is at least as illusory as the ontical finitude of the body. We may ask why it is that, although Vaughan distinguishes "spiritual" and "mental" forms of intuition, the bracketed terms have often been taken as synonymous. I suggest it is because the "pure positivity" of God the Father is a strictly human projection of the intuitive *mind.* I venture to say that, at bottom, there *is no* "unadulterated wholeness," but only the *(w)holeness* of the ontological mothers, and the *absolute holeness* of their original midwife. The upshot is that, whereas Vaughan views the "highest" form of intuition as entailing "disembodied spirit,"[4] we must speak instead of the *deepest* form of intuition as the *null* intuition of the "holey" *minera.* Far from being disembodied, this dark maternity is the *densest* of bodies, associated as she is with the "eternal sway of matter" (Jung 1967, 238), with the diamond-like hardness of the *lapis philosophorum.*

In view of the foregoing, let me take the liberty of revising Vaughan's account of fourfold intuition for our purposes by subsuming "spiritual" intuition under *mental* intuition and offering the null intuition of the *minera* as the basis of all higher-dimensional intuitive forms. Since Vaughan evidently equates "physical intuition" with the *sensory* realm, the sphere of "bodily sensations" (1979, 66), let us rename it "sensuous intuition." Finally, we retain Vaughan's term "emotional intuition," defined simply as that mode of intuitive functioning that is "associated . . . with feelings" (66).

Turning to the question of phylo-functional development, the unique character of zero-dimensionality is plainly in evidence once more. Whereas vegetative touch, animal fear, and human archaic thinking are all embryonically implied in S1, the same cannot be said for the intuition of the mineral sphere. In no positive sense is there intuition here, not even in an embryonic one, since there is no *minera* in any such sense: no self or subject, no object, no Being. Yet the utterly selfless, meontological *minera* does *give* intuition. Whereas the *anima,* functioning as selfless midwife to the cognitive mother, helps distill mental emotion in S4; and whereas the *vegeta* facilitates hearing and sight for the S3 *anima* and *cogito* (respectively); the *minera* assists with three gifts in S2: sensuous intuition goes to the *vegeta,* emotional intuition to the *anima,* and mental intuition to the *cogito.* Beyond noting that the mental intuition accompanying magical thinking in S2 must be a more concrete mode of functioning than its S3 and S4 functional isotopes, can we say any more about it? Let us first consider magical thinking itself in a little more detail.

Recall that, for Gebser, magical awareness is completely absorbed in the point-like details of its one-dimensional environment. And yet, according to the principle of *pars pro toto* that governs magical participation in the world, every point of experience is inseparably united with every other point and with the whole. Thus, "magic man" inhabits a "point-like unitary world" (1985, 48), a "point-related unity in which each and every thing intertwines and is interchangeable" (48). The complete equality of each point or part with every other, and with the whole, "brings with it what we may call thinking by analogy or association. . . . Magic man feels things which seem to resemble one another as 'sympathetic to,' or 'sympathizing with,' one another [cf. 'sympathetic magic']. He then proceeds to connect them by means of the vital nexus—not the causal nexus" (50). The term "vital nexus" alludes to "the realm of vegetative energy," to the "vegetative intertwining of all living things" (49). Thus magical thinking, embedded as it is in the sensuous matrix of the *vegeta,* does not function discursively. It does not operate by an if/then procedure in which propositional elements assumed to be distinct from one another are externally linked by causal relations acting locally in a cognitive space ("if X is true and Y is true, then Z must be true"). Rather, magical thinking forges direct associations based on the *indistinguishability* of elements (on their "intertwining"). But we *are* dealing here with a form of thinking, not with a mode of sensation per se. Though it is "sleep-like" and "only dimly conscious" (Gebser 1985, 251) compared with more mature forms of thinking, magical thinking is, in Jung's sense, still a "rational" function. However weakly, it involves representation; reflection upon and evaluation of externally experienced reality from the inner perspective of a subject. Sensation as such, on the other hand, is "irrational"; it is more *presentational* than representational, entailing a more immediate (less reflective) reaching out to, grasping, and taking in of its field of experience. But is the S2 human being not more plantlike than human, more sensuous than cognitive? The S2 operations of the human evidently do "recapitulate" the primal sensuality of the *vegeta.* In this stage, humankind is presumably well immersed in the largely undifferentiated world of olfaction discussed in the previous subsection. Nevertheless, we are seeing that, *in addition* to the potent sensuality of early human experience, the S2 human being, *as* a human being, engages in an activity that, while being plant*like,* is distinctly *cognitive:* magical thinking. And, by courtesy of the *minera,* this immature form of thinking is accompanied and aided by an initial form of mental intuition.

We have found that *Cartesian* intuition (mental intuition III) supports the deductive mediations of the *cogito* by providing an unmediated apprehension that "is more certain . . . than deduction, because it is simpler. . . . Thus, anybody can see by mental intuition that he himself exists, that he thinks, that a triangle is bounded by just three lines, a globe by a single surface, and so on" (Descartes 1628/1954, 155). Since the Cartesian subject's intuition of its own distinct existence is the *primary* intuition (as discussed above), the implication of the passage is that this "irrational" self-apprehension is what essentially undergirds the subject's rational operations upon its world. In like manner, the magical subject's immature kind of self-intuition affords a measure of support for its incipient rational activities. But *does* the magical subject really intuit its own distinct existence? Is it not governed by the principle of *pars pro toto* in which it is inseparably united with its world? Without a differentiated apprehension of itself, how could the magical subject engage in *any* kind of rational activity? We know from Gebser that this immature subjectivity is not *entirely* incognizant of itself as a distinct entity. To reiterate Gebser's characterization of "magic man,"

> he is distinguishable above all by his transition from a zero-dimensional structure of *identity* to one-dimensional *unity.* . . . The man of the magic structure has been released from his harmony or identity with the whole. With that a first process of consciousness began; it was still completely sleep-like: for the first time, not only was man *in* the world, but he began to face the world in its sleep-like outlines. . . . The more man released himself from the whole, becoming "conscious" of himself, the more he began to be an individual. (1985, 46)

Thus "magic man" is not without self-consciousness. On this basis, s/he can engage in a form of self-intuition that functions in the service of magical thinking.

Now, to avoid the possibility of confusion, let me reinforce the contrast drawn above between magical thinking and primal sensation. We have just learned that magical thinking is only minimally self-conscious, weakly reflective. How, then, are we to clearly distinguish it from an "irrational" function like primary olfaction, about which we must apparently draw similar conclusions? Magical thinking is indeed a weakly reflective, highly concrete activity when compared to mythic and fully rational thinking. But compared to rudimental sensation and other noncognitive modalities, it is in fact qualitatively—no, *ontologically*—more abstract. By the same token, whereas mental intuition—being "presentational" or "irrational"—functions more concretely than the "rational" members of the cognitive family (thinking and mental emotion), it is decidedly more abstract than noncognitive forms of action. In general, we can forestall conceptual confusion by keeping in mind the distinction between (1) relations among members of a given ontological family (vertically arrayed cells of table 7.1), and (2) relations among members of different families (horizontally arrayed cells). Whereas the former entail the same lifeworld, the same order of wild Being, the latter involve different orders of Being. This is not to say that the several orders of the flesh are simply on separate tracks. They intimately overlap one another, as we have seen. The human being is thus not a purely cognitive creature but can feel anger, smell the fragrance of flowers, and so on. Still, these overlapping modes of functioning must be recognized as not merely quantitatively or qualitatively distinct, but as *ontologically* so.

Let us now turn to *mythic* thinking. Gebser tells us that the transition from the "sleep-like" magic structure to the "dream-like" mythic structure brings with it a more differentiated order of consciousness. In the process, a new and more mature form of thinking is made possible. Gebser calls it "oceanic" (1985, 252). Oceanic thinking is wavelike or circular: it "posits and postulates—immediately modifying the posited by a new postulation, and ultimately returning to the point of inception" (252). Here there is a "cyclic motion of thoughts beginning with one 'concept' to which others are added before the return to the initial concept" (252). As an example, Gebser offers a fragment from Heraclitus: "'For souls it is death to become water; for water it is death to become earth. But from the earth comes water and from water, soul'" (252). Gebser augments his commentary on oceanic thinking with a quote from Dionysius Areopagita (ca. 500 CE): "'The souls also have rational knowledge by their discursive endeavor in a circular motion to achieve the truth of things'" (253). The nonlinearity of mythic thinking bespeaks the fact that it is embedded in the emotional matrix of the *anima*. Of course, we *are* dealing with a form of thinking here, not with a mode of primary emotion. Though it be "dream-like" (Gebser 1985, 72), mythic thinking entails an ontologically more abstract manner of functioning than that of animal emotion. It remains true, however, that the S3 human, submerged as s/he is in the lower-dimensional lifeworld, is more emotional than cognitive. In this stage, humankind evidently is steeped in the passionate affiliations of animal life. Still, *in addition* to the powerful emotionality of S3 human experience, the human being, *as* a human being, engages in an activity that, while being animal-*like*, is distinctly cognitive: mythic thinking. And, by courtesy of the *vegeta* and the *minera*, this still immature form of thinking is aided by isotopic forms of "irrational" apprehension, viz. visual perception I and mental intuition II (respectively). Thus the circular operations of mythic thinking are undergirded by dreamlike imagery and by the self-intuition of mythic subjectivity.

To be sure, the *vegeta*'s offer of sensuous assistance is not limited to the *cogito*. The *vegeta* supports the *anima* as well. We know that, like the *cogito*, the *anima* functions "rationally," in Jung's sense. It, too, acts by way of reflecting upon and evaluating its experience, though this valuation is ontologically more concrete than its cognitive counterpart, proceeding by means of primal feeling. To the "rational" *anima* the *vegeta* presents the "irrational" gift of *hearing*, which, when accepted by the *anima* in S3, emerges from S2 latency to undergird the *anima*'s affiliative passion. The *anima* is also assisted by the *minera*, of course. This midwife tenders the gift of *emotional intuition*, which, when accepted by the *anima* in S2, grounds the *anima*'s primal anger in a mode of unmediated self-apprehension more concrete than any form of mental intuition. Next, in passing to S3, a sublimated isotope of emotional intuition (emotional intuition II) forms to furnish support for animal love. So, just as human thinking is facilitated by the gift of mental intuition, animal emotionality is aided by emotional intuition. For the "level of intuition [that] comes into consciousness through feelings," Vaughan gives the example of "instances of immediate liking or disliking with no apparent justification," or an overall sensitivity to the emotional "'vibes'" of a situation (1979, 69). In the first instance, however, emotional intuition should involve the emotional subject's apprehension of its own existence.

Finally, we take note of the *minera*'s gift to the *vegeta*, accepted by the latter in S2. Here the largely "irrational" workings of sensuous consciousness (primal smell and taste) are undergirded by the more complete "irrationality" of *sensuous intuition.* This is what Vaughan calls "physical intuition," and associates with "a strong body response . . . bodily sensations" (1979, 66). More basically, sensuous intuition should entail the sensuous subject's apprehension of its own concrete existence.

3. GYRATIONS OF CO-EVOLVING FLESH

Having identified the members of each of the four quaternities in phylofunctional terms, we are now prepared to examine in greater detail the co-evolution of the several orders of wild Being. As in the more abstract treatment provided in chapter 5, we view the coordinated circulations of the flesh as topodimensional birthing processes, self-nativities carried out by dimensional mothers with the aid of their midwives. In the cycles of self-Appropriation (the "clockwise" nativities), there is dimensional divergence, whereas those of Proprioception (the "counterclockwise" rebirthings) bring dimensional convergence, a synchronizing or harmonizing of dimensional gyrations. The synchrony may be thought of as a "dance," the harmony as a "song," or as a "symphony" of the lifeworlds.

3.1. Cycles of Divergence: Self-Appropriation

To begin, we return to the primal nativity scene. It is S1, the opening stage of Ontogeny. The higher-dimensional mothers are embedded in the matrix of the zero-dimensional *minera* and thus are but nascently implied. They are embryos that have not yet given birth to themselves. All embryos are "negatively contained"; they rest in the blackness of the mineral womb, the "black hole" of absolute nothingness; of timelessness, spacelessness, boundlessness. The nascent ontological tactility of the primordial *vegeta* is topodimensionally embodied by the fractionally one-dimensional lemniscatory body (shown as 0d/1d in table 7.1); the undifferentiated fear of the *anima* finds its topological flesh in the fractional two-dimensional Moebius body (0d/2d); and the archaic thinking of the *cogito* is realized in the three-dimensional Kleinian embryo (0d/3d). It is not only that the S1 *vegeta*'s sense of touch is but *implicit,* being retroactively differentiated in S2 from an S1 lack of differentiation among the senses; not only that the *anima*'s feeling of fear is actualized in this way from an initial "emotional synaesthesia"; and not only that the *cogito*'s archaic thinking arises thus from a "cognitive synaesthesia." More than that—given the absence of spatiotemporal boundaries in S1 that the compartmentalizing linearity of table 7.1 fails to reflect—the *vegeta, anima,* and *cogito* are undifferentiated from each other. All dimensions of the flesh are inchoately intertwined in the opening stage of Topogeny.

Advancing to stage two, dimensional divergence commences. To "score" this first movement of the "topodimensional symphony," we use the "musical notation" provided in tables 5.1, 5.2, and 7.1. In S2, the 1d/1d lemniscatory *vegeta* completes her first self-nativity in a process whereby tactility is sublimated into olfaction, with the tactility of S1 being retroactively differentiated. The passage is facilitated by the *vegeta*'s acceptance of the encouragement that had been extended to her in S1 by the sub-lemniscatory *minera.* Musically speaking, the sensuous mother of S1 is a tactile "undertone," 0d/1d, which receives midwifely support from the *minera*'s sensuously intuitive "overtone," 1d/0d (see chapter 5's discussion of overtones and undertones). Responding in S2 to the midwife's selfless act of Appropriation on her behalf, the mother engages in an act of self-Appropriation in which the midwife's latent sensuous intuition becomes manifest as the mother's own. There is thus the fusion of 1d/0d and 0d/1d enantiomorphs wherein the *minera* is introjected by the *vegeta,* absorbed by it in such a way that overtone-undertone opposition is "annihilated," the appearance being created of subject-object opposition internal to the 1d/1d *vegeta.* With the matrix thereby reduced, the quasi-mature S2 operations of the *vegeta* now consist of two subject-object differentiated functions: (1) *olfaction,* which is the "eigenfunction" of the *vegeta,* its principal mode of autonomous motherly action; and (2) *sensuous intuition,* the gift of the midwife introjected by the mother so as to undergird her olfactory sensuality.

The *anima* and *cogito* enact similar self-nativities in passing to S2. In the former case, fear is sublimated into anger, the process being catalyzed by the emotional mother's acceptance of support from the intuitive midwife. The *anima*'s self-Appropriation accordingly entails a fusion of 2d/0d and 0d/2d enantiomorphs such that this mirror opposition dissolves into the 1d/2d emotional sphere, whereupon we have the animal eigenfunction of anger undergirded by emotional intuition I. In the case of the *cogito,* self-Appropriation involves transforming archaic thinking into magical thinking, the sublimation being facilitated by the fusion of 3d/0d and 0d/3d enantiomorphs wherein the mental intuition latent in S1 is now manifested as a support function internal to the new, 1d/3d form of thinking.

We know, of course, that the S1 state of affairs is not simply left behind in passing to S2. Instead, consciousness becomes layered or stratified in such a manner that the primordial situation of S1 is partially eclipsed, persisting as a penumbra. I pointed out above that this way of reading the Ontogenetic phase transition is *topological,* inasmuch as it mirrors what happens in going from one side of the Moebius strip to the other, or from one perspective of the Necker cube to the other, where the initial side or perspective "persists in the background." But we must understand this "background" to be *interior,* to lie within, since Topogenetic advance essentially involves a process of *containment.* What we are presently rediscovering is that higher topodimensional bodies introject or internalize lower topodimensional bodies, swallow or contain them. It is true that, in the *opening* stage of Topogeny, it is the zero-dimensional midwife who provides containment. Yet the absolute negativity of the *minera,* her total lack of internal structure, precludes her from actually containing higher topodimensional bodies in any positive sense of the word. All she can do is contain them *negatively,* i.e., support the *self*-containment of the higher-dimensional bodies via her selfless enantiomorphic offerings. Then, with passage to S2, containment switches from "negative" to "positive" as the meontological midwife is now herself contained by the dimensions of motherly Being. Through the absorption of the sub-lemniscatory midwife's three overtones by the higher-dimensional mothers, she is taken into them, "ingested" in such a way that her contributions (the three respective forms of intuition latent in S1) are manifested on their surfaces, and she herself is rendered subliminal, occlusively subordinated or repressed, relegated to an interior penumbra of awareness. Or—returning to that quintessential fairytale of alchemy—we can say that, in S2, the Mercurial *minera* becomes "bottled up." In fact, the whole primal S1 scene—with its undifferentiated tactility, fear, and archaic thinking—is "bottled up" in moving to S2. In attempting to read table 7.1 in a nonlinear, topological fashion, it might help to regard the phase transition from S1 to S2 as one in which the matrix of the latter becomes superimposed upon that of the former, creating a translucent layer through which light from S1 filters through.

Let us focus our attention on the S2 relationships among the *vegeta, anima,* and *cogito.* In this stage, the first has reached quasi-maturity whereas the latter two have not. To be sure, the *anima* and *cogito* have gained a measure of individuation inasmuch as they have extricated themselves from the zero-dimensional matrix of the *minera.* Yet they are now embedded in the one-dimensional matrix of the *vegeta.* Unlike the *minera,* the *vegeta* is an order of Being. Its relation to the *anima* and *cogito* is therefore not purely midwifely. Functioning meontologically, the *vegeta* does selflessly contain animal and human self-containment. However, in the *ontological* aspect of its functioning, the S2 *vegeta* also *threatens* that self-containment. We shall soon see what this means, but let us first consider more closely the *vegeta*'s midwifely role.

The layout of table 5.1(b) suggests that, before the lemniscatory *vegeta* can serve as midwife to the Moebial *anima* and Kleinian *cogito,* it must complete its own development. Only then does the *vegeta* feature the enantiomorphic overtones—2d/1d and 3d/1d—that are necessary for the negative (selfless or meontological) containment of animal and human undertones—1d/2d and 1d/3d, respectively. I have acknowledged, however, that table 5.1(b) separates the dimensional windings in an artificial fashion that neglects their temporal overlap. Tables 5.1(a) and 7.1 rectify this limitation by bringing out a sense in which lifeworlds interweave. Nevertheless, while dimensional gyres are indeed temporally interwoven, we know that they are not simply contemporaneous, since each gyre possesses its own "time scale," its own order of temporality. How, then, are we to read the timing of the S2 *vegeta*'s meontological facilitation of animal and human Ontogeny? If it does not simply occur *after* the *vegeta*'s own development or simply simultaneously with it (as tables 5.1(a) and 7.1 seem to suggest), when does it take place?

The midwifely functioning of the S2 *vegeta* may be clarified by enlisting a curious concept from contemporary physics. The theory of electromagnetism includes the possibility of energy potentials or waves that are *advanced.* An "advanced potential" is defined as "any electromagnetic potential arising as a solution of the classical Maxwell field equations, analogous to a retarded potential solution, but lying on the future light cone of space-time; the potential appears, at present, to have no physical interpretation" (Lapedes 1978). According to philosopher Mary Hesse (1965), the application of advanced solutions of Maxwell's equations "leads to the apparently paradoxical result that one event affects another . . . at a time . . . *before* it occurs" (279–80). Applying this notion to the *vegeta*'s S2 midwifely influences upon the *anima* and *cogito,* I suggest the effects are exerted as "advanced waves." In this regard it should be helpful to keep in mind that the *vegeta, anima,* and *cogito* are indeed wavelike dimensional actions, topodimensional circulations that are vortical in nature (see chapter 5, section 2). What I am proposing is that one-dimensional cyclonic action "carries" the higher-dimensional vortical waves, i.e., provides them with enantiomorphic midwifely support, at a time that is "before" the one-dimensional vortex has matured to the point of being enantiomorphically equipped to do this. The enantiomorphic "carrier waves" (i.e., overtones) by which the *vegeta* assists the *anima* and *cogito* are thus "advanced waves."

But is the notion of the "advanced wave" not an outright contradiction? How would it be possible for the *vegeta*'s midwifely influence to occur at a time that predates her acquisition of the capacity to exert such an influence? It would *not* be possible were relations between dimensional waves governed by a single order of temporality. An effect cannot precede its cause on the same scale of time. The "advanced" midwifely influences of the *vegeta* upon the *anima* and *cogito* derive precisely from the fact that these dimensional waves constitute *different* orders of temporality. It is because the vegetative temporal gyre is more "tightly coiled" than its higher-dimensional counterparts that it can appear to be "ahead of itself" in relating to them.

By way of clarification, consider again the analogy of the clock (chapter 5). In the relationship between the second and minute hands, we have a simple model of different time scales. All sixty ticks of the second hand occur within the same tick of the minute hand. Might we not say, then, that the faster, more compressed cycle is temporally undifferentiated relative to the slower time scale? Are not the distinct ticks of the second hand all the same tick to the minute hand, which cannot make such fine distinctions? In classical physics, of course, this relativistic effect is obviated by the fact that the two time scales are entirely commensurable; they are incorporated as inter-convertible subdivisions of a single metric. On the other hand, what *we* are dealing with are not quantitative temporal differences but *ontological* ones: incommensurable orders of temporality. As a consequence, relativistic effects cannot be eliminated. Being unable to reduce the temporalities of the *vegeta, anima,* and *cogito* to a unitary metric, we must come to grips with the idea that, while the different stages of Ontogeny are indeed distinct within the relatively tight temporal coiling of the *vegeta,* in relation to the more "loosely wound" circulations of the *anima* and *cogito* the moments of vegetative individuation do lose their distinctiveness; they all occur at the "same time." It is this temporal "compression" of the one-dimensional winding relative to higher-dimensional windings that enables the *vegeta* to function as midwife in "advance," to negatively contain animal and human Ontogeny at the "same time" that it has not yet completed its own Ontogeny. In its latter "moment" of unfulfilled development, the *vegeta* affects the *anima* and *cogito* in a manner that is contrary to the mature *vegeta*'s midwifely influence. Instead of encouraging animal and human self-containment, the *vegeta* menaces it. How so?

To contain is to enclose within boundaries. While the self-containment of animal and human maternal dimensions receives nothing but "encouragement" (negative containment) within the matrix of the utterly boundless mineral midwife, in the matrix of the *vegeta* the blessings are mixed. Insofar as the S2 *vegeta* is not only functioning "in advance" as a boundless midwife but also as a mother who is still concerned with her own self-binding, her eigenfunctioning poses a danger to the functioning of the higher-dimensional mothers. Why is that so? In S2, the self-containment of the *vegeta* has gone further than that of the *anima* and *cogito.* The latter two are but weakly self-contained vis-à-vis the former; being largely undifferentiated from the *vegeta,* they are wide open to its influences. Those influences are "positive" in the quasi-mature moment of the *vegeta.* That is, rather than deriving from midwifely selflessness, they stem from the *vegeta*'s ongoing motherly efforts to contain herself. Thus they threaten to swamp the weaker self-containment capacities of the still embryonic *anima* and *cogito.* It is because they are flooded by the potencies of the lower-dimensional mother that the two- and three-dimensional mothers are more vegetative than animal or human (respectively), more sensuous than emotional or cognitive. In sum, the S2 *vegeta* influences the *anima* and *cogito* in both a "positive" and a "negative" way: in her motherly moment, the *vegeta* overpowers the still immature higher-dimensional mothers; in her midwifely moment, she gives them selfless support. To complete the account of S2 vegetative functioning, let us add what we have already noted about the *vegeta*'s relation to the *minera:* with the passage from S1 and concomitant fusion of 1d/0d and 0d/1d enantiomorphs, the *minera* is "positively contained" by the self-Appropriative *vegeta,* repressively "bottled up" by it.

The next phase transition brings a further divergence of topodimensional windings. With movement to the third stage of Appropriation, anger is sublimated into love and the differentiation of S1 fear and S2 anger is completed retroactively. By this process, the 2d/2d Moebial *anima* concludes her initial self-nativity. The birthing is facilitated by the *anima*'s acceptance of the encouragement that had been extended to her in S2 by the lemniscatory *vegeta.* The emotive mother of S2 is the undertone of anger, 1d/2d, which receives midwifely support from the *vegeta*'s auditory overtone, 2d/1d (here the *vegeta* operates in "advance" of completing her own motherly individuation). Responding in S3 to the midwife's selfless act of Appropriation on her behalf, the mother engages in an act of self-Appropriation in which the midwife's latent hearing function becomes manifest as the mother's own. There is thus the fusion of 2d/1d and 1d/2d enantiomorphs wherein the midwifely *vegeta* is introjected by the *anima,* absorbed by it in such a way that overtone-undertone opposition is "annihilated," the appearance being created of subject-object opposition internal to the 2d/2d *anima.* With this matrix reduction, emotional intuition I is carried forward in sublimated isotopic form as emotional intuition II. The quasi-mature S3 operations of the *anima* consist, then, of three subject-object differentiated functions: (1) *love,* which is the eigenfunction of the *anima,* its principal mode of autonomous motherly action; (2) *hearing,* the gift of the sensuous midwife introjected by the mother so as to undergird her affiliative activity; and (3) *emotional intuition II,* the refined functional isotope of the intuitive midwife's gift to the emotive mother that serves to further ground the mother.

The *cogito* enacts a similar self-nativity in passing to the third stage of Appropriation. Magical thinking is transformed into mythic thinking and the differentiation of S1 archaic thinking and S2 magical thinking is enhanced retroactively. The process is facilitated by the cognitive mother's acceptance of assistance from the sensuous midwife, with 3d/1d–1d/3d enantiomorphic fusion leading to the distillation of visual perception I as the primary support function of the new form of thinking (2d/3d). Also, the intuitive midwife's gift of mental intuition is refined into the isotopic functional form, mental intuition II.

The movement to the third stage of Appropriation entails a further stratification of consciousness such that the S2 state of affairs is now partially eclipsed and the S1 situation is eclipsed more deeply than it was in S2. The occluded stages are repressively contained or "bottled up," creating "bottles within bottles." If we need another metaphor to clarify said penumbral relationships (and perhaps we do, given the recondite nature of these matters!), we may think of the nesting of Chinese boxes or Russian dolls: the S2 matrix is contained within S3, and the S1 matrix within S2. It can also be said that, in KB S3, the *cogito* is contained by the *anima*. But, in this case, the containment is *negative*. That is, the still immature 2d/3d cognitive mother is not eclipsed by the *anima* but given midwifely encouragement by her via her 3d/2d overtone, which operates as an "advanced potential." Yet it is also true that the MB S3 *anima,* in the quasi-mature *motherly* moment of her functioning, *threatens* the self-containment of the *cogito*. Being flooded by overwhelming animal potencies, the still embryonic *cogito* is more animal than human, more emotional than cognitive.

With the final phase transition in the process of clockwise Appropriation, the divergence of topodimensional gyres reaches its highest degree. In passing to stage four, mythic thinking is sublimated into rational thinking and the differentiation of S1 archaic thinking, S2 magical thinking, and S3 mythic thinking is completed retroactively. In this way, the *cogito* concludes her first self-nativity. The transformation is catalyzed by the fusion of 3d/2d and 2d/3d enantiomorphs. Here the midwifely *anima* is introjected, absorbed by the *cogito* in such a way that overtone-undertone opposition is "annihilated," the appearance being created of subject-object opposition internal to the 3d/3d *cogito*. With the matrix thereby reduced for the last time, the mental emotion that was latent in S3 now becomes manifest within the cognitive sphere. In addition, visual perception and mental intuition are carried forward in isotopic forms that are more rarefied than those they took in S3. So the quasi-mature S4 operations of the *cogito* consist of the four subject-object differentiated functions initially examined in section 1 of this chapter: (1) rational thinking, the eigenfunction of the *cogito;* (2) mental emotion; (3) visual perception II; and (4) mental intuition III. Consciousness has become maximally stratified in this stage. It is now constituted of four layers, the penumbra shading back from S3 to the dark recesses of S1.

3.2. Cycles of Convergence: Proprioception

The stratification of consciousness in the clockwise cycles of Ontogeny speaks to the divergence of the topodimensional gyres. Dimensional mothers birth themselves, bring about their own individuation, by extricating themselves from the lower-dimensional wombs. It is in this process that lower-dimensional matrices become backgrounded or "bottled up." In the layering of consciousness that this entails, dimensional gyres are partitioned from one another. Of course, these initial nativities only bring *quasi*-individuation. To give birth to herself in earnest, each mother must undergo a second nativity, one requiring that she "shift gears," switch her orientation from forward to backward, clockwise to counterclockwise. Once the transition is made from the posture of self-Appropriation to that of Proprioception, the mother can proceed to move "backward through the birth canal." As she consciously retraces her steps to her point of origin, the lower-dimensional matrices from which she emerged are no longer repressively occluded but thankfully acknowledged. In this way, the "genies come out of their bottles" and the several dimensional spheres harmonically converge. In Gebser's terms, a mode of integral consciousness is realized in which dimensional spheres that had become obscure during the stages of dimensional divergence now become diaphanous. The shift from stratified consciousness to integral consciousness can be grasped with topo-experiential precision by returning to the Necker cube model (figs. 6.1(b) and 6.2).

Ordinarily, the Necker cube is viewed in such a way that, at any given moment, a single perspective is dominant, the other being relegated to the background (though not rendered opaque). The eclipse of one perspective when viewing the other models the stratified condition of consciousness prevalent in the clockwise, forward-oriented, Appropriative stages of dimensional development. With the help of figure 6.2, however, we saw that it is possible to go beyond the customary manner of viewing the cube and apprehend its perspectives as dynamically integrated. In counteracting the figure-ground tendency to project one perspective to the fore and occlude the other, the paradoxical interpenetration of perspectives symbolizes the shifting of gears to counterclockwise, to the retrograde act of Proprioception. Accordingly, if clockwise phase transitions are understood as Necker-cube-like passages that relegate previous stages to a background penumbra, we may regard counterclockwise transitions as Proprioceptively reversing that forward thrust, and, in the process, bringing about an integration in which all stages, matrices, and dimensional spheres become diaphanous to one another. Evidently, then, whereas the clockwise matrices are to be read as superimposed upon one another in translucent layers, we are to read the counterclockwise matrices as *transparent*—not in the classico-modernist sense of static simultaneous visibility, but in the topo-paradoxical sense of dynamic interpenetration.

In advancing through the stages of Proprioception, the matrix reductions carried out in the interest of self-Appropriation are reversed. Matrices now expand, with enantiomorphs reappearing in a new capacity. No longer do they express the relationships between lower-dimensional midwives and immature higher-dimensional mothers, since the mothers have now achieved their individuation. Instead of being the means by which containment is offered one-sidedly to undeveloped orders of dimensional flesh, the enantiomorphs serve as vehicles for the mutual containment of individuated dimensionalities, for their harmonic attunement. But we must bear in mind the *asymmetry* that is involved here. In any given inter-dimensional relation, the higher-dimensional member contains the lower ontologically ("positively"), whereas the lower contains the higher meontologically ("negatively"). It would be a mistake, however, to treat the respective containment contributions of synchronized dimensions as if they could simply be teased apart. For there is such close harmony between these dimensions that the containment of one by the other flows into its converse without a break.

In previous pages, partial indications were given of topodimensional convergence. We have now reached the point where a more comprehensive account is required, one that rehearses the stages of counterclockwise coevolution in greater phylo-functional detail.

We know that, in the Kleinian winding, the initial shifting of gears to the backward orientation occurs in stage five. This entails the Proprioception of rational thinking and its three concomitant support functions: mental emotion, visual perception II, and mental intuition III (the experiential meaning of this quaternary Proprioception will be elaborated upon in the next chapter). These Proprioceptions are essential for the self-acknowledgment of the cognitive mother, and for her acknowledgment of the contributions to her self-birthing made by the noncognitive midwives. Mental emotion is the functional distillate that had resulted from the matrix reduction of KB S4, when the *cogito* had accepted the *anima*'s offer of assistance by appropriating the midwife's emotional overtone as her own. In this process, the 3d/2d and 2d/3d enantiomorphs of KB S3 had been annihilated as such, gaining new expression within the sublimated cognitive sphere. Now, with the KB S5 *Proprioception* of mental emotion, the cognitive mother receives the midwife's emotional offering in a fully conscious way. In like manner, the mother's Proprioceptions of visual perception and mental intuition constitute her thankful acknowledgments of the midwifely gifts that had been proffered to her by the sensuous *vegeta* and the intuitive *minera* (respectively). Of course, the Proprioceptions in question are only a prelude to inter-dimensional synchrony, since the cognitive mother's recognition of noncognitive assistance remains primarily *cognitive.* Although the Proprioceptions enacted in KB S5 pave the way for the emergence of the noncognitive dimensions, the hegemony of human thinking is maintained. Nonhuman lifeworlds continue in eclipse and the divergent relation among topodimensional gyres is upheld.

With the passage to the second stage of counterclockwise action, dimensional convergence begins in earnest. The matrix expansion attendant to this phase transition signifies that the Kleinian mother—moving "backward through her birth canal" in this second self-nativity—is Proprioceiving her mythic nascency in KB S3, the stage wherein the then immature 2d/3d mother had been emotionally flooded by the quasi-mature 2d/2d Moebial mother, yet had received midwifely containment from the Moebius, as well, via the 3d/2d "advanced potential." The KB S6 reprise of this earlier situation is clearly no mere regression to it. For, with the new Proprioception, the KB S5 realization of 3d/3d rational thinking is not just left behind, nor does it merely persist as a background penumbra. The individuated rationality of the previous stage is wholly transparent in KB S6, so that the now-mature *cogito* is able to *hermetically contain* the *anima*'s surging potencies rather than be overwhelmed by them. For its part, the *anima* has also undergone development. By MB

S4, it has advanced beyond its stage of quasi-maturity and enacts its own Proprioception, one entailing a self-apprehension of the function of love. Should we say that the *cogito* and the *anima* attain independence in KB S6–MB S4? What happens is that these orders of wild Being become individuated *in relation to one another.* Therefore, it is not so much *in*-dependence that marks their maturation but *inter*dependence. With the matrix expansion of KB S6–MB S4, 3d/2d and 2d/3d enantiomorphs assume the "musical" role of "mirror fifths" that channel the mutual attunement of emotive and cognitive (w)holes. These individuated spatiotemporal orders of the flesh presently contain each other asymmetrically. Moebial and Kleinian topodimensional vortices now gyrate in synchrony.

The harmony of dimensional spheres is by no means complete with the notes that are struck in KB S6–MB S4. Though human cognition and animal feeling are distilled with respect to each other in this stage, neither has yet become fully attuned to the sensuous and intuitive lifeworlds. Here is where additional preparatory Proprioceptions are necessary. Since the KB S5 Proprioception of visual perception II (visual experience in abstract Cartesian space) is not enough to set the stage for the liberation of the *vegeta,* there must be another Proprioception of this kind in KB S6. Thus the Proprioception of mythic thinking that aligns it with animal love includes a repeated Proprioception of visual experience. A felt sense is obtained of visual perception I, of the dreamlike world of concrete images. In this way, the contribution of the sensuous midwife is cognitively acknowledged for a second time, though the *vegeta* remains in eclipse. The KB S6 *cogito* similarly Proprioceives mental intuition II, paying additional cognitive tribute to the *minera* while not yet releasing it from repressive containment. In the same fashion, the MB S4 *anima* pays emotional homage to the *vegeta* through the Proprioception of hearing, and gives thanks to the *minera* via Proprioception of emotional intuition II.

The third stage of counterclockwise circulation brings a further convergence of topodimensional spheres. With the matrix expanding once more, the 3d/3d Kleinian mother presently Proprioceives the even more primal 1d/3d magical situation, in which she had been overpowered by the sensuality of the 1d/1d lemniscatory mother, yet had also received midwifely containment from the lemniscate through the 3d/1d "advanced potential." Of course, the KB S7 *cogito* is mature now and thus can wholly *contain* the *vegeta*'s burning sensuality rather than being consumed by it. The liberated *vegeta,* for its part, has undergone its own Ontogeny. By LB S3, it has gone beyond its clockwise stage of quasi-maturity and currently enacts a Proprioception of primal olfaction. Thus attaining individuation in synchrony, the cognitive and sensuous orders of fleshly Being become attuned to each other, with the asymmetric mutual containment of these (w)holes being channeled by the newly functioning 3d/1d and 1d/3d enantiomorphs.

In this third stage of counterclockwise action, it is not only the *cogito* that harmonically converges with the *vegeta,* but the *anima* as well. The 2d/2d Moebius mother, moving backward through her own "birth canal," Proprioceives the state of affairs in which her nascent rage, 1d/2d, had burned in the sensuous flames of the 1d/1d lemniscatory mother, though the inchoate Moebius body had also received midwifely support from the lemniscate through the 2d/1d "advanced potential." Now, in MB S5, the *anima* is mature and can ontologically contain the *vegeta,* who, also being mature, contains the *anima* in a meontological way. The asymmetric mutual containment of lemniscatory and Moebial bodies is channeled by 2d/1d and 1d/2d "mirror octaves."

Nevertheless, the harmony of dimensional spheres is still incomplete in KB S7–MB S5–LB S3. Although human cognition, animal feeling, and vegetative sensation are all distilled with respect to each other, none are wholly attuned to the mineral world. Yet each enacts a preliminary Proprioception that paves the way for this ultimate attunement: the *cogito* Proprioceives mental intuition I, the *anima* Proprioceives emotional intuition I, and the *vegeta* Proprioceives sensuous intuition.

The fourth and final stage of counterclockwise circulation carries to completion the convergence of topodimensional gyres. With the full-blown expansion of the Ontogenetic matrix, the wholly individuated KB S8 mother Proprioceives the negative containment of her archaic cognition within the "black hole" of the mineral midwife. This happens in such a way that the three-dimensional Kleinian mother and zero-dimensional sub-lemniscatory grandmother now embrace one another—the former positively, the latter negatively. Their asymmetric mutual containment is mediated by 3d/0d and 0d/3d "mirror whole notes." Similarly, the fully mature, passionately affiliative MB S6 mother Proprioceives the mineral containment of her embryonic fear in a manner that brings about the asymmetric mutual containment of the two-dimensional *anima* and zero-dimensional *minera,* a conjunction that is channeled by 2d/0d and 0d/2d enantiomorphs. Lastly, the individuated olfactory mother of LB S4 Proprioceives the *minera*'s meontological containment of her nascent tactility so as to achieve asymmetric mutual containment of the one-dimensional *vegeta* and zero-dimensional *minera,* the nexus occurring via 1d/0d and 0d/1d. Now all topodimensional wheels are turning in asymmetric synchrony. The symphony of the lifeworlds has reached its crescendo. The alchemical circle has finally been squared.

At the end of section 2.1 of this chapter, I intimated the following:

When all four dimensional windings have been brought into harmony, when all four of "Ezekiel's wheels" are spinning in synchrony, when the alchemical circle has fully been squared, the dimensional spiral will then expand in logarithmic fashion, with the dialectical matrix now growing from 4^2 to 5^2. In this wider turning of the spiral, a new and more dialectically intricate, *four*-dimensional order of the flesh will come into play beyond the Kleinian—a "meta-Kleinian" onto-topological action pattern laid out in 5 × 5 matrices. In this novel lifeworld, there should be an entirely new mode of functioning beyond thinking, feeling, sensing, and intuition, and each of the five functional families should consist of five sub-functions. Indeed, there should be a *whole new order of nature* that surpasses the *minera, vegeta, anima,* and *cogito.*

What is the task that confronts us with regard to this higher dimension of Being? Should we seek to extend our conceptual analysis so as to elucidate the fifth order of the flesh as we have the first four? Clearly the answer is no, for nothing that would be limited to conceptualization—to thinking or writing—could provide a positive grasp of the unique functionality that transcends the cognitive mode. It would be an exercise in futility to attempt to encompass the meta-cognitive via a cognitive analysis aspiring to even greater heights of mental abstraction. Evidently, what we must do instead is continue the process of *questioning* the abstract manner in which we have been operating. Advancing that process means concretely realizing the Proprioceptions about which we have thus far been speaking abstractly. In enacting this *coagulatio,* cognitive Ontogeny will reach completion in relation to the Ontogeny of lower-dimensional orders of the flesh. It will be from this topodimensional harmony, from this synchronization of backwardly spinning fleshly mothers, that the chrysalis will be spun for the meta-cognitive mother.

CHAPTER EIGHT

DIMENSIONAL SELF-SIGNIFICATION

ROUND ONE: COGNITIVE SELF-SIGNIFICATION

In chapter 2, the third dimension of wild Being was fleshed out by the self-signification of this Kleinian text. Developmentally situated, this act of embodiment is the opening stage in the process of Proprioception (stage five of Kleinian Topogeny, denoted KB S5 in table 7.1). In subsequent chapters, we turned to the lower dimensions of the flesh. After adumbrating those lifeworlds abstractly in chapters 3 through 5, their concrete contents were specified in chapters 6 and 7. This did not bring them to full realization, of course. For, in analyzing their contents as we did, we implicitly maintained the underlying threefold division among content, container, and uncontained (object, space, and subject). Thus we signified the lower dimensions in such a way that they did not yet signify *themselves*. The question of lower-dimensional self-signification remains to be considered in concluding this book. But before turning to that, let us attempt to go further with the self-signification of the cognitive dimension.

Recall the first step in the process of self-signification indicated in chapter 2, section 3: I had to remove my cloak of anonymity and stand present as the author of the text. This was necessary because, as long as I maintained the customary posture of authorial detachment, whatever I signified was naught but an other; no concrete *self*-signification was possible. In the earlier chapter, I went on to demonstrate that self-signification additionally requires the *topodimensional upgrading* of the physical signifiers themselves: the one-dimensional typographic bodies of ink of which this written text is composed need to be recast in terms of three-dimensional Kleinian signification. I now want to deal more extensively with the *experiential* aspect of self-signification. To that end, we will take a closer look at what it means to *stand present.*

I am here, in my apartment in Vancouver, present at this keyboard, working on this text. At the moment, it is 8:49 a.m. on October 18, 2004. Do I *truly* stand present by acknowledging in this way that I, Steven Rosen, am here and now writing this text? Consider the posture that is implicitly assumed in this way of regarding "Steven Rosen." He is viewed as a particular person situated in a particular place at a particular moment in chronological time. Thus he is approached in the "forward" or "clockwise" orientation, regarded as an ontical being. This manner of presenting him is in fact but a *re*-presentation. Steven is signified as an object cast before a subject that itself remains anonymous. For, in this alleged self-signification, Steven actually has divided himself into *subject qua object*—a certain person who can be defined in terms of certain objectifiable characteristics—and *subject qua subject*—a still elusive anonymity. How, then, can we say that the author truly stands present here and now? Objectifying himself in this way, the "here" and "now" that he signifies is actually *elsewhere and otherwhen* relative to a resituated here and now, an originating presence whose locus is the new anonymity established by the split. In rendering Steven an object, the here and now is displaced to the nameless subject qua *subject.*

It is certainly true that ontological self-signification could not have been realized as long as the author continued to write in a largely anonymous fashion. The person behind these words had to make his presence felt. Only by dropping his cloak of anonymity and addressing the question of Being in a more *personal* way could he have approached Being. But a more complete approach to Being requires a fuller presencing, and this, in turn, means that the forward thrust by which Steven objectifies himself must be drawn back in.

This is far easier said than done, of course. Although the author of this text may wish to counteract his self-objectification by Proprioceiving his pre-objective bodily source, he is confronted with the compelling sense that the body behind these words is only that of the objectified ontical being, *Steven Rosen.* In typing these abstract words, no doubt I could gain a measure of bodily awareness simply by bringing my attention to my hands and fingers as they touch the keyboard. The proprioceptive sense I would thus obtain of the working of my muscles is readily achievable. But there is one thing I normally do *not* proprioceive, namely, the *"I" itself*—"that unique touch that governs the whole tactile life of my body as a unit, that *I think* that must be able to accompany all our experiences" (Merleau-Ponty 1968, 145). This "I" is what Burrow called the *"I"-persona* (see chapter 7). It is the prime "identity operator"[1] upon which all of "my" specific operations are based; it is the "master word" that lies behind all these particular words. The "I think" is what must be Proprioceived in order to surpass Steven's finite particular body and gain access to the generic thinking body (fourth flesh). That is, cognitive Being must think itself.

The matter at hand could not be more salient since it concerns the very authorship of this text. The author must stand present here and now. But if the objectifying displacement of the here and now is to be avoided, it is *Being* that must stand present, not just the particular being, Steven Rosen. We should not forget that this ontological presence is paradoxical: it is also an *absence*. As Heidegger put it, "Being . . . ambiguously names what is presently present and also . . . what is absent" (1946/1984, 37). Or, to repeat, Being possesses a *Kleinian hole;* it entails *(w)holeness,* not simple wholeness. The Proprioceptive movement that brings this "absent presence" to realization discloses the "onto-biographical" nature of this text: three-dimensional Being emerges as its author.

Exactly what sort of experiential practice is required for the Proprioception that is called for? In what specific way is our apprehension of this text to change in order that we may consciously engage three-dimensional Being per se? Can anything more be said about just how we are to enter the generic body? In Gendlin's experiential approach to the pre-objective, we are advised to enter the body through the *middle:* "Let your attention refer inside, directly, physically, to the comfort or discomfort in the middle of your body" (1991a, 45), say, in your "chest [or] stomach" (1996, 1). The alternative I suggest so as to facilitate the Proprioception of the thinking body that governs this ontological writing is that we enter the body through the *head.* My proposal is based on the work of Trigant Burrow.

We know that, according to Burrow, the "I"-persona has a distinct site of operation within the human organism. It is found in what Burrow termed the "cerebro-ocular" region (Burrow 1953, 526), that is, in the cerebral cortex of the brain and its associated organ of vision. Burrow pointed out that it was through the phylogenetic development of the cerebral cortex that language and symbolic activity first arose. Therefore, to gain an immediate sense of this activity, it seems one would have to "enter the body" through the cerebrum. But this conclusion was informed by more than a simple logical deduction. Burrow claimed to have had a spontaneous experience of the "I"-persona's bodily base, one that profoundly influenced all his subsequent research. After a prolonged period of interpersonal strife involving the members of the group that he had established to investigate such "I"-based conflict, he began to notice a distinctive pattern of tension around his eyes and forehead. Burrow recognized in this the concrete expression of the "I"-persona.

I have proposed that, to enter the KB S5 stage of Ontogeny, we must engage in Proprioception, which means consciously receiving the gift of thinking that organic Being has given to itself. In Burrow's terms, this entails bringing attention not so much to sensations in the chest or stomach, or in the hands, but to the ocular-facial or "cephalic segment" (1953, 249–54), viz. the area of the body around the forehead and eyes. Burrow would caution us not to confuse the "I"-persona that resides therein with the ego of the allegedly isolated individual. We might say that this persona is the *species-wide* "subject" that lies behind the appearance of individual subjectivity—the subjectivity of "Steven Rosen," for example. But while it is through the "I"-persona that we, as a species, create the impression of ourselves as merely isolated, disembodied subjects, the generic "I" itself is no disembodied subject. Rather, it is the *bodily process* that is central to human functioning as a whole. Therefore, when Burrow became attentive to the "I"-persona rather than continuing to be unwittingly governed by it, he experienced this palpable pattern of tension around the eyes and forehead against the "tensional pattern of the organism as a whole" (Galt 1995, 31). He was thus presumably able to apprehend in an immediate way what he called the "solidarity of the species" (Burrow 1953, 71)—what Merleau-Ponty called the "flesh of the world."

Following his first spontaneous Proprioception of the generic organism, Burrow sought to cultivate the experience in a systematic practice he named "cotention" (Burrow 1932). He described his procedure as one of setting aside daily experimental periods in which he "adhered consistently to relaxing the eyes and to getting the kinesthetic 'feel' of the tensions in and about the eyes and in the cephalic area generally" (1953, 95). Burrow might say that, behind "Steven Rosen's" work with this text, behind the typing of these very words, the *master* word or "I"-persona operates, and that this generic "I" manifests itself concretely in the muscular activity of "Steven's" *eyes,* continually engaged as they are in minute acts of binocular convergence that serve to objectify his world (chapter 7). It may well be easier to proprioceive movement in other parts of the body than in the eyes; since ocular activity is subtler, a great deal more practice may be necessary. Nevertheless, if Burrow is right, it is proprioception of the activity in and around "my" eyes that is required for generic Proprioception. So it seems that, if "I" am to go beyond functioning exclusively as an "abstract head," "my" practice evidently has to include obtaining a bodily sense of the very "head" that presently directs this writing.

Now, in the previous chapter we learned from Burrow that the "cerebro-ocular" region of the brain is closely linked to "affecto-symbolic" functioning, and to the intuitive certainty of our own "rightness." Restated in the language of the present volume, visual perception, mental emotion (affect), and mental intuition operate jointly to undergird the cognitive activities of the thinking subject in the fourth stage of Kleinian Ontogeny. Therefore, in advancing to stage five, the felt sense of our visual objectifications should be coupled with a proprioceptive awareness of our affective reaction tendencies and of the concomitant experience of self-certainty, the sense of the absolute "rightness" of any particular apprehension we may have, which is rooted in the mental intuition of our own existence.

In this regard, let me offer a personal acknowledgment of the motivational underpinning of this book. My impetus for writing is far from pristine. Though I truly hope to contribute to human understanding and to the betterment of our species, behind this intention there is also my desire to reinforce my self-image, to win the recognition of my colleagues, to show you how "brilliant" I am, and to prove to you that I alone have gained mastery over some of the most challenging and significant issues that human beings have ever faced! (Perhaps you will be inclined to make your own acknowledgments of the motives that underlie *your* endeavors, including your current reading of this book.) So while there are moments when I might like to think myself capable of proceeding in a purely dispassionate way that is detached from my "baser" feelings and self-centeredness, the fact is that every word I have written, however rarefied and cerebral, and these very words I am now writing, are colored by an emotional, self-possessed undertone.[2] Does this recognition of the mental emotion and mental intuition that undergird the present text constitute a *Proprioception* of that text? Clearly it does not. To engage in Proprioception, it will not suffice to take back the particular self-oriented emotions I experience. Beyond that, the "I" itself must be taken back. It is in topodimensionally withdrawing the projection of *Steven Rosen* as the prime container of visual perception, mental emotion, and mental intuition that thanks are given for these gifts afforded by the topodimensional midwives. In this enactment, the perceptions, emotions, and self-intuitions that accompany "my" writing in fact become the experiences of humanity at large—that is, of Kleinian Being. Apparently, then, "I" must obtain a palpable internal impression of the region of "my" head from which originates "my" present feeling of frustration at the difficulty of conveying these ideas—and the subsequent feeling of "righteous self-satisfaction" when success seems achieved! Proprioceptions of this kind, insofar as they lead backward (through "Steven Rosen") to the "I"-persona as such, open a pathway to the ontological.

Now, I observed in the previous chapter that the set of Proprioceptions carried out in KB S5 is a forerunner of more condensed orders of Proprioception. In KB S6–MB S4, with the derepression of the *anima,* emotional signification coagulates. How is this to be realized as *a self-signification of the text* enabling the Moebial dimension of Being to acknowledge her co-authorship? Once again a *standing present* is called for.

ROUND TWO: COGNITIVE-EMOTIONAL SELF-SIGNIFICATION

Although the (quasi-)mature rational *cogito,* operating through the vehicle of Steven Rosen (fig. 8.1, left), governs the text in its KB S4 and KB S5 modes of expression, by no means is the immature mythic *cogito* merely absent from the scene. Beneath the abstract emotions of Steven Rosen, noted above, lie the subtextual feelings of a less developed personage. Let us call him "Stevie" (the name my parents gave me as a child) (fig. 8.1, right). Steven's desire to prove himself to the community of thinkers whom he hopes will read this book is rooted in pre-Oedipal Stevie's longing to be loved. It is long-repressed Stevie who must now stand present.

Figure 8.1. Steven (left) and Stevie (right)

November 12. Yesterday, a few hours after writing the previous sentences, Stevie did make his presence felt to me in an unprecedented way.

Late in the afternoon, my wife and I became embroiled in a highly charged emotional exchange. It left me with a sense of deflation, a feeling of total inadequacy. I had disappointed the woman I love so dearly. The pained anguish that appeared in her eyes was there because of my failings.

Marlene went to the bathroom and I into the unlit living room, my heart still heavy with the feeling that I had failed miserably to meet her expectations of me. Gazing out of the window I could see the twinkling lights of West Vancouver on the far side of the Burrard Inlet, and the image I could not shake was the hurt expression on Marlene's face.

I began to weep. To sob bitterly. My shoulders heaved in spasms of uncontrollable emotion. The feelings that burst from my throat echoed loudly through the room. This bellowing lament was a breakthrough for me. Not since childhood had I experienced and expressed my emotion so fully. It isn't that I'd never cried since then. But on the rela-

tively rare occasions when this would happen, the feeling seemed to get "stuck in my throat," as if the upsurgence of emotion reached a bottleneck there that prevented full vocalization, choking off further expression. Yesterday it was different. Yesterday the surge of emotion flowing up from my heart broke out of the blockage in my throat and continued into my head. My heart was opened to a wider awareness.

When Marlene heard my howls of grief and came in to comfort me, she could hear a mournfully repentant voice say, "I tried. *I* tried." *It was the voice of* Stevie.[3]

Here, beneath the layer of Steven's mental emotion, lies a more concrete motivational subtext: Stevie's need to be loved. The penumbra is brighter now. The child's feelings shine forth more clearly on the page. Still, the descriptive containment of these feelings given by the journal entry does not constitute a *hermetic* containment; the disclosure of the emotional subtext does not add up to a *Proprioception* of the text. For the latter, it will not be enough to gain cognizance of Stevie's particular feelings. *Stevie himself* must be taken back in. The projection of Stevie as a particular being whose particular emotions are contained in his particular body must be withdrawn. Thus moving backward, the feeling-saturated cognitions of the child at play beneath the surface of the text come to be recognized as the *generic* cognition of mythic humanity. This second sealing of the three-dimensional Kleinian vessel will contain the immature *cogito* so as to bring a greater degree of authentic maturity than had been achieved in the first sealing. Further individuation will be realized. But something more will happen with the Proprioception of Stevie. The "genie will be let out of the bottle." The noncognitive *anima* will be released from cognitive containment to enact her own Proprioception, that which corresponds to the initial hermetic sealing of the two-dimensional Moebial vessel. In Proprioceptive conjunction, Kleinian and Moebial authors of the text will then collaborate on their joint "ontobiographies."

Now, we may restate what we have found about KB S5 Proprioception by saying that, for the ontological circle of self-signification to be closed, (1) the particular signifiers printed on the page need to be traced backward to their master signifier, the "I"; (2) the "I" must be realized paradoxically as "I am not-I" (lest it be merely an *objectified* "I"); and (3) this paradoxically self-signifying "I" has to be fleshed out topodimensionally via

Kleinian embodiment. In the course of the Proprioception, the authorship of the text passes from ontical Steven Rosen to wild Being. For the *second* sealing of the Kleinian vessel that is presently at hand, the Proprioceptive circle cannot be closed without first distilling Steven's written text to the denser subtextual medium of Stevie's concrete images and spoken words.

It is not that imagery was not used in the initial KB S5 Proprioception carried out in chapter 2. Here illustrations of the Moebius strip and Klein bottle were enlisted in the realization of the Kleinian text. But the images in question, being employed in the service of the rational *cogito,* were sublimated; they were processed through the auspices of visual perception II (see table 7.1). That is, the illustrations were viewed as figures in Cartesian space; they were experienced as distinctly circumscribed objects cast before the perspective of a viewer who regarded them with a considerable degree of detachment. Stevie's KB S3 mode of imaging evidently is different.

Philosopher David Lavery (1983) helps us better appreciate the distinction between visual perception II and visual perception I. He draws the contrast between "focal vision" and "peripheral vision" (27–28). The former gained supremacy in the aftermath of the Renaissance with the rise of perspective. It was because the Cartesian observer detached himself from the world that he was able to "put things in perspective" (25). Thus standing aloof, he could bring the objects under his visual scrutiny into sharp focus by the process of binocular convergence upon them. Lavery intimates that the eyes worked differently prior to the Renaissance. The preperspectival person, participating more fully in the world, being immersed in it, did not primarily focus upon objects held off at a distance. Rather, the sense of being *encompassed* by one's environment led one to experience things and people in a "peripheral" fashion, i.e., as being *all around one.* In formulating his account of this earlier way of seeing, Lavery draws from the writings of philosopher Owen Barfield, who is subsequently quoted as follows: The "'earlier kind of knowledge . . . was at once more universal and less clear. We still have something of this older relation to nature when we are asleep, and it throws up the suprarational wisdom which many psychoanalysts detect in dreams'" (24). The old peripheral vision is *oneiric* in character: it possesses a dreamlike quality, as Gebser said of mythic imagination.

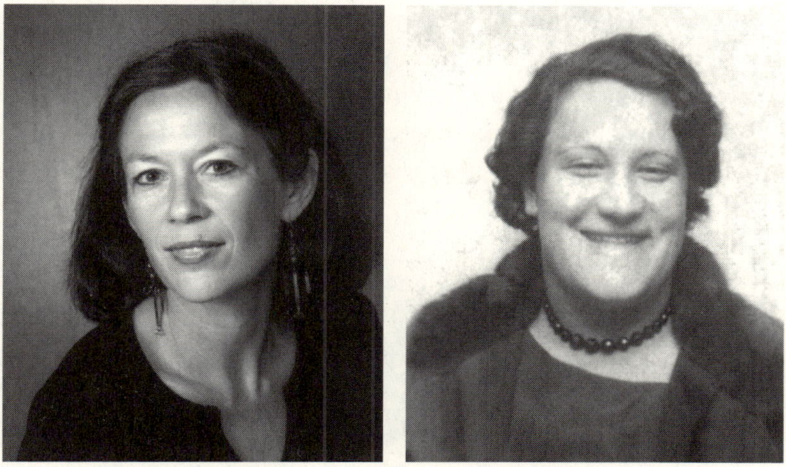

Figure 8.2. Marlene (left); and Stevie's mother (right)

Standing present with a text now distilled, Stevie may picture Marlene (fig. 8.2, left). *Her image* is presently the text. And he views the picture diffusely. Since he does not bring her face into sharp relief, Marlene's visage can fluidly transmute into that of another (fig. 8.2, right), much as in a dream. It is not a *written* word that accompanies such peripheral imaging but one that is *spoken*. Stevie does not inscribe Marlene's name. He soulfully voices it. And when he says to her, "I *tried*," we hear his plaintive intonation.

The transition from the child to the ontical adult, or, phylogenetically, from mythic to rational culture (KB S3→KB S4), in fact is marked by the passage to the written word from a word that was spoken. The change is brought out in Gebser's portrayal of how mythic consciousness, characterized by "utterance" and "voicing," gave way to rational consciousness, where "seeing and measuring" (1985, 146) were more prominent (the written word is seen, of course, not voiced and heard). In *Interfaces of the Word* (1977), Walter Ong attributes great significance to this transformation of the means by which humans communicate. Ong hypothesizes that, over the past five thousand years, consciousness has evolved from being governed primarily by aural perception, sensitivity to sound, to being controlled by vision. The cultural concomitant of this has been the transition from orality to script, from social transactions based on the spoken word to those dependent on writing.

Following the inception of script around 3500 BCE, says Ong, "the world of primary orality was torn to pieces by writing and print, which then created, agonizingly, a new kind of noetic and a new kind of culture based on analysis and self-conscious unification" (1977, 10). What, specifically, was the primary experience like? Ong offers the following proposition:

> The psyche in a culture innocent of writing knows by a kind of empathetic identification of knower and known, in which the object of knowledge and the total being of the knower enter into a kind of fusion, in a way which literate cultures would typically find unsatisfyingly vague and garbled and somehow too intense and participatory. To personalities shaped by literacy, oral folk often appear curiously unprogrammed, not set off against their physical environment, given simply to soaking up existence, unresponsive to abstract demands such as a "job" that entails commitment to routines organized in accordance with abstract clock time (as against human, or lived, "felt," duration). (18)

And of the spoken word, Ong says:

> [It] is of its very nature a sound, tied to the movement of life itself in the flow of time. Sound exists only when it is going out of existence: in uttering the word "existence," by the time I get to the "-tence," the "exis-" is gone and has to be gone. A spoken word, even when it refers to a statically modeled "thing," is itself never a thing. . . . No real word can be present all at once as the letters in a written "word" are. The real word, the spoken word, is always an event . . . an action, an ongoing part of ongoing existence.
>
> Oral utterance thus encourages a sense of continuity with life, a sense of participation, because it is itself participatory. Writing and print, despite their intrinsic value, have obscured the nature of the word . . . for they have sequestered the essentially participatory word . . . from its natural habitat, sound, and assimilated it to a mark on a surface, where a real word cannot exist at all. (20–21)

Despite Ong's overdrawn contrast between the "real word" and the written word, and his nostalgic preference for the former, he brings home some important distinctions between these modes of interaction. Of course, while the written word has gained supremacy over the spoken word, we still do speak to each other. Nevertheless, spoken language, though still being transmitted through the concrete medium of sound, has taken on, in its abstract pattern of organization, the character of written discourse. We have come to speak as though we were writing. The clue Ong provides as to the nature of *original* speech—that it is "tied to the movement of life itself in the flow of time"—is reminiscent of what Gebser says of mythic consciousness in general: it is "oceanic" (1985, 252), as exemplified in the circular rhythm of the fragment from Heraclitus: "'For souls it is death to become water; for water it is death to become earth. But from the earth comes water and from water, soul'" (252). How different this is from the prosaic linearity of rational expression later to gain sway. For the text of Heraclitus to be properly conveyed, it must be recited, voiced aloud in the rhythmic tempo of a song.

The Jungian analyst Mary Lynn Kittelson provides further insight into the nature of original hearing and speech in her essay, "The Acoustic Vessel" (1995). She begins by taking note of the *repression* of the old sonority that has occurred in our society: "Mostly . . . we work visually. Our words are heard primarily as content. We pay scant heed (consciously, at any rate) to how things sound" (89). Thus:

> Through the [domination of the] eye, the vibratory and participatory aspects of experience fall into the shadows. To be "ear-minded" is to be resonant, layered, slower, sensing things out before the light. . . . Unlike light, whose vibratory nature is a less immediate experience, sound and silence reverberate in a palpable way. . . .
>
> Our collective experience in modern society has supported auditory inattentiveness and misuse of sound. . . . Chatter, complaint, jargon, interminable "how-to's," and the hyped-up headline style of broadcast news are . . . incessant. . . . As adults we have lost our ear-minded center, which was vibrant in previous generations and remains so in much of infant and animal life. According to one neurolinguistic programming study, "Most people in the U.S. do not actually hear the sequence of words

> and the intonation patterns of what they, or other people, say. They are only aware of the pictures, feelings and internal dialogue that they have in response to what they hear" (Bandler and Grinder 1979, 124). (Kittelson 1995, 90–91)

Using terms like "song," "melody," "prelude," "overture," and "opera" (96), Kittelson emphasizes, both metaphorically and literally, the rhythmic musicality of uninhibited sonance. *Imagination* surely plays a role in such soundings, as Gebser says of mythic "utterances" and "voicings." The images may be concretely visual (visual perception I), or acoustical, "as in [the images of] a poem or song" (103). Like Gebser, Kittelson relates primal sonority to myth in an explicit way. After observing that soundings of this kind are "noticeably present in many creation myths" (91), she focuses on the Greek myth of Echo, which voices "longing, dreams, loneliness, pain, and poignant search" (101). "The mythic image of Echo," says Kittelson, "suggests a potential for resonance in a way that 'neurosis' or mindless 'parroting' does not" (102). In the passionate murmuring of Echo, there is a "calling back, and calling back, and calling back" (101) that is "full of meaning" and "evokes the kind of listening that poignantly discovers the significance in Echo's sounds" (102).

When Kittelson goes on to describe echo experiences as "the singsong offerings of the psyche" (102), Stevie comes to mind. "I *tried,* I *tried,*" he echoes in his mournful cadence, with the image of Marlene's face flickering behind his softly focused eyes. For Stevie truly to stand present in the text, his words must echo here and now and his oneiric images gain presence. The words must sound forth beneath the mute graphic marks that Steven has printed on the page; the dream-light of the images must radiate beneath the cold abstraction of these skeletal bodies of ink.

It is clear, however, that while such a distillation of the text is necessary for the KB S6 Proprioception now required, it is not sufficient. More is needed than Stevie's ontical presence. His voiced words and preperspectival images must be *retracted* in realization of the ontological. We saw in the previous chapter that the KB S5 Proprioception is neither left behind in passing to KB S6, nor is its presence filtered translucently. Rather, the individuation achieved in the earlier stage is fully transparent in the stage that succeeds it. Evidently, the transparency of the KB S5 Proprioceptive adult is required to support the Proprioception of the KB

S3 child. Only by sustaining the light of mature cognitive self-comprehension can a regression to KB S3 be averted. But exactly what is it that happens in KB S6 beyond the ontical presencing of Stevie? Just how is Stevie's ontical text ontologically transformed?

Again, for the initial Proprioception enacted in KB S5, Steven enters his body through his head to obtain a bodily sense of the head that directs his writing. In the process, he turns his awareness to the tensions in his eyes behind which lies the master word, the generic "I" constituting the innermost core of Steven's particular "I." In the KB S6 Proprioception of *Stevie,* the head must be entered once more, but now on a different trajectory. We might even say that a "different head" must be entered.

Recall that the distinct anatomical locus of the rational "I" is the "cerebro-ocular" region, that is, the cerebral cortex of the brain and the organ of vision associated with it. The KB S5 Proprioception entails drawing attention backward into this cerebral brain (Burrow's practice of "cotention"). I suggest that, in KB S6, the "master signifier" is not the "I"/eye of the cerebral cortex that is linked to the written word, but a *sub*cortical, "synaesthetic I," that which lies behind Stevie's voiced word and preperspectival image.

(Before going into further detail on the brain and other organ centers, I feel it necessary to add a caveat similar to the one with which I opened the previous chapter. The following analysis by no means aspires to do justice to the vast body of scientific research on these organs. In the present context, only those aspects of somatic functioning relevant to ontological phenomenology will be considered. I must also acknowledge that many of the relationships I propose depart from mainstream thinking to the point of sounding "esoteric." However, as far as I know, none of my submissions are inconsistent with established fact. In hypothesizing a connection between original emotion and the heart, for example, I do not deny the evolution of certain *brain* centers concerned with emotion, though I do not deal with them explicitly in my selective account.)

The brain research of Burrow was primarily concerned with activity in the cerebral cortex. Almost two decades after Burrow's death, a theory emerged that brought greater emphasis to *lower* centers of the brain. The neurophysiologist Paul D. MacLean (1968) proposed that the human brain is *triune*—that it actually consists of *three* brains, one arising from the other in the course of phylogeny (see fig. 8.3). MacLean identified the cerebrum as the neomammalian brain, hypothesizing that it evolved out of an older, paleomammalian structure associated with animal functioning. Anatomically, the new brain formed *around* the older brain so that the latter came to be contained within it. The structural containment is also functional. Whereas the cerebrum is the layer of the brain in which cognitive activity is consciously processed, the old mammalian brain or limbic system is linked to operations of a subconscious nature. Recent research using sophisticated imaging techniques (PET studies) confirms that while neocortical activity prevails during waking consciousness, in dreaming the limbic system gains control (Braun et al. 1997, 1998; Maquet et al. 1996). The paleomammalian brain evolved out of and came to contain an even older structure: the *reptilian brain.* Lying inside the base of the skull, this primal core functions in a largely unconscious, vegetative fashion (the way reptiles appear to function).

I propose that while the KB S5 Proprioception entails a movement into the neomammalian brain, in KB S6 it is the paleomammalian brain that is reentered. Figure 8.3 provides us with a rough idea of the spatial relationship among the three brain centers, but the diagram is of course an objectification. Here we view the representation of the brain as appearing *in front of* us, when what is required is a *backward* movement

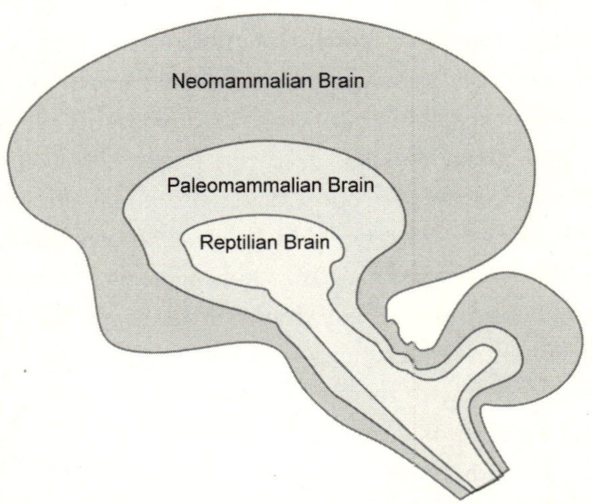

Figure 8.3. Schematic diagram of triune brain (from Paul D. MacLean, "Alternative Neural Pathways to Violence," in *Alternatives to Violence,* ed. L. Ng [Alexandria, VA: Time-Life, 1968])

into the brain itself. Entering the brain in reverse via Proprioception brings a realization of the "brain-as-lived" (Leder 1990, 113); of the onto-phenomenological, generic brain; of the brain as *flesh of the world.*

To pave the way for the retrograde passage into the inner recesses of the old mammalian brain, Stevie first moves from the image of others (fig. 8.2) to the image of himself (fig. 8.1, right). Then, proceeding in reverse through this master signifier, he works his way to the ontological core of his ontical self-image. Whereas Steven reenters his head through the "I"/eye governing the written text and moves backward into the cerebrum, Stevie proceeds in relation to the *coagulated* text, retreating from his imaginal "I" (the bodily image of himself) into the "dream brain" constituted by the limbic system. When Stevie stands present, his softly focusing eyes move in the saccadic rhythm of a dream. The limbically based eye movements (REM) thus generated are to be Proprioceived. Guided by the KB S5 Proprioception of Steven that remains transparent in KB S6, Stevie's KB S3 dream then becomes *lucid;* he awakens in it to obtain a consciously felt sense of the limbic eyes behind which lies the ontological "I" of the mythic *cogito.*

As in KB S5, the KB S6 Proprioception is indeed about achieving a certain lucidity; in fact, it is about *seeing the light.* In my previous work (Rosen 2004), I explored the special role of light in the human lifeworld: light is no mere object that is seen but is that *by* which we see. In discussing the implications of the famous Michelson-Morley experiment on light that gave impetus to Einstein's theory of relativity, I demonstrated that the phenomenon of light—instead of lending itself to being treated as an object open to the scrutiny of a subject that merely stands apart from it—must be understood as entailing the *inseparable blending of subject and object,* of seer and seen. For his part, Heidegger adumbrated an intimate linkage among lighting, thinking, and Being: to think Being is to think light (1964/1977, 370–92). Presently I am proposing that the linkage is tangibly realized via the Proprioceptive self-lighting of the phenomenologically lived brain. Through such Proprioception, we see that *by* which we see, see our own seeing, rather than continuing to see only the objects from which light is reflected. And the self-seeing in question goes beyond the ontical activity on the optical surface of the face to encompass an *ontological self-thinking* that penetrates the depths of the lived brain. This ontological action surely is not limited to cognitive-visual functioning. The brain does not just think itself by seeing itself; in KB S5 it does this by lucidly feeling and clearly intuiting itself, as well; that is, mental emotion and mental intuition III are also Proprioceived here. In all cases, *clarity* is gained; illumination or enlightenment is realized. What happens in KB S6? With the waking Proprioception of Stevie's dream eyes (visual perception I) and his self-intuition (mental intuition II), the *limbic* brain illuminates itself, bringing about the more concrete sealing of the Kleinian vessel. The self-illumination now involved is that of mythic humanity at large.

But, again, while rational Steven can reflect on mythic self-signification through the medium of this written text, for the self-signification of the mythic text to actually take place, Stevie must stand present with the *oneiric image*. His *image* (fig. 8.1, right) must be the text.

Let us say that KB S6 self-signification involves a special kind of *mirroring*. When I comb my hair or shave my face, the classical default setting normally operates: the face I see in the mirror is taken simply as object, and the subject who views this face is relegated to the background. Thus rendered invisible, the subject enjoys the comforts of anonymity. But then there are those moments when gazing in the mirror can be disturbing (and not just because the "object" I see has too many wrinkles and too large a nose!). I look at myself and realize that it is *I* who am doing the looking. There is something unnerving about the experience, intimating as it does an odd commingling of viewer and viewed. Gazing at myself in this way, I am likely to be struck by a sense of alienating derealization. My natural inclination is to shake off the feeling and resume my normal way of looking in the mirror, that in which the comfortable split is upheld between the one I view—Steven as the object in the glass—and he who does the viewing—the anonymous subject ensconced in the background.

But suppose I sought to *sustain* the uncanny perception of myself instead of brushing it aside or letting it go. I have tried to do this and found it difficult. It is much like attempting to hold both perspectives of the one-dimensional Necker cube without just blurring them. In the present case, however, the difficulty is compounded by the fact that it is the *three-dimensional subject and object* that are to be held together. Still, it might seem that if I could somehow stay with the paradox for a long enough time, a clearer glimpse could be gotten of the uroboric blending of the one who looks into the mirror and the one who is seen therein. On such an occasion, I would neither simply be inside my body looking out, nor be outside of it looking back upon it from a disembodied vantage point. Rather, I would align myself with the Kleinian circulation of inside and outside. In the process, "I" would literally be *beside* "myself," ecstatically present to "myself." On a moment like this the circle of Proprioception would be closed and Steven Rosen and his anonymous other would flow unbrokenly together, penetrate one another in mutual transformation. In such a moment of ecstasy, seer and seen would gain expression as the *seeing process,* the transitive gyration by which wild Being sees itself (recall from the preface Heidegger's allusion to "man's ecstatic sojourn in the openness of presencing"; 1964/1977, 390).

However, in the case of the concrete "text" constituted by the bodily mirror image, the fact is that Steven cannot consummate the ecstatic marriage to Being while operating on his own; the union can be achieved only in conjunction with *Stevie.* To be sure, Steven has already consummated the marriage in *abstract* terms. In his KB S5 enactment, he has closed the paradoxical circle of Proprioception; the ecstasy of cerebral self-lighting has been realized insofar as the text has come to signify itself (this is the work of chapter 2, elaborated upon experientially in the previous section of the present chapter). Yet this is an ecstasy of the *written* "I," not of the image one sees in the mirror. Setting the text aside, I may surely look into the glass and gain a momentary sense that it is *I* who am looking. Generally, however, the abstract "I" that presides over the written text also holds sway in the world at large. I noted above that we have come to speak as if we were writing. It also seems true that we now tend to *see* as if we were *reading.* That is why the uncanny mirroring experience cannot be sustained—because Steven *reads* his mirror image (interprets it abstractly), rather than viewing it concretely. Steven is the "I" through whom mirror perception normally takes place, and he does not correspond literally to the bodily image he perceives; rather, Steven's identity is centered in the more abstractly embodied, non-imaginal, linguistic "I." Since the "master signifier" or "I" that lies at Steven's core is the written word and not the image, as long as this "I" maintains its detached control in the sphere of concrete perception, it will *objectify* the image it sees in the glass (the fleeting experience of being "beside myself" may indicate a momentary lapse in said control). It is true that, for the KB S6 Proprioception of the image to take place, the KB S5 Proprioception of the word is necessary. Yet it is not sufficient. For, while the circle of self-signification may close in the written text to encompass Steven's cerebral identity, it cannot close with respect to the mirror image without the presence of the being whose core identity corresponds to that image, viz. Stevie.

Now, whether perceiving the text-as-concrete-image or the written text, the "I"/eye is the source of perception, and it is to the "I"/eye that perception returns in realizing that *I* am the one who is regarding the text. In *holding* the paradox, the bodily source is made more explicit. Thus moving backward into the body, a felt sense is obtained of the optical activity by which the "I"/eye composes itself as object (image or written word). Going further with this "self-reversal" (chapter 2) in which the "I"/eye draws back in upon itself, a palpable glimpse is obtained of the "brain-as-lived" (Leder 1990, 113). The proposition is that *different* brains are involved in the respective Proprioceptions of written and imaginal texts. Whereas the Proprioceptive pathway of the former leads us backward into the neomammalian cerebrum, it is the paleomammalian limbic brain that is Proprioceived in the case of the latter. To focus more definitively on just what is entailed in entering these distinctive spheres of generic flesh, let us return to Burrow.

For Burrow, experiencing the "phyloörganism[4] as a whole" (1953, 445) requires the practice of *cotention:*

> Our aim was to maintain a steadfast internal sense of the balance and tension connected with the eyes. . . . Through the kinesthetic balance of the ocular muscles it became possible, following long-continued practice, to maintain the eyes in a state of equilibrium for increasingly longer periods. In thus holding to an internal awareness of the tensional balance of the muscles controlling the eyes, there was . . . the coincident occurrence of a sensation of stress or tension in the region of the eyes and in the anterior part of the head. With the maintenance of this more steadfast and centralized position of the eyes, it was disclosed that the sense of physiological strain or tension was due to the opposition of the affecto-symbolic pattern (ditentive thinking) to the primary pattern of cotention. (1953, 372–73)

In other words, by "bringing about a kinesthetic awareness of the eyes through the consistent maintenance of a steadfast balance in oculomotor tensions" (444), eventually a sense was cultivated of the "solidarity of the species" (71) that lies beyond the delusion of the isolated individual (e.g., Steven Rosen).

A detail of Burrow's method is relevant to our purposes: "The first step in the method . . . consisted in closing the eyes. While in one's attempt to restore cotention the eyes need not necessarily be closed, it seemed helpful . . . to employ this procedure" (372). Naturally, I cannot close my eyes if I am to engage in an actual mirroring of the text. In cotentively mirroring the *written* text, my eyes do not light upon an image of myself but upon a word—the "I" that governs all these particular words. Here I do not merely *view* the sign but *read* it. My cotentive oculo-cerebral processing of this sign supports the conceptual reading of it that transforms it into a *Kleinian* sign. In this way, the neomammalian brain thinks light; as ontological author, it uroborically thinks itself. What about the ontological mirroring of the more concrete text? Suppose I attempt to steady my eyes in Burrowian fashion while I look at my image in the glass. What would be the result?

Around 1980, when I first began describing to my students the perspectival integration of the Necker cube, one fellow[5] said it reminded him of an eye exercise he had read about and had later come to practice. He and his partner would gaze at each other pupil to pupil, with pairs of corresponding pupils operating synchronously but separately. This mirroring procedure evidently counteracts the natural tendency *either* for one participant simply to focus on both of the other participant's eyes at once so that they appear simultaneously, *or* for one participant's eyes to come into focus on but a single eye of the other participant at a time. The effect of the special technique is that participants view each other's eyes *both* together *and* separately! When I tried the procedure myself with several partners, on some occasions we momentarily experienced an eerie sense of penetration, as if we were moving *through* each other. I understood why my student associated this exercise with the Necker cube. In viewing the cube in the integrative way, we neither bring our eyes into focus on but a single perspective at a time, nor view the perspectives in juxtapositional simultaneity. Instead, the perspectives are grasped as *flowing through each other* in a manner that thoroughly blends succession and simultaneity. Thus we experience the *paradoxical interpenetration* of perspectives (fig. 6.2). After hearing from my student about the eye exercise and seeing its relationship to the Necker cube, I incorporated it into a novel (Rosen 1985), referring to it here as the "Moebius gaze": the eyes of the participants interact as do the separate sides of the Moebius strip that nonetheless twist into each other as one.

It should be clear that when we look at each other in the Moebius way, though our eyes are not focusing as they normally do, neither is the binocular convergence of our eyes simply being *prevented,* since that would only result in a loss of focus. In such a case, we would look at each other and see naught but a blur. In the "Moebius gaze," rather than merely obstructing the focusing process, we are *counter-acting* it, moving backward against the action of its habitual forward thrust. In chapter 4, such a retrograde movement was illustrated through the example of handling a piece of fabric whose fibers are arranged in a certain direction. By running one's fingers against the grain of the material, its direction becomes more clearly discernible. Similarly, in moving against the objectifying impetus of binocular convergence, awareness is no longer limited to the resultant object that is focused upon, but encompasses the entire focusing process. So, in practicing the Moebius gaze, participants do not just desist from perceptually objectifying each other; evidently, they obtain a felt sense of the whole course of objectification, including the *pre*-objective phase from which the objectification first arises. In a word, the Moebius gaze is *proprioceptive.*

I propose that the so-called Moebius gaze—which, in fact, is more accurately described here as *Kleinian*—is, in effect, a form of cotention practiced with the eyes open. Its counteractive gesture brings an "awareness of the eyes" (Burrow 1953, 444) that involves "a steadfast . . . sense" of their "tensional balance" (372)—i.e., a sense of the dynamic process by which the eyes converge to create their objects. Suppose, then, that, gazing into the mirror, I view myself in the cotentive Kleinian way. Counteracting the tendency for my eyes to focus *either* upon both mirror counterparts at once *or* upon a single one at a time, these modes of focusing are blended dialectically so as to disclose the focusing process. Mirroring my eyes in this fashion, would I not *see my own seeing* rather than merely see a pair of objects in the glass? There is a way to facilitate such special seeing.

In the field of visual research a technique has been developed known as *stereoscopic viewing*. Perceiving a figure by this method, disparate im-

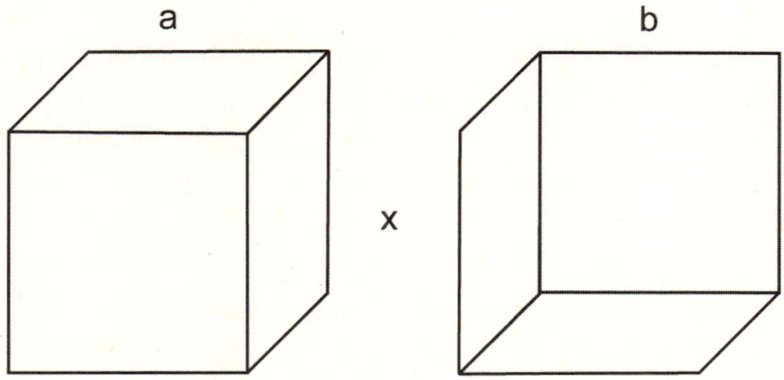

Figure 8.4. Stereoscopic construction of Necker cube

ages of it are fused to create a heightened experience of depth.[6] As with the perspectival integration of the Necker cube, stereoviewing requires the percipient to counteract the habit of binocular convergence. We can in fact construct the cube stereoscopically, and this can help us with its integration.

Figure 8.4 parses the cube into its component perspectives. Instead of focusing on either perspective alone, direct your view to the X marked halfway between the perspectives and defocus your eyes, allowing them to cross. After a while, a third image should appear between the ones printed on the page. The new image fuses original perspectives, a fact you will be able to confirm by observing that the labels "a" and "b" have become superimposed. The image you have created stereoscopically is of course the Necker cube. Working with the cube in this manner should make it easier to work against the habitual tendency to fixate on but one of its perspectives, since the viewer is now able to gain a tangible sense of how both perspectives are integrally involved in the dynamic process by which the cube is produced.

In much the same way, stereoviewing my mirror-reflected eyes should feed back to me the very process of my viewing them. Gazing into the glass, I direct my view to the bridge of my nose and allow a loss of focus. A "third eye" will soon appear between the other two. Continuing to peer at myself in this manner, flanking eyes disappear and the third eye constitutes itself as a "cyclopean" eye. But the cyclops thus evinced is not simply a transcendent unity. For I am able to see in the third eye the ongoing interplay and interpenetration of opposing eyes. In this autostereopsis, I can therefore obtain a sense of the way my two eyes, entailing disparate visual perspectives, converge to form a single perception in depth. In so stereoviewing my eyes, what I am seeing in the glass is no mere object upon which my eyes converge but a reflection of the binocular convergence process. At the same time, I can sense this optical process from *within* via Burrowian kinesthesia ("muscle sense").

The critical question, however, concerns what it will take for this proprioception to become Proprioception. It all depends on which "I" does the viewing, Steven or Stevie. When Steven stereoscopically views his eyes in the mirror, he can surely gain cognizance of the viewing process as it occurs on the optical surface of his body. But we know that he can do *no more* than this. Though Steven's surface visual proprioception does overcome the simple objectification of his mirror image, the "*central* vision"[7] or thinking subject housed in his cerebrum, remaining aloof, objectifies said proprioception. Because Steven's core identity is rooted in the written word and not the image, the autostereopsis of his image succeeds in Proprioceiving neither word nor image, for it can encompass neither cerebral nor limbic subjectivity. Stereoviewing his image in the glass, Steven cannot help but turn the viewing process into a *product,* an object that is cast before his non-imaginal subjectivity. Necessarily proceeding in this manner, Steven stereoviews his mirror-reflected eyes in a detached and clinical way, being unable to realize the ontological ecstasy by which his own subjectivity blends with the object that he views. What is required for imaginal Proprioception is backward entry into the paleomammalian, limbically lived brain. Steven cannot do this alone, since he is essentially a cerebral creature. Only with the assistance of Stevie can the KB S6 Proprioception of the text be enacted. Thus, for the ontological mirroring of the text-as-image, Stevie must participate in the autostereopsis. Waking up within his dream and standing present in the imaginal text, Stevie is to regard himself autostereoscopically. Unlike Steven, Stevie's core identity *is* rooted in the image. Therefore, rather than remaining aloof from the image of himself that he encounters in his stereoscopic mirroring, he can enter into an ecstatic exchange with that cyclopean image that effects the self-lighting of the limbic brain.

Having focused exclusively on the imaginal aspect of Stevie's text over the last several pages, the dual character of the text must now be reconsidered. The child's sub-cerebral lifeworld does not just involve the visual image but also the voice; it includes sound as well as light. Evidently, then, a *dual* Proprioception is called for.

Though Stevie's manner of functioning may be simpler than that of Steven, the structure of Stevie's identity is actually more complex. Whereas the ego of Steven is centered in a unitary core, Stevie possesses an *alter* ego: the presence of "another" is found in his midst. This schismatic feature is to be expected of the young child, since, at this early stage of development, identity has not yet crystallized into the singular form it will later assume. Young children often have "imaginary" playmates or companions, a "departure from reality" that adults tend to discourage, or tolerate condescendingly. Quite frequently, the companion takes the form of an animal, whether a creature that is invisible, or an inanimate object that is personified, such as the proverbial teddy bear. The undeniable importance of animals to children recapitulates their great significance in ancient mythic culture, as discussed in the two previous chapters. This is consistent, of course, with the basic ontological analysis set forth in those chapters: while the rational *cogito* stands alone, its mythic counterpart is paired with the *anima* (see table 7.1). We have found that, in the second stage of Proprioception, the KB S6 backward movement to the ontological core of the *cogito* is synchronized with the MB S4 backflow of the now-liberated *anima* (the "genie has been let out of the bottle"). Understood in terms of self-signification, we may presently say that the KB S6 Proprioception of Stevie's imaginal "I" is coupled with the MB S4 Proprioception of Stevie's derepressed animal familiar, his alter-"I." I suggest, moreover, that whereas the KB S6 aspect of the dual Proprioception follows a pathway from eye to limbic brain and culminates in *thinking light,* the trajectory of the MB S4 *anima* is from ear to heart and climaxes in *feeling sound.*

The world of primary feeling that "primitive man shares with the animal" (Neumann 1954, 329) is "resonant, layered, slower" (Kittelson 1995, 90) than the vibratory world of light; it is "'ear-minded'" (90); i.e., it is a world of sound. And the pervasive influence of animals upon the mythic human being evidently occurs through this sonorous channel. Recall Neumann's observation that the gods that presided in the earliest Egyptian dynasties were animals. According to the psychologist Julian Jaynes, the Egyptian deities controlled mortal human beings through the power of the *voice.*

In his book, *The Origin of Consciousness in the Breakdown of the Bicameral Mind* (1976), Jaynes hypothesizes that mythic awareness was radically different from the rational consciousness later to emerge. Mythic individuals "do not sit down and think out what they do. They have no conscious minds such as we say we have, and certainly no introspections" (72). Jaynes contends that, for the mythic person, "the gods take the place of consciousness. . . . The beginnings of action are not in conscious plans, reasons, and motives; they are in the actions and speeches of gods" (72). The early human being thus found himself or herself in a largely passive position, at the disposal of external forces personified by the deities: the divinities spoke and mortals listened, acting accordingly. Jaynes goes on to ask:

> Who . . . were these gods that pushed men about like robots and sang epics through their lips? They were voices whose speech and directions could be as distinctly heard . . . as voices are heard by certain epileptic and schizophrenic patients. . . . The gods were organizations of the central nervous system. . . . The gods are what we now call hallucinations. . . .
>
> [In antiquity,] volition, planning, initiative is organized with no consciousness whatever and then "told" to the individual in his familiar language, sometimes with the visual aura of a familiar friend or authority figure or "god," or sometimes as a voice alone. (73–75)

Later, in a section on the "Authority of Sound," Jaynes raises the "profound question of why such voices are believed, why obeyed" (94–95). It is because they possess for the listener an overpowering sense of reality. The listener is "somehow face to face with elemental auditory powers, more real than wind or rain or fire, powers that deride and threaten and console, powers that he cannot step back from and see objectively" (95). Viewing the mythic experience of sound as closely akin to the auditory hallucinations of schizophrenics, Jaynes quotes one schizophrenic as reacting to an invisible voice "'as though all parts of me had become ears, with my fingers hearing the words, and my legs, and my head too'" (96). Toward the end of his book, Jaynes notes that the "phenomenon of imaginary companions in childhood . . . can be regarded as another vestige" (396) of the primal state of affairs. As with the schizophrenic, the unseen companion exerts a highly potent influence on the child through the medium of sound.

To confirm the *animal* origins of the "elemental auditory powers," we turn to the historian of religion Mircea Eliade. In *Myths, Dreams, and Mysteries* (1960), Eliade underscores the significance of animals in early human culture: "animals are charged with a symbolism and a mythology of great importance for the religious life. . . . The prestige of animals in the eyes of the 'primitive' is very considerable; they know the secrets of Life and Nature, they even know the secrets of longevity and immortality" (63). Eliade goes so far as to hypothesize that, in the earliest form of religion, "the Supreme Being may very well [have] manifested himself in the shape of an Animal" (177).

According to Eliade, archaic mythology is often characterized by nostalgia for a bygone "paradisiac epoch" in which people "understood the language of animals and lived at peace with them" (59). If the original intimacy with the animals and their language could be recovered, this would be tantamount to "ascension into Heaven," to meeting "with the gods" (60). We learn from Eliade that the essential role of the *shaman* in mythic societies is precisely to reestablish contact with the "paradisiac" sphere of the animal divinities. Eliade observes that "during his initiation, the shaman is supposed to meet with an animal who reveals to him certain secrets of the craft, or teaches him the *language of the animals,* or who becomes his *familiar spirit*" (62).[8] Once initiated, the shaman can practice his craft by using the "secret language":

> The shaman imitates, on the one hand, the behaviour of the animals and, on the other, he endeavors to copy their cries. . . . The shaman produces mysterious sounds. . . . You think you are hearing the plaintive cry of the peewit, mingled with the croaking of a hawk, interrupted by the whistle of a woodcock: it is the shaman who is making these noises by varying the intonation of his voice. . . . A good many of the words used in a shamanic session have their origin in the imitation of the cries of birds and other animals. (1960, 61–62)

So it is sound that is the means by which the shaman taps into the realm of animal potency, and this suggests that sound was indeed a primary channel—if not *the* primary channel, through which mythic humanity felt the impact of the animal lifeworld. That world, I propose, is the world of "elemental auditory powers." Ontologically, the sound wave plays a role for the *anima* similar to that which the light wave plays for the *cogito*. Sound is the archetypal waveform of animal action. It is the form the dimensional vortex takes in the sphere of animal Being. Furthermore, while the luminosity of the *cogito* is centered in the brain, the *anima*'s sonority finds its rightful place in the heart. For we know from chapter 6 that the world of animal feeling is rooted in the heart. Therefore, if the KB S6 Proprioception entails a backward movement from the oneiric eyes in which the mythic *cogito* sees/thinks the light of the limbic brain, thereby resealing the Kleinian vessel more concretely than in KB S5—then the accompanying MB S4 Proprioception involves a retrograde circulation from the ears in which the loving *anima* hears/feels the sound of the heart, and thus seals the Moebius vessel in hermetic fashion. With these topodimensional vessels spinning synchronously, they contain one another in the asymmetric manner described in previous chapters. Human and animal lifeworlds now pulsate in onto-ecological harmony.

The question for the present chapter, of course, is whether this abstractly allusive account of dual Proprioception can be tangibly distilled via the self-signification of the text. For that, Stevie must stand present. Given the dual nature of his identity, he will be accompanied by his alter ego, his animal counterpart. Whereas the Proprioception of Stevie per se deals with tracing his ontical visual image back into the generic limbic system of the lived human brain, the concomitant Proprioception of Stevie's "double" is essentially auditory: it is concerned with moving through the plaintively voiced "I" back into the sonorous animal heart.

The "I" plays the same role in the written text as does the mirror image in visual perception: it reproduces the identity of the subject in an externalized form, establishing the possibility of the uncanny realization that *I* am the one who regards myself. In the case of the *spoken* "I," the mirroring becomes an *echoing,* allowing awareness that I am the one who *hears* myself. To *sustain* the ecstatic blending of subject and object in the MB S4 act of Proprioception, the body must be entered through the system of speech. Voicing his plaintive "I," Stevie's vocal cords vibrate to produce that word of himself whose sound echoes in his ears. The echo is surely not to be heard in a simply objective way, as if the sound were merely emanating from an external source. What is required instead is what philosopher David Applebaum (1983) calls "hearing-immediately" (117). Whereas the "direction of ordinary listening is toward the object, a word-sound . . . hearing-immediately is directed exclusively toward itself, the activity of audition. It thus occupies itself with the inner relations of audition" (117). Emphasizing the processual nature of hearing-immediately (it "pertains to activity, unfolding, doing"; 118), Applebaum relates it to phenomenology:

> [The] attributes of immediate-hearing . . . remind one of Marcel's analysis of the "body-as-mine." Their commonality allows us to conclude without the shadow of a doubt that it is the management of the lived-body experience, through the means of immediate-hearing, that produces its special ontologizing possibilities. Immediate-hearing converges with the general field of sensation underlying all organic movement, willed or no. The experience of this general field Marcel calls *sentir;* Husserl derives his notion of kinaesthesia from it; others have spoken of it as *proprioception.* (1983, 119; emphasis added)

So, with immediate-hearing, we redirect "sensate perceptivity back to the naked bodily whole" (120). In so doing, rather than listening with the "expectation of a result external to our listening . . . we listen to our own listening" (121).

Now, just as the visual Proprioception of KB S6 is facilitated by autostereopsis, the auditory Proprioception of MB S4 may be aided by what might be called *autostereophony.* This sonorous self-signification can be related to shamanic practice. Consider Eliade's account of the primary steps in the shaman's journey:

> A shamanic session generally consists of the following items: first, an appeal to the auxiliary spirits, which, more often than not, are those of animals, and a dialogue with them in a *secret language;* secondly, drum-playing and a dance, preparatory to the mystic journey; and thirdly, the trance . . . during which the shaman's soul is believed to have left his body. The objective of every shamanic session is to obtain . . . ecstasy. (1960, 61)

To be sure, the Stevie of KB S3 does not function as a shaman. Though accompanied by an animal counterpart, he does not intentionally speak the *language* of animals. He voices the word "I," not an animal sound. By so articulating the "I," Stevie asserts the human side of his identity that has been developing through his childhood, and this keeps his animal companion on its "leash." To the extent that Stevie has been socialized into the world of human signification, his animal double has become "domesticated." Of course, in bouts of emotional regression, the child's "I" can quickly dissolve into an animal howl. But the Stevie of old must not be confused with the one who stands present in the text for the purpose of Proprioception. With KB S5 Steven standing behind him to lend support, Stevie has now come of age and can function not merely with the innocence of the child but with the wisdom reminiscent of the shaman. And the self-signification of his "neo-shamanic text" unleashes the "elemental auditory powers" of animal vocalization in a non-regressive fashion.

In ritual practices, drumming is often accompanied by repetitive chanting. Our neo-shamanic self-signification commences with a chanting of the "I" that is not just an echoing of human identity but involves an appeal to the alter-"I." Rather than attempting to "tame" the animal double by imposing the human word upon it, as happens in KB S3, in the second stage of Proprioception the other is invited to express itself fully in its own language, which is the quintessential language of sound. It is in transforming the spoken human word into a primary animal vocalization that the MB S4 Proprioception of sound is enacted.

We have found that the KB S6 Proprioception entails experiencing the underlying process by which opposing visual perspectives converge to form a perception of oneself as an object in space. The auditory counterpart of this autostereopsis is *autostereophony.* The drumming practice so central to shamanic ritual can play a key role here. For example, repeated chanting of the "I" can be accompanied by a synchronized drumbeat, with the drummer moving his instrument (or his head) rhythmically from side to side so that left and right ears alternate in receiving primary exposure to sound. In effect, this establishes mirror-opposed auditory standpoints, creating the impression of listening stereophonically to the "I" that is intoned. As the player beats the drum faster and faster and moves the instrument accordingly, opposing auditory aspects converge and interpenetrate to form a Moebial body of sound that is "ecstatic," since the echoing standpoints are both distinct from one another and merged (i.e., they are outside and inside of each other). And the neo-shamanic practitioner—caught up in the sonorous transpermeation of "I" as object and "I" as subject—is herself ecstatic, is beside herself, both out of her body and within. Thus Eliade can say that the drum is a "specifically shamanic instrument [that] plays an important part in the preparation for the trance: the shamans both of Siberia and Central Asia say they travel through the air seated upon their drums. We find the same ecstatic technique again among the Bon-po priests of Tibet" (1960, 102). Where do these shamans travel in their ecstatic voyages? They pass into the world of primary sound, the realm of echoing animal spirit that they have unleashed. The animals that reside here are hardly docile companions subservient to the human word; they are godlike numinous presences to the shaman of old. Similarly, with the neo-shamanic transformation of the voiced "I" into the basic animal vocalization, the "genie" is unfettered. The sonorous animal into which the chanting author is transmuted upon striking the animal hide that covers the drum is a "transpersonal, spiritual being. He is transpersonal because, although an animal . . . he is such not as an individual entity, not as a person, but as an idea, a species" (Neumann 1954, 145). In other words, the animal invoked by the neo-shaman's drum-attuned self-chanting is *onto-dimensional.* It is the two-dimensional Moebius *anima* that signifies itself in the MB S4 Proprioception of the text-as-sound.

Moreover, the self-signification of sound reverberates in the *heart.* In many a shamanic practice, the fundamental link between the drumbeat and the heartbeat is acknowledged and accentuated (see Braine 1995). As Lisa Sloan puts it in recounting the experience she had with drumming while writing her doctoral thesis on shamanism, "the drum beat simulated the rhythm of a heart beat. . . . The sound of this beating drum stimulated and resonated with my own heart, suggesting to me that it is indeed through the heart that the shaman sees into the other world" (1999, 83).

Summing up the requirements for the synchronized self-significations of the KB S6 and MB S4 texts, Stevie must stand present, and, with the support of KB S5 Steven, he must engage in autostereopsis and autostereophony (respectively). For the Proprioception of the text-as-visual-image, he must gaze into the mirror in the cyclopean manner that allows him to see his own seeing. For the Proprioception of the text-as-sound, Stevie's animal double must be invoked as Stevie-become-shaman chants the "I" to the stereophonic drumbeat by which he can hear his own hearing. Let us take stock of just how far we have gone with all of this.

In the midst of the written text, Stevie has indeed stood present with his more concrete text, his images (figs. 8.1 and 8.2) and his spoken words (words inscribed in the manuscript but intended to be *voiced*). Nevertheless, the child's appearance here has been short-lived. By and large, the written word of *Steven* has held sway. This is hardly surprising when we consider how long the abstract "I" has reigned in our culture, and the profundity of its influence. In order for the denser self-significations to be enacted in earnest, the supremacy of Steven's word must continue to be challenged so as to permit Stevie's presence to be felt more consistently and fully.

The coagulated self-significations require that, once ontical Stevie can appear more reliably, he must be *ontologized.* It helps that the autostereopsis evidently entailed in this is something that Steven, for his part, seems able to do, at least in Stevie's absence. Steven can more or less stereoview his mirror image so as to view his own viewing process. But he does not find this easy to do. Apparently, he needs to become more proficient at the task of counteracting the normal tendency toward perceptual objectification. Of course, when Steven alone performs the autostereopsis, at the ontological level he still engages in objectification. To carry out the KB S6 Proprioception that ecstatically blends object and subject, it is *Stevie* who must be present for autostereopsis.

What of the autostereophony that should be necessary for the MB S4 co-Proprioception? Can Steven do with sound what he more or less can do with sight? Can he "listen to his own listening"? Does he possess enough proficiency in the techniques of chanting and drumming (or equivalent procedures) that he can hear himself stereophonically? Not yet. And even if the answer eventually becomes *yes,* Steven, functioning as neo-shamanic author of the sonorous text, would then need to subdue his own voice in favor of Stevie's and *ontologize* that voice, inviting Stevie's powerful animal companion to assume authorship, to voice its "elemental auditory powers" (Jaynes 1976, 95), to speak its "*secret language*" (Eliade 1960, 61).

It seems, then, that a great deal remains to be done for the self-signification of the "companion texts" accompanying this written corpus.

And the very fact that there *are* companion texts—whole bodies of work to be carried out in close synchrony with the written work, yet enacted in their own distinctive media—appears to define the natural limits of the present volume, which is, after all, primarily a *written* work. Evidently, in explicitly recognizing the need to author denser texts that parallel the current one, we have reached a bifurcation point. The text has undergone a dehiscence; it has burst open like a seedpod to yield "seedling texts" that will require much nurturance and attention. It is clear that this cannot happen solely within the framework of the present volume. Companion "volumes" (not to be confused with *written* books) will indeed be needed for the bona fide realization of the denser self-significations. It will be in those opera that the lower-dimensional orders of Being will assume their authorial roles beside the *cogito* in the ontobiographical collaboration. In the present opus, we must be content with merely launching those works, with setting their wheels in motion.

ROUND THREE: COGNITIVE-EMOTIONAL-SENSUOUS SELF-SIGNIFICATION

In the previous section, we explored the heartfelt mythic subtext of Stevie that lies beneath the rational abstractions of Steven's written text. It is now necessary to acknowledge an even deeper textual substratum: the sensuous magical realm of the still younger child. The text of the adult is not without its intimations of that sphere.

Steven certainly hopes that his readers will find his work well reasoned and logically compelling, in spite of its abstruseness! But the fact of the matter is that his logical constructions are erected on an *ana*logical base. In other words, the rational thinking that governs Steven's text is a sublimation of a *magical* thinking operating below the surface. Recall Gebser's way of expressing the magical: "magic man" inhabits a world "in which each and every thing intertwines and is interchangeable" (1985, 48), and this "brings with it what we may call thinking by analogy or association. . . . Magic man feels things which seem to resemble one another as . . . 'sympathizing with' one another. He then proceeds to connect them by means of the vital nexus—not the causal nexus" (50). That is, magical thinking—participating in the "realm of vegetative energy," the "vegetative intertwining of all living things" (49)—does not build its conclusions in the discursive linear manner whereby distinct propositional elements are linked to each other one step at a time; rather, it forges direct associations based on the indistinguishability of elements. Has the modern individual now put "magic man" behind him or her and gone on to function in a purely rational way? Clearly s/he has not. Even rigorous thinkers such as logicians and mathematicians will acknowledge the loose associative substratum that underlies their tight constructions, especially if they possess some insight into their own thinking processes.[9] It goes without saying, then, that magical thinking is at play beneath the surface of the *present* text.

Can a form of *emotion* be detected in Steven's work that ties in with magical cognition? Stevie's need for love is not the only primal feeling working from below to spur Steven's labors on this text. The foregoing chapters contain a great many sharply phrased, pointed critiques of the ideas of other thinkers. Reexamining what has been written, we may recognize in places a certain caustic undertone, even an attitude of irritability and dismissiveness. However valid the argumentation here, however helpful it may be in shedding light on the limitations of contemporary thought, does it not at the same time betray a tendency toward aggression on the part of its author? Let me not beat around the bush. Let me admit that, beneath the surface, childish anger certainly is present, though the sublimated form that it assumes in the text may indeed serve a useful purpose.

The sometimes acerbic language of Steven's text may be described as *pungent,* a word that derives from the Latin, *pungere,* to prick or puncture. In his etymological study, Eric Partridge (1958) relates the term to "pugnacious" and "pugilistic" (535). *Webster's New Twentieth Century Dictionary* defines "pungent" as "sharp and piercing . . . sharply penetrating; expressive; biting; as, *pungent* language . . . keenly clever; stimulating." But the primary definition *Webster's* provides relates the word to *olfaction:* "producing a sharp sensation of taste and smell; acrid."[10] Is the "acrid smell" of this text purely figurative, or is there actually a subtextual level at which olfaction operates more literally, just as we have seen that an ear-based text underlies the visual? Though the olfactory text is certainly well obscured when vision is dominant, the phylofunctional analysis carried out in the previous chapter does seem to point to the existence of such a corpus. We must consider, then, a substratum of primordial symbolic operations in which meaning is signified by "sniffing it out," picking up its scent. This accords, in fact, with anthropological literature that indicates the prominence of olfaction in the magical significations of Paleolithic hunter-gatherers. For example, Montague (1974) and Classen (1993) describe indigenous peoples (the Ongee and the Inuit, respectively) whose cultural worlds largely revolved around the function of smell, and Kohl and Francoeur (1995) propose that the earliest linguistic markings consisted of acrid traces of urinary or sexual scent left on environmental objects.[11]

Overshadowed as it is by rationality, by no means does the magical thinking that may find expression in Steven's text possess the primal vitality of pre-mythic thinking, or that of the infantile cognition that recapitulates pre-mythic phylogeny. Similarly, Steven's sublimated anger hardly approximates the intensity of infantile rage. Even less does the pungently sensuous "undertaste" of his work approach the potency of undifferentiated smell that operated primordially. Still, the magical *cogito* of early childhood (KB S2) is not simply absent. Like its mythic counterpart, it resides beneath the surface. And whereas the mythic *cogito* shares its identity with a loving "animal companion," the even more diffuse identity of the magical *cogito* is shared with *two* nonhuman "companions": a more primal, fury-filled *anima,* and the "scent-uous" *vegeta.* What we now want to do is invite the *self-signification* of the coagulated *anima* and *vegeta,* but the invitation to those lower dimensions of the flesh cannot properly be extended with Steven's three-dimensional mode of rational signification continuing to hold sway, or even with the prevalence of Stevie's mythic signification. If the two-dimensional *anima* is to appear for the second sealing of the Moebial vessel (MB S5); and if the one-dimensional *vegeta* is to stand present for the first hermetic closure of the lemniscatory vessel (LB S3); then a further "regression in the service of the text"[12] must be enacted. An even more deeply repressed child must make its presence felt. We may call this nascent being "Baby Stevie." Once the infant is brought to presence, he must be Proprioceived. In the backward movement that is called for, the baby's cognition will come to be recognized as the *generic* cognition of magical humanity. And, with the Proprioception of Baby Stevie (KB S7), both the *anima* and the *vegeta* will be free to carry out their own Proprioceptions. In threefold Proprioceptive conjunction, Kleinian, Moebial, and lemniscatory authors of synchronized texts will collaborate on their joint ontobiographies.

However, getting Baby Stevie to stand present is much easier said than done. The appearance here of the older child was difficult enough. His arrival seemed to hinge on the curious conjunction of November 11, when Steven's call for the presencing of Stevie was followed the same evening by a bout of regression in which the child actually materialized in an unprecedented way (see above following fig. 8.1). This odd serendipity seemed to make it possible for Stevie's presence to be felt in the text, to the limited extent that it has been. The presencing of *Baby* Stevie is another matter, since the raging infant is far more alien to the coolly cerebral Steven than is Stevie. Nevertheless, Baby Stevie *will* have to show up if work is to proceed in the "scentuous" companion opus to this written volume. Required here will be a distillation of Steven's written text into a subtextual medium that is even denser than the concrete images and spoken words of Stevie: the medium of *intuition.*

While the mythical image is bathed in dream-light (visual perception I), the world of magic is darker, entailing as it does a deeper state of sleep (Gebser 1985, 121). Magical functioning is mediated by the most concrete form of cognitive intuition: mental intuition I. The magical subject has neither an abstract concept of himself, nor an image of himself, but only the dim "gut-level" intuition of his own being. Since this dark apprehension is all that supports the elemental operations of consciousness here, it must suffice as the basis for the text. Therefore, when Baby Stevie stands present to engage in self-signification, the "master signifier" of his magical text will be a bodily self-intuition, not a self-image. And this rudimental intuition of his own existence is what Baby Stevie must Proprioceive (with Steven's assistance, of course). (I noted above that magical signification can have an *olfactory* quality, but this only reflects the *vegeta*'s influence on the *cogito,* not the *cogito*'s signification of itself.)

If Steven's Proprioception involves the passage backward through the abstract "I" to the ontological core of the neomammalian lived brain; and if Stevie's Proprioception entails the retrograde movement through his self-image to the paleomammalian brain; then Baby Stevie's ontological realization should require moving in reverse through the intuition of himself to the generic *reptilian* brain. We found above that the so-called reptilian brain is the most primitive member of MacLean's triune grouping. This deeply embedded cranial structure is largely unconscious.

Since Baby Stevie is more deeply repressed than Stevie, more will have to be done by way of encouraging his appearance. But while it may be more difficult to coax the infant's presence, once that happens it should be easier for him to surpass the ontical. Is that because the infant is more ontological? On the contrary, it is because he is *less* ontologically individuated than are his comparatively mature counterparts. The dialectic of Being is such that its degree of concealment in the ontical division of subject and object is directly proportional to its strength. The projection of subject-object opposition is relatively weak for the weakly ontological infant, so the opposition is easier to overcome. In the contrasting case of Steven, by KB S4 a powerful inclination to split subject and object has become deeply engrained. This makes "counterclockwise" movement *against* the grain more challenging. Steven's ontical orientation is indeed reversed with the KB S5 realization of ontological paradox, but only with the substantial effort necessitated by the strong resistance of a well-developed grain. Steven's backward movement from the optical periphery to the lived cerebral cortex thus proves to be a daunting process. Proprioception should be somewhat less of a problem for Stevie, since, in his case, the division of subject and object enacted in KB S3, being weaker than Steven's KB S4 division, is not as deeply engrained. The challenge is to get Stevie to stand present in the first place, so that he can then carry out the KB S6 retrograde movement through his self-image to the limbically lived generic brain. Though an even greater challenge must be met in bringing Baby Stevie to presence, the subject-object division magically enacted in KB S2 is so feeble that its KB S7 counteraction should involve little effort. With barely any split between subject and object to act against, not much resistance should be encountered in passing to the ontological. Whereas Stevie's ontological self-signification requires reentering his head via his oneiric eyes and moving backward through his self-image from the optical surface of his body to his limbic core, Baby Stevie evidently would have a "shorter trip," since there would be no well-defined body surface to move back from (despite the baby's appearance to the ontical adult). Given the rudimentary differentiation of ontical body surface and ontological core, the self-signification of Baby Stevie's text should simply involve the immediate intuition of the species-wide reptilian brain. Here there would be little need for the kind of paradoxical mirroring carried out in the Proprioceptions of Stevie and Steven.

When Baby Stevie arrives on the scene, it will be with his two nonhuman "companions." It is not surprising that Baby Stevie should possess one more alter ego than does Stevie, given the fact that the infant's identity is less cohesive than the older child's. While the latter can express to us in words and deeds the nature of his "imaginary friend," the baby is not so articulate. Upon appealing to the ontological analysis of the previous chapters, two things are suggested in this respect. First, the magical human infant is accompanied by an animal counterpart that is more primitive than the companion of the older child: whereas the alter ego of the latter is loving, that of the former is filled with rage. Second, the infant's other companion is not an animal but a vegetable. The transformation of Baby Stevie and his nonhuman "triples" that is now called for corresponds to the threefold Proprioception already adumbrated. In the third stage of Proprioception, the KB S7 movement to the ontological core of the magical *cogito* is synchronized with the MB S5 backflow that follows the unleashing of animal rage, and the LB S3 backflow following the release of vegetative "scentuality."[13] Put in terms of self-signification, the KB S7 Proprioception of Baby Stevie's intuitive "I" is coupled with the MB S5 Proprioception of his animal alter-"I," and with the LB S3 Proprioception of his vegetable alter-"I." I propose, moreover, that while the KB S7 aspect of the triple Proprioception entails the cognitive self-intuition of the reptilian brain, in MB S5 there is an emotional self-intuition of the animal heart, and in LB S3 a sensuous self-intuition of the vegetable "groin." What we are seeing is that the magical *cogito*'s noncognitive companions also function through the medium of intuition. Whereas the MB S3/S4 *anima* expresses itself via heartfelt vocalization, the soundless fury of the MB S2/S5 *anima* is conveyed by means of emotional intuition I (table 7.1). In a similar fashion, the primal olfaction of the LB S2/S3 *vegeta* is mediated by sensuous intuition (see previous chapter for discussion of the several types of intuition).

If it still seems odd to speak of vegetable olfaction, recall (from chapter 7) that while plants do not have specialized organs of smell, their operations are guided by "chemoperception" (Boller 1995), short-range chemical interactions with their environment that are the equivalents of olfactory activities in animals and humans. Since the *vegeta* holds sway in S2, humans and animals alike presumably function here in a plantlike manner, with smell serving as the predominant sense modality. We have seen that the lifeworld of the *anima* prevalent in MB-KB S3 is a sonorous milieu that pulsates more slowly and is denser than the human world of light. Denser and slower still is the *vegeta*'s S2 world of smell.[14] The proposition that this is first and foremost a primordially *sensual* life-world is supported by the well-established connection between smell and sexuality (e.g., Wright 1964, 1982). Of course, plants do not have groins any more than they have noses, but they certainly possess organs of reproduction! What was brought out in chapter 6 is that the vegetative sphere of "vital forces" (Gebser 1985, 270) is essentially concerned with the force of life itself, of sexuality and sensuality. It is this arche-erotic domain that is to be revisited non-regressively in the threefold Proprioception at hand. We may gain further insight into this task by resuming our exploration of shamanism.

We know that the shaman's goal is an ecstatic return to the bygone "paradisiac epoch" (Eliade 1960, 59). While the journey is often guided by animal spirits or deities, animals are not the only beings who serve in this capacity. According to Eliade's *Shamanism* (1964), "In South America, as everywhere else, helping spirits can be of various kinds: souls of ancestral shamans, spirits of plants or animals" (91–92; see also Neumann 1954, 145). In fact, plants commonly play a significant role in many aspects of shamanic practice. Rituals quite frequently involve the burning of plants like juniper and thyme (Eliade 1964, 118n) as incense. The pungent odor of the smoke can have an intoxicating effect; it can help generate the "magical heat" (474–77) that facilitates the shaman's entry into the state of ecstasy. Eliade notes that many practitioners "eat highly aromatic plants; they hope thus to increase their inner 'heat'" (475). Related to this is the "ecstatic function of the [hot] vapor bath, combined with intoxication from hemp smoke, among the Scythians" (475). "There is every reason to believe," says Eliade, "that the use of narcotics was encouraged by the quest for 'magical heat.' The smoke from certain herbs, the 'combustion' of certain plants had the virtue of increasing 'power.' The narcotized person 'grows hot'; narcotic intoxication is 'burning.' . . . Often the shamanic ecstasy is not attained until after the shaman is 'heated'" (476–77).

Would we not expect that the state of "boiling ecstasy" attained by the shaman should be accompanied by the arousal of sexual desire? Eliade confirms this in detail (1964, 71–81). In one account, a Siberian shaman tells of how he is possessed by erotic spirits: "whether big or small, they penetrate me, as smoke or vapour would" (72–73). Is it possible that this "smoke or vapour" that stirred the shaman's loins had a *vegetable* origin; that it was occasioned by the burning of potent incense? Typically, shamans do become *incensed* in the course of their ecstatic journeys. They go into violent paroxysms reminiscent of epileptic convulsions (24ff.). In his frenzied agitation, the shaman may submit himself to torments: "He gashes himself with knives, touches white-hot iron, swallows burning coals" (476–77). In this heated condition, the shaman literally fumes. Joseph Shipley (1984) establishes some interesting etymological links in this regard. The Indo-European root "dheu I" is generally related to "smoke; dust . . . strong smell," from which derives the Greek "*thuos:* incense. *Thyme:* to cause to smoke; used in sacrifices" (70). The same root gives the Latin "*fume, fumigate; perfume . . . fury,* from the excesses after inhaling incense at sacrificial festivals" (70). What is the upshot of all this? It is that, beyond the shaman's entry into the sonorous lifeworld of animal feeling, s/he enters an even denser, more primordial vegetable domain of powerful scents and sensuality, a realm of primal fury not unlike the sulfurous underworld so frowned upon in the patriarchal scriptures.

But the shaman does not merely descend. Her journeys to the "nether world" are not mere bouts of regression, for there is a "method to her madness." She is deliberately seeking enlightenment. Eliade emphatically distinguishes shamanism from psychopathy. He notes that, "for all their apparent likeness to epileptics and hysterics," shamans "control their ecstatic movements" (1964, 29). To take one example, at the same time that a certain Yakut shaman "'gashed himself with a knife, swallowed sticks, [and] ate burning coals,'" he "'bubbled over with intelligence and vitality'" (29). In general, the shaman demonstrates an "astonishing capacity to control . . . ecstatic movements," and, "'intellectually, he is often above his milieu'" (30). It is because the shaman maintains the light of intellect, preserves reflective consciousness during her excursions, that her descent is at once an *ascent.*

Eliade has much to say throughout his work about the shaman's "magical flights" to "Heaven" or the "celestial spheres." The critical importance of the *vegeta* in these trips aloft is symbolically expressed by the fact that the ecstatic journey is most often enacted by climbing "the Cosmic Tree [that] is supposed to be situated at the Centre of the World and that connects Earth with Heaven" (1960, 65). Eliade describes one such ritual at length (1964, 190–200). The Altaic shaman sets up a young birch tree in the middle of the yurt (ceremonial hut), positioning it so that it protrudes through the smoke hole in the top. Notches are cut in the tree as footholds for the ascent to Heaven. The ritual fire is lit, and, after sacrificing a horse, the shaman begins his ecstatic climb. As he goes up, he becomes more and more heated. With his agitation mounting, he beats his drum violently, loudly intoning nonhuman sounds to announce the presence of accompanying animal spirits. Whereas the skin stretched over the drum comes from the hide of an animal, its barrel is fashioned from one of the branches of the Cosmic Tree. As Eliade explains in his more abbreviated account of the Altaic ceremony, since "the frame of his drum [is] made of the actual wood of the Cosmic Tree, the shaman's drumming transports him magically near to the Tree; that is, the Centre of the World, the place where there is a possibility of passing from one cosmic level to another" (1960, 65). Contributing to this "vegetable magic" is the periodic burning of incense (juniper). With the ceremony reaching its climax, the shaman encounters "Bai Ulgän, the supreme God" (1960, 64) and exits rapturously via the same hole atop the yurt through which the intoxicating smoke drifts out.

What bearing does all this have on the matter of self-signification? Evidently, the ontological self-signification of the text now at hand entails a shamanesque transformation in which we "turn up the heat." The "regression in the service of the text" that brings Baby Stevie to presence will indeed be incendiary. For the infant will stand present with all his rage and "scentuality." To facilitate Baby Stevie's appearance, pungent incense may be burned that helps trigger the connection with early childhood.[15] Yet the "descent" will also be an "ascent," for the presencing of Baby Stevie will be no *mere* regression, since Steven will be there with his reflective intelligence to assist in containing the process. Thus Baby Stevie, come of age and now functioning in the manner of a shaman, will ecstatically enact the KB S7 Proprioception; the text-as-mental-intuition will be taken back in through the direct apprehension of the reptilian brain. In the process, a "supreme God" will be encountered, or a Goddess, perhaps: the Kleinian Mother, whose serpentine generic body will be uroborically sealed for the third time.

But Kleinian Proprioception will not be the only ecstasy to occur here. For we know that, when Baby Stevie shows up, he will not be alone. His animal and vegetable alter egos will accompany him. The rage of the magical child reflects the influence of the *anima;* the infant's potent scentuality (his "polymorphous perversity," as Freud would say) derives from his relation to the *vegeta*. In commencing the KB S7 Proprioception, Baby Stevie-become-shaman unleashes his nonhuman companions, who are now free to perform Proprioceptions of their own. Thus it is that the MB S5 *anima*, more "heated" now than in MB S4, can enact its second Proprioception, the self-signification of the text-as-emotional-intuition that entails an immediate apprehension of the furious animal heart. In the course of this rapture, the generic body of the Moebius Mother is hermetically sealed for the second time. Beyond that, the LB S3 *vegeta* carries out a Proprioception of the text-as-sensuous-intuition that establishes a direct link to the vegetative organs of reproduction and hermetically seals the generic body of the lemniscatory Mother. In the threefold Proprioception that transpires, the Mothers of the flesh gyrate backwards in synchrony as they give birth to themselves.

Now, while the process of self-signification has been related to shamanism in the present chapter, previously we related this work of individuation to *alchemy*. Eliade acknowledges the parallels between shamanism and alchemy (1960, 170; 1964, 410–11), and, in a book devoted to the latter (1962), he suggests we view it as a later form of shamanism, one that developed independently, growing out of such enterprises as mining, toolmaking, and the like. Jung, for his part, in considering the alchemical significance of the "philosophical tree," cites Eliade:

> The shaman climbs the magic tree in order to find his true self in the upper world. Eliade says in his excellent study of shamanism: "The Eskimo shaman feels the need for these ecstatic journeys because it is above all during trance that he becomes truly himself: the mystical experience is necessary to him as a constituent of his true personality." The ecstasy is often accompanied by a state in which the shaman is "possessed" by his familiars or guardian spirits. By means of this possession he acquires his "'mystical organs,' which in some sort constitute his true and complete spiritual personality." This confirms the psychological inference that may be drawn from shamanistic symbolism, namely that it is a projection of the individuation process. This inference, as we have seen, is true also of alchemy. (1967, 341)

In the same paragraph, Jung emphasizes the importance to individuation of the *coincidentia oppositorum,* the paradoxical union of opposites: "The process [symbolic of individuation] usually consists in the union of two pairs of opposites, a lower (water, blackness, animal, snake, etc.) with an upper (bird, light, head, etc.), and a left (feminine) with a right (masculine). The union of opposites . . . plays . . . a great and indeed decisive role in alchemy" (341). It seems reasonable to suppose, then, that alchemy and shamanism alike were concerned with an approach to psychospiritual or psychophysical development that is essentially dialectical. Of course, the paradoxical union of opposites is a central theme of the present volume, if not *the* central theme. Here the conjunction has been expressed in terms of ontological phenomenology, and fleshed out via topology; ontological individuation (Ontogeny) has been seen as a process that climaxes with a series of topo-dialectical Proprioceptions. Having focused on relating Proprioception to shamanism in this chapter, I would now like to return to the equivalent relationship with alchemy. Correlation of the stages of Proprioception with the conjunctions of alchemy was already partially indicated in earlier chapters. Before I attempt to complete the account, let me add a word of caution.

Elsewhere (Rosen 1995), I brought out the distinction between modern alchemy and the alchemy of old. Ancient and medieval alchemy were for the most part prerational. They involved a certain absence of discrimination between *psyche* and *physis* or subject and object, an inability to distinguish the two sharply. It was in this deficient sense that, in old alchemy, "the physical and the psychic [were] . . . blended in an indissoluble unity" (Jung 1968, 279). Just this fundamental confusion was reflected in pre-Renaissance alchemy's notorious difficulty with sealing the "Spirit Mercurius" into the bottle. In my essay, I demonstrated the promise of the modern topological approach to the problem. I suggested that there is a greater likelihood of hermetic closure in the new alchemy because it benefits from advances in science, mathematics, psychology, and philosophy—that is, from a heightening of reflective consciousness. What the modern alchemist can grasp that the ancient alchemist could not is the *onto-topodimensional* nature of the uroboric vessel, entailing as it does the self-permeation of three-dimensional Kleinian Being. The new incarnation of the alchemical bottle, being made of "perfected glass" (Rosen 1995, 137)—i.e., constructed in terms of the conceptually mature, highly differentiated idea of mathematico-ontological dimension—can ecstatically contain the "Mercurial genie" in a way the old bottle could not.

To be sure, it is not that the alchemist or shaman of old was merely engaged in an act of regression. No doubt s/he possessed a high degree of native intelligence and brought it to bear in the ecstatic rituals s/he performed. However, given the cultural context in which s/he was embedded, s/he could not have possessed the order of reflective awareness that emerged later, with the rise of modern science, mathematics, and philosophy. Therefore, if alchemy most essentially entails *solve et coagula* ("dissolve and coagulate"), the alchemist of old could not yet have achieved the *solutio* necessary for sealing the vessel in a fully hermetic fashion. In correlating the stages of Proprioception with the alchemical conjunctions, then, I do not mean to suggest a simple return to the earlier, prerational form of alchemy. Relevant in this regard is the observation of the analytical psychologist Marie-Louise von Franz. Following her cogent critique of Western rationalism, she noted: "The rationalism of the 17th century . . . had one advantage after all: it drove father-spirit and mother-matter so far apart that now we can reunite them in a cleaner way" (1975, 42).

Now, in chapter 6, we associated the initial Proprioception, the one enacted in KB S5, with "the first stage of conjunction or individuation" in alchemy (see Jung's *Mysterium Coniunctionis,* 1970, 475). In referring to this stage as the *unio mentalis,* Jung cites the sixteenth-century alchemist Gerhard Dorn: "The *unio mentalis,* the interior oneness which today we call individuation, [Dorn] conceived as a psychic equilibration of opposites 'in the overcoming of the body,' a state of equanimity transcending the body's affectivity and instinctuality" (471). Here,

> the mind (*mens*) must be separated from the body. . . . By this separation (*distractio*) Dorn obviously meant a discrimination and dissolution of the "composite," the composite state being one in which the affectivity of the body has a disturbing influence on the rationality of the mind. The aim of this separation was to free the mind from the influence of the "bodily appetites and the heart's affections," and to establish a spiritual position which is supraordinate to the turbulent sphere of the body. (1970, 471)

Therefore, in achieving the intellectual maturity that brings mental development to fruition, one must surmount "everything bodily, sensuous, and emotional . . . the soul's appetites and desires" (472).

But Dorn well recognized that the *unio mentalis* is only the *opening* stage of conjunction. Stage two of the process occurs when the mind, having effected its separation from the body (*solve*), is then reunited with it (*coagula*). "It is significant for the whole of alchemy," says Jung, "that in Dorn's view a mental union was not the culminating point but merely the first stage of the procedure. The second stage is reached when the mental union, that is, the unity of spirit and soul, is conjoined with the body" (465). Jung goes on to comment that the "reuniting of the spiritual position with the body obviously means that the insights gained should be made real. . . . The second stage of conjunction therefore consists in making a reality of the man who has acquired some knowledge of his paradoxical wholeness" (476).

Note that, when Dorn speaks of overcoming the body in the first stage, two spheres of embodiment are specified: "the bodily appetites and the heart's affections." Jung apparently associated these spheres with "sensuous" and "emotional" (472) functioning, respectively. Presumably, in the second stage, when the mind reunites with the body, the bodily centers in question, having been subdued in the *unio mentalis,* would now be reanimated. In keeping with our Topogenetic analysis, I propose that what Jung (following Dorn) identified as the second stage of conjunction be seen as actually entailing *two* stages, the first of these involving emotion and the second sensuality. Such a formulation permits the alchemical stages to be coordinated with the Topogenetic ones. On this account, restoring contact with the "heart's affections" and the "bodily appetites" corresponds respectively to the dual Proprioception of KB S6–MB S4 and the triune Proprioception of KB S7–MB S5–LB S3. This leaves one stage to be accounted for in the process of individuation.

ROUND FOUR: EMBRYONIC SELF-SIGNIFICATION IN THE *UNUS MUNDUS*

Still following Dorn, Jung adumbrates an ultimate stage of conjunction beyond the union of mind and body. For Dorn, "the third and highest degree of conjunction was the union of the whole man with the *unus mundus*. By this he meant . . . the potential world of the first day of creation, when nothing was yet 'in actu,' i.e., divided into two and many, but was still one" (1970, 534). Thus, the union with the world constituting the third stage of conjunction—which for us is the fourth stage—is "not with the world of multiplicity as we see it but with a potential world" (534). Can any more be done to conceptualize this *unus mundus* in positive terms? According to Jung, "there is little or no hope that the unitary Being can ever be conceived, since our powers of thought and language permit only of antinomian statements" (538) with respect to it. Nevertheless, "we do know beyond all doubt, that empirical reality has [such] a transcendental background" (538).

The *unus mundus* or "potential world" evidently corresponds to the inchoate sphere of absolute nothingness. This utter negativity prevailing on "the first day of creation"—S1 of Topogeny—is met with once again on the "last day": KB S8–MB S6–LB S4 ("the first must come last"; Heidegger 1946/1984, 18). When Jung subsequently likens the "antinomian" encounter with the *unus mundus* to the "ineffable . . . experience of *satori* in Zen" (1970, 540), we are reminded that only the logic of Zen can do justice to the zero-dimensional realm. Consequently, if we follow this logic, we cannot allow the implication, as Jung did, that the *unus mundus* is "one" *as opposed to* being "divided into two and many." We must say instead that it entails *neither* unity *nor* diversity. And when we apply the neither/nor of Zen to the "potentiality" of the *unus mundus,* we realize that the bracketed word cannot signify an *Aristotelian* potential. Though "potentiality" surely is inherent in the embryonic sphere, it cannot mean nascent positive being, any more than the mere negation of said being. The "potentiality" of the *unus mundus* must refer to *absolute* negation, in the sense of Tanabe.

Now, what must the alchemist do to achieve union with the *unus mundus?* The final stage "of the opus alchymicum was indubitably the production of the lapis or *caelum*" (Jung 1970, 535). The philosopher's stone ("which corresponded to . . . Dorn's *caelum*"; 536) is a "symbolic prefiguration of the self" (539). To prepare for the "consummation of the . . . *unio mystica* with the potential world" (537) one must realize oneself as "'living stone,'" a "'stone that hath spirit'" (539). Accordingly, "Dorn sees the . . . highest degree of conjunction in a union or relationship of the adept, who has produced the *caelum,* with the *unus mundus*" (539). In Topogenetic terms, the "production of the *caelum*" corresponds to the fourth and final sealing of the Kleinian vessel. In KB S8, the three-dimensional Kleinian Mother solidifies her individuation to the point of making herself into "living stone" so that she may enter non-regressively the stony world of the zero-dimensional Grandmother. Indeed, the "potential matter of the first day of creation" (Jung 1970, 537) corresponds to the "prima materia" (534n) or Prime Mother, which, elsewhere, Jung correlates with the lapis (1967, 235). Jung also relates the stone to the "dark sphere" associated with the "*massa confusa*" or original chaos of the world (1968, 325), to the "darkness of inanimate matter" (1968, 304), and to "the mysteries of cosmic matter" (1967, 96). At the end of chapter 6, I proposed that the zero-dimensional *minera* be grasped as the "holiest" of dimensional bodies, since it is "*all* hole." It is this "absolute holeness," nothingness, or negativity that the *unus mundus* symbolizes. In a word, it betokens *death*. In the fourth stage of conjunction, the *cogito*, along with the *anima* and *vegeta,* must steel themselves for the confrontation with death, so that they may enter this most primordial sphere without merely succumbing.

Although Jung does not relate the *unus mundus* to death in an explicit way as far as I know, his colleague and collaborator, Marie-Louise von Franz, does just that in *Number and Time* (1974). Recognizing the *unus mundus* at play in the religions of West Africa, ancient Egypt, and China, von Franz notes that, in each case, "the *unus mundus* is identical with the realm of the dead, the spirit land" (271). In each, moreover, the ultimate aim is resurrection, which means the awakening of complete self-awareness in the very midst of death. Certain West African medicine men, for example, venture into the underworld guided by a divinity called "Gba'adu," who "represents the most powerful magic, and 'the highest possible degree of self-knowledge a man can attain'" (269). Associating "Gba'adu" with "the alchemical Mercurius, who was considered to personify the *unus mundus*" (269), von Franz notes that, "According to certain ideas of the alchemists, the individuated human being who has become unified must join himself to this mercurial spirit" (270). This is of course the crowning stage of conjunction identified by Jung (1970), as von Franz well knows. For her part, she wishes to emphasize that the final *coniunctio* must entail an encounter with death:

> In the experience of death the alchemist . . . hoped to discover . . . an exalted and final step in the achievement of oneness. . . .

> This experience is particularly beautifully extolled in an ancient alchemical text which can be traced back to the Egyptian liturgy of the dead. In this text the *unus mundus* is pictured as a transcendental experience of wholeness attainable only in the resurrection mystery after death. (1974, 270)

Von Franz explains that the climactic phase of individuation is to be enacted by means of a radical transformation in which the erstwhile rule of the ego is surpassed in realization of the universal Self. This is another way of saying that the conjunction with the *unus mundus* requires attainment of the philosopher's stone, since the "precious stone . . . clearly represents . . . an image of the Self" (276). So "the symbol of the Self" is "the end-image of the individuation process" (283), and this symbol is embodied in the stone. But how, specifically, is the domination of the ego to be surmounted in a way that brings the Self to concrete presence?

In seeking to address this question, von Franz is guided by the ancient "Chinese concept of life after death, as described by Richard Wilhelm" (279). Egoic identity is "bound to the physical body" (280)—which means the finite particular body, the body construed as an object-in-space. It is this objectification of the body that grounds one's ontical sense of oneself from early infancy (see discussion of the "mirror stage" in chapter 4). The particular ego, operating alone, cannot survive the decimation of the physical body brought on by the ravages of death. Only with the assistance of a "universal ego" (280), viz. the Self, is survival possible. To make this generic identity a concrete reality, it is necessary to construct a "subtle body," a "'body of a spiritual kind'" (279): "By building up the spiritual body through meditation exercises, the Chinese attempted in this life to disengage the energies attached to one's ordinary body and thus to endow the seminal power, the entelechy—or, translated into our modern terms, the Self—with a new body" (280). Once this subtle body was fashioned, one was purportedly "able to retain an individual identity after death within the . . . *unus mundus*" (280).

Importantly, the "meditation exercises" leading to the creation of the "subtle body" involved conscious recognition of a "retrograde movement of life energy (as we also saw indicated in the 'heavenly orders')" (280).[16] According to von Franz,

> The Self, as a psychophysical monad or ultimate nucleus of the personality, does not merely engender the ego consciousness emanating from it at birth and during the growth of the individual's personality. At death it also draws the ego back into itself and contracts. . . . The moment of death forms the decisive shock . . . as the ego plunges into the inner monad and unites with it. When an individual consciously participates in the individuation process, and thereby prepares himself for this moment by exerting himself to experience it as consciously as he can, he will succeed in experiencing the ego's transposition into the Self knowingly. But when he remains, as it were, hemmed in by floating psychic contents which are autonomous and unintegrated, consciousness becomes deflected and slips into a state of unconsciousness, which the ancient texts symbolized as being imprisoned by underworld demons. (280–81)

In the language of Jungian psychology, the meditation enabling the individual to consciously follow the retrograde movement of life energy entails the *withdrawal of projection.* This is what it means to "integrate autonomous psychic contents." In the first stages of life, when the movement of energy is "forward," myriad identifications are formed that serve the development of the particular ego. The contents thus projected are free-floating or autonomous because the individual is unaware of the process of projecting them. For the individual to gain the degree of self-knowledge necessary to sustain her or him in the *unus mundus,* the projections must be retracted, drawn back in, proprioceived.

However, in the confrontation with death it will not suffice merely to retract the projections of particular contents, personal identifications that have hardened into "objective truths" so as to meet the needs of the particular subject. At a deeper level, it must be recognized that—along with the objects—*subjectivity as such* has been projected. When this happens, the primary projection that is withdrawn is *object-in-space-before-subject.* Now it is Self or Being per se that is engaged in the backward movement, and, instead of simple proprioception, we have Proprioception.

The individuation at stake in the process of alchemical conjunction clearly does not just involve the development of the particular individual, but of the Self, of Being as such. We may say, with von Franz, that the Self engenders ego consciousness at birth, and add that it does this in order to facilitate *its own* development (chapter 4). Then, in the climactic stages of Ontogeny, the ontological Mother reorients herself. It is by shifting into "reverse," by engaging in Proprioception, that the Mother endows herself "with a new body" (von Franz 1974, 280), a generic one that surpasses the finite particularity of the body-in-space. In so doing, she reaches fulfillment.

Von Franz implicitly deals with this theme of Ontogeny in her interpretation of the Egyptian liturgy of the dead:

> When the One (the Self) acquires its new, adequate, perfect body, namely, the supernatural resurrection body; then it destroys its former body and permeates all other bodies. Translated into modern psychological language this would mean: when the Self, in a state of "becoming" within earthly man, fully attains its "body," i.e., its goal—the mandala of the *unus mundus*—it exerts an annihilating effect on man's earthly existence, because it has attained a state of being as the all-permeating one-continuum. (271)

Completing the "resurrection body" in this way means "entombing" it, sealing it in stone, and this, in turn, corresponds to the final hermetic sealing of the alchemical vessel.

Now, at the end of the chapter in which she deals with death and resurrection as the climax of individuation, von Franz touches on shamanism: "young shamans among the circumpolar peoples sleep by the graves of great shamans, in order to receive their teachings" (1974, 284). Presumably, the "great shamans" are those who have gone through all the stages of conjunction, climaxing with the stage of union with the *unus mundus*. Does Mircea Eliade have anything to say on the theme of death and resurrection? He says a great deal about this. Examining at length the patterns of initiation by which shamans in various parts of the world take up their calling, Eliade (1964, 53–59) finds evidence of a universal motif. In cases of self-initiation, the candidate undergoes a spontaneous crisis. Typically, s/he falls ill, retreats from normal everyday life, and suffers an emotional and/or physical breakdown. The crisis reaches its climax in a symbolic death (the initiate may be on the verge of actual death, have a near-death experience, descend into the "underworld," etc.). This is followed by regeneration. The candidate returns to life miraculously and is transformed. S/he now possesses extraordinary powers, such as the ability to heal, to foresee the future, to fly, to communicate with the dead, and to converse with the gods. Often shamans do not elect themselves in this way but are chosen for initiation by the tribe. In these cases, the crisis is deliberately imposed on the neophytes. They are isolated from ordinary group life and subjected to torturous ordeals that bring them to the brink of death. The result, however, is much the same. The candidates return from their symbolic death and are spiritually transfigured. In considering the key elements in the election of the shaman, Eliade thus states: "We have seen that the choice of the shaman . . . involves a quite complex ecstatic experience, during which the candidate is believed to be tortured, cut to pieces, put to death, and then to return to life. *It is only this initiatory death and resurrection that consecrates a shaman*" (1964, 76).

It seems plausible to suppose that, in the ecstatic rituals described earlier (rounds 2 and 3), the shaman is essentially reenacting the death and resurrection first experienced during initiation. On this interpretation, the shaman's ascent, her climb up the Cosmic Tree, her flight into the sky, may be understood as the ecstasy that comes from gaining consciousness in the realm of the dead.[17] Eliade confirms unambiguously that shamanic ascent is equivalent to resurrection: "ascents = resurrection" (1964, 236). To achieve the full measure of ecstasy, it appears that the light of consciousness would have to be directed backward toward itself in such a way that the radiance created would not *objectify* the afterlife. In alchemical terms, this retrograde self-lighting is what is needed to complete the formation of the "subtle body," thereby making the Self or "universal ego" into a concretely conscious, stonelike reality. A realization of this sort would indeed be "heavenly," since it would involve an ecstatic conjunction in which one would be fully beside oneself—inside oneself and outside at the same time. Let me presume to inquire into the extent to which the shaman or alchemist actually realized such a *coniunctio*.

We have found that a full-fledged *coagulatio* must be preceded by a complete *solutio,* which means that the consciousness required for the ultimate conjunction would have to include the highly abstract, reflective awareness only achieved with advances in modern thinking. If ecstasy entails standing outside of oneself as well as within, I venture to propose that the traditional shaman or alchemist could not have achieved full-fledged ecstasy because he could not have stood outside of himself to the same extent that the abstracted, self-alienated modern individual does. What I am suggesting, then, is that the "subtle body" can be completed in earnest, Self or Being can be realized in full, only by bringing to bear the more sharply discriminatory topological consciousness of the "*neo*-shaman." For the vessel to be sealed hermetically, it must be sealed in a *Kleinian* manner.

Eliade (1964) substantiates the limitations of the older shamanism. In examining his account of the archaic notion of the underworld, we find evidence that heaven and hell were confused. He discloses, for example, that the Altaic shamans begin their descent into the underworld by first *climbing a mountain.* After noting "several incongruities" in this odd journey, he concludes that, "Probably there is contamination between the themes of ascent and descent" (204). Eliade observes similar contradictions among the Yakut shamans (206), and goes on to speak of the "confusion of their religious and mythological conceptions" (235). In the Altaic case, it is especially interesting that the shaman's descent (following his ascent) is through "'the smoke hole of the earth.'" (202). For we found above that the Altaic ritual of climbing the Cosmic Tree into heaven reaches its climax when the shaman exits the yurt through the smoke hole in the *top.*

Perhaps we can say that the early shaman's confused identification of the hole to heaven with the hole into the underworld reflects his underlying inability fully to differentiate hole and *whole.* Wholeness (the "heavenly light above") and holeness (the "darkness below") were confounded. This is certainly not to suggest that the shaman functioned in a totally undifferentiated manner. If he had not brought a sufficient measure of reflective awareness to the undifferentiated underworld, he could never have experienced his ecstatic resurrections. In contrast, the Western rationalist experiences anything but embodied ecstasy, given his thoroughgoing repression of the *unus mundus.* The point I wish to make, however, is that the consciousness of the archaic practitioner of ecstasy was not differentiated *enough* to fully contain the ontological turmoil of the *unus mundus.* It is when the "world of the dead" is accessed but not fully contained that its inherent ambiguity becomes manifested in the sort of projective objectification that Jung attributed to the medieval alchemist. "In an age when there was as yet no empirical psychology," says Jung, "everything unconscious, once it was activated, was projected into matter—that is to say, it approached people from outside. It was . . . a concretization such as we frequently encounter in the psychology of primitives" (1968, 277–78). Clearly, then, when the alchemist or shaman of old underwent ecstasy, this was not expressly understood as a differentiated fusion of mind and matter, subject and object, hole and whole. Because the psyche and the material world were still being *con*fused, for the most part the ecstatic union of inside and outside was seen to be happening externally. The tendency was to interpret it literalistically, as a "trip to heaven"—or was it to *hell?* Once again, to fully contain the "genie" of the *unus mundus,* to enclose her hermetically, the vessel must be sealed in the paradoxical manner wherein wholeness and holeness are just as completely differentiated as they are fused. I submit that only a consciousness that has suffered the repressions of rationalism, that has gone through the modernist stage of alienated self-consciousness and come out on the other side, is equal to this task.

Now, what of the *phylo-functional* nature of the *unus mundus?* Table 7.1 tells us that, in this S1 domain, the *minera,* with its "null intuition," negatively contains (via its dimensional overtones) the primitive tactility of the *vegeta,* the primal fear of the *anima,* and the archaic thinking of the *cogito.* As noted in the previous chapter, the *minera*'s "null intuition" refers to the fact that, while this zero-dimensional origin serves as the source of higher-dimensional intuition, it does not "itself" function intuitively, simply because the *minera,* as the ultimate midwife, is utterly selfless. Possessing neither Self nor Being, it entails absolute negation. Within this sphere of neither/nor, the archaic *cogito* functions in but an embryonic fashion; its capacity to think is a bare potential, an *implication,* in Gendlin's (1991b) sense of this term. The same must be said for the animal fear that pervades the "land of the dead,"[18] and for the undifferentiated tactility of the *vegeta.* Moreover, in this inchoate realm it is not only the case that differentiation is lacking *within* the functional families, blurring the distinctions among the several forms of thinking, of feeling, and of sensing. It is also true that the distinctions *among* the families are only incipient. Functioning is thus ubiquitously nebulous within the zero-dimensional matrix.

Although there is hardly a hint of zero-dimensionality in Steven's written text, it follows from our analysis that such a substratum must exist. Beneath the magical thinking that itself is but weakly evidenced here lies an embryonic mode of thinking; beneath the angry subtext there is primeval fear; beneath the faintly detectible "acrid odor" of the text, there is a global tactility. Can we even speak of these deepest strata as "subtexts"? A text entails signification, symbolic mediation of some kind. Is the term appropriate for what transpires within the zero-dimensional matrix? To be sure, the zero-dimensional *minera* per se, being utterly boundless, is purely non-symbolic; it functions in a totally unmediated way. But we cannot say the same for the higher orders of dimensional action that, in S1, are under the *minera*'s sway. While the significations in which they engage are by no means articulated, neither are they simply absent. Again, they are *embryonic* (as indicated by the dashed lines that compose the S1 clockwise rotation circle of table 7.1). So even the lowest strata of Steven's text can be described as "subtexts," albeit ones that are merely *implied*.

In the KB S8–MB S6–LB S4 encounter with the *unus mundus* that is required to complete the several processes of individuation, the embryonic significations of S1 must be counteracted by embryonic *self*-significations (denoted by the dashed rotation circles directed counterclockwise). A final round of Proprioceptions is in order to bring about the climactic closures of the topodimensional vessels. For this to happen, Steven evidently must enact a still further "regression in the service of the text." The most primitive part of himself must be brought to presence and be Proprioceived. Concomitant to this Proprioception in which nascent cognition is recognized as the generic cognition of archaic humanity, the *anima* and *vegeta* will carry out their own concluding Proprioceptions. And the threefold Motherly Proprioception occurring within the *unus mundus* will liberate the *fourth,* viz. the Selfless mineral Grandmother. In this final *coniunctio* of dimensional Mothers, all topodimensional bodies will engage one another in asymmetric mutual containment.

However, if getting the baby to stand present is more difficult than doing this with the child, securing the presence of the embryo will be more difficult still. As Steven sits here coolly drafting this text, how far removed he is from the undifferentiated tactility and primal fear presumably operative in the prenatal matrix. Be that as it may, it *will* be necessary for him to gain access to his embryonic core if work is to proceed in the densest companion opus to this written volume. The requisite movement "backward through the birth canal" is not unlike the process of rebirthing described by the transpersonal psychiatrist Stanislav Grof in his essay, "Modern Consciousness Research and Human Survival" (1985).

Grof ascribes great significance to the "perinatal" stages of development, those occurring around the time of birth. He claims that "the perinatal process transcends biology and has important psychological, philosophical, and spiritual dimensions. It would be an oversimplification to interpret it in a mechanistic and reductionistic fashion" (30). Grof associates the initial stage in the birthing process with an experience of "overwhelming fear. . . . The very onset of biological birth is experienced as imminent vital danger and threat of enormous proportions" (30–31). The "second clinical stage of childbirth" entails a "gradual propulsion of the fetus through the birth canal" involving an "enormous struggle for survival" (31). "From the experiential point of view," this stage includes "a fight of titanic proportions, sadomasochistic experiences, intense sexual arousal . . . and encounter with fire" (31). Finally, there is "the actual birth of the child," the stage wherein "the agonizing process of the birth struggle comes to an end" (31). Here the negative emotions of the preceding stages are transformed into a "flood of positive emotions toward oneself, other people, and existence in general" (33). It is undoubtedly true that Grof's analysis does not conform in every detail to the foregoing phylo-functional analysis of what happens in the first three stages of Appropriation. And yet—in the transition from cold primordial fear to the heat of sexuality and aggression to a sublimated stage of loving affiliation—an overall correspondence is evidenced. Moreover, Grof calls for a *reliving* of the birthing process. As a psychiatrist influenced by Freud, he understands that, when a person is in crisis, the difficulty can only be fully addressed by returning to its "psychological roots" (37). In his view, however, we are currently embroiled in a multitude of crises the dimensions of which are not only personal, but also "transpersonal" (37), since the dilemmas besetting us extend beyond the individual sphere to directly encompass humanity at large. For Grof, addressing this "global crisis" means returning to the "perinatal level of the unconscious and the dynamics of the death-rebirth process [which] represent a repository of difficult emotions and sensations" (37). In effect, then, Grof is urging a movement of awareness that takes us backward through the birth canal in a manner that is transpersonal as well as personal.

How is this to be achieved? According to Grof, "There exists a wide spectrum of ancient and Oriental spiritual practices that are specifically designed to facilitate access to the perinatal and transpersonal domains" (28). Among these are "various shamanistic procedures, aboriginal rites of passage and healing ceremonies, death-rebirth mysteries, and trance dancing in ecstatic religions" (28). Grof goes on to emphasize the *phylogenetic* aspects of such rituals: "one can transcend the limits of the specifically human experience and identify with the consciousness of animals, plants, or even inorganic objects and processes" (34). From the outset he makes it clear that experiences like these cannot simply be dismissed as hallucinatory epiphenomena of the individual human brain. Although the "traditional model of the human psyche that dominates academic psychiatry is personalistic . . . observations of the last few decades have drastically changed our understanding of . . . the psyche" (27). This has resulted in a greater appreciation of the world beyond the person as an isolated unit of matter (i.e., a finite particular object-in-space). In a concluding remark, Grof comments, "All we can say is that somewhere in the process of confrontation with the perinatal level of the psyche, a strange qualitative Moebius-like [!] shift seems to occur in which deep self-exploration of the individual unconscious turns into a process of experiential adventures in the universe-at-large" (36).

In preceding sections, we explored alchemico-shamanic procedures for facilitating movements back to earlier stages of ontical action, and for bringing about their ontological counteractions. Though it is only in the present section that we have begun to consider explicitly the theme of death and resurrection, is it not clear in retrospect that each of the processes described above can be regarded as involving such a motif? For in each case it appears we can say that ontical being is confronted with its essential limitations, thrown into crisis, and brought face-to-face with its death. And in each, death can be survived, the ego can be resurrected, by a retrograde act of self-signification that transforms it into a "universal ego," an order of Being or Self.

Actually, the *initial* death is the purely cognitive one implicitly confronted from the outset of this book. Passing beyond KB S4, the rational intellect is brought to wrack and ruin, or, in the words of Tanabe, reason is "torn to pieces by antinomies and cast into the pit of contradictions" (1986, 28). However, "when the critique of reason that takes place in philosophy progresses to the point of an absolute critique and thus reaches the end of its tether, a way to the suprarational 'death-and-resurrection' of reason is necessarily thrown open" (28). The alchemical term for the death of conventional rationality is the *nigredo* ("blackening"), often symbolized by decapitation. This severance of the head leads to the *unio mentalis,* the KB S5 rebirth of reason as the paradoxical[19] self-signification constituting the first manifestation of the "subtle body" or Self. We know that the next Proprioception entails an *emotional* crisis. There is a "trial by water," an ordeal precipitated by a regression to MB-KB S3 wherein ontical human being (here in the person of Stevie) and its animal alter ego are flooded by feelings and drown. In the hoped-for KB S6 and MB S4 resurrections, the outpouring of feeling is to be hermetically contained via the retrograde acts of self-signification that concretely close the Kleinian and Moebial ontological vessels. Next comes the ordeal by fire. Regressing further back into the birth canal to S2, things dry out and heat up and we are brought to that sulfurous, scentuous place of burning passion and rage that consumes the ontical egos ("Baby Stevie" and his companions). After perishing in the flames, the egos can rise Phoenix-like from the ashes if the incendiary ontical thrusts accompanying their demise are Proprioceived in the LB S3, MB S5, and KB S7 realizations of universal flesh. If this happens, hermetic containment is achieved once again, the self-signification being even more concrete than before.

The ultimate confrontation with death comes with the movement all the way back through the birth canal to its source, the *unus mundus* of S1. If regression into the scentuous realm involves turning up the heat, in the final ordeal there is an exposure to *cold.* The crisis the ego now confronts does not merely concern its cognitive, emotional, or sensuous life, but its very being. The primitive emotion triggered by this existential endangerment is cold fear. More tangibly still, one experiences a suffocating sense of closeness at the prospect of being entombed. With the icy fingers of death tightening their hold, there is no longer any possibility of maintaining even a modicum of egoic detachment, as one could do in the more abstract visual, auditory, and olfactory realms. So now, in this densest and blackest of underworlds, tightly imprisoned within the cold bowels of the *minera,* one is really "up against it." Grof portrays this as an "experience of cosmic engulfment" (1985, 31). Returning to the deepest recesses of the birth canal, "the individual faces a situation that can best be described as no exit or hell. He or she feels stuck, encaged and trapped in a claustrophobic nightmarish world" (31). In the alchemical literature, this final ordeal is depicted by the burial of the "king and queen" following their *hierosgamos* (the holy wedding that symbolizes the penultimate *coniunctio*).

If one can consciously confront the demons of the S1 netherworld, then the ultimate resurrection becomes possible. Let me put it in terms of the work that will be necessary in the most concrete companion volume to the written text. To repeat, a regression in the service of the text will be required that will take Steven beyond the spheres of the child and the infant to the first stirrings of the embryo. We know that the deeper the regression, the more difficult its enactment. It will be harder for Steven to summon up and fully confront the unbridled instinctuality of Baby Stevie than the feelings of Stevie. Gaining conscious access to the embryonic sphere of "null intuition" will surely be most difficult of all. The ego death that Steven will need to face and accept here will be *physical,* not just emotional or sensual. On the other hand, we are also aware that while regression to earlier stages of egoic functioning is indeed more challenging the deeper we attempt to go, once these ontical modes of action are genuinely engaged, they are increasingly easier to counter-act via Proprioception. In S1, the ontical thrust that is to be counteracted is exceedingly weak; in fact, the projection of subject-object opposition is only a bare implication. Given the inchoate character of the clockwise ontical "grain," the ontological movement against this grain required in KB S8–MB S6–LB S4 will be inconsequential. Therefore, if awareness *can* be brought to the merely implicit significations that transpire in the black void of the *unus mundus*—which means nothing less than accepting one's total demise as an ontical being—then the self-signification that brings resurrection can occur with no more than an embryonic effort. Of course, for the hermetic vessel to be resealed in this way, it must previously have been sealed. That is to say, the effortless Proprioceptions that are to take place in KB S8–MB S6–LB S4 must be preceded by the more arduous Proprioceptions (the latter remaining transparent during the enactment of the former). Should regression to an earlier stage occur in the absence of previous Proprioception, the concrete reenactment of Proprioception would be precluded; in that case, the return movement would be *no more than* regressive. The containments provided by the rational *cogito* ("Steven"), the loving *anima,* and the scentuous *vegeta* are therefore indispensable to the more primal acts of containment. Supporting the self-signification of the embryo is that of the infant; supporting the infant's Proprioception is the child's; and the child's Proprioception is backed by that of the mature adult.

Now, in the foregoing sections, Proprioceptive movements were tracked through various organ centers of the body. The self-signification of KB S5 entails a retrograde movement through the neomammalian brain. In KB S6 and MB S4, there is a dual Proprioception channeled backward through the paleomammalian human brain and animal heart, respectively. Then, in the triadic Proprioception of KB S7, MB S5, and LB S3, the synchronized retrograde movements pass through the reptilian human brain, the animal heart,[20] and the vegetable reproductive center. What anatomical trajectories are followed in the final set of Proprioceptions?

Although MacLean postulated that the human brain is triune, he acknowledged that there actually exists a "fourth brain" serving to ground the other three: the *neural chassis.* According to MacLean, the basic neural machinery required for self-preservation and the preservation of the species is built into the lower brain stem and spinal cord. This neural chassis is comparable to the chassis of an automobile. If, by itself, "the neural chassis . . . [is] likened to a vehicle without a driver" (MacLean 1990, 23), evidently there evolved three different drivers for the same vehicle, "each radically different in structure, chemistry, and organization" (23). The "three different drivers" are of course the neo- and paleomammalian brains and the reptilian brain. In considering MacLean's model, Jantsch observes that the "neural chassis . . . is older than the three drivers and may be traced back to an early phase of the appearance of multicellular organisms" (1980, 165–66). If this primeval brain "were not driven by three 'drivers,'" comments Jantsch, it "would resemble an empty and unguided vehicle" (165). Phylo-functionally, the neural chassis is the brain associated with the *cogito*'s archaic thinking. In the KB S8 Proprioception of this embryonic form of cognition, the original brain is apprehended in a transpersonal and unmediated fashion.

Given that the regression of the *cogito* to the embryonic level discloses *three* nonhuman companions—the *anima,* the *vegeta,* and now the *minera*—it might appear at first glance that the closing set of Proprioceptions should be fourfold. But this proves not to be the case. The KB S8 embryonic Proprioception of the neural chassis is synchronized with the retrograde embryonic movements of MB S6 and LB S4, the latter two actions being channeled respectively through the fearful animal heart and the primitive tactility of the "vegetable groin." Yet no Proprioception of the "mineral gut" is included, not even an embryonic one. We are well aware of the reason for this. Since the zero-dimensional *minera* is *totally* undifferentiated, since it possesses no ontical grain whatsoever, there can be no movement against the grain in realization of the ontological.

With respect to the possibility of a complete regression through the birth canal to the moment of conception, Grof notes that "many people experience very concrete and realistic episodes which they identify as fetal and embryonal memories. It is not unusual under these circumstances to experience (on the level of cellular consciousness) full identification with the sperm and the ovum at the time of conception" (1985, 34). For the "cellular consciousness" of the zygote, the S1 lack of functional differentiation noted above is coupled with a lack of differentiation among the organ correlates of the four functional families. In this germinal milieu, the separation of brain, heart, groin, and gut is purely nascent. Moreover, the functional and anatomical lack of differentiation holding sway in the *unus mundus* should be accompanied by a *phylogenetic* absence of differences that bespeaks the initial entanglement of evolutional windings explored in the last chapter. But, once again, the *Proprioception* of this primal state of affairs entails more than a mere regression to it. With the differentiations carried out in previous stages now shining diaphanously in the zero-dimensional sphere, topodimensional vessels are not merely wide open to each other but also hermetically closed. The mutual containment of uroboric bodies will be asymmetric, of course. In the dance of the dimensional Mothers, the higher-dimensional will encompass the lower in a positive (ontological) way, whereas the lower will embrace the higher negatively (meontologically). So shall turn the wheels of Ezekiel in the final round of Proprioceptions.

But will it really be the *final* round? The conclusion previously reached on this ultimate question now bears one more repetition: When all four dimensional windings have been brought into harmony, when all four of "Ezekiel's wheels" are spinning in synchrony, when the alchemical circle has fully been squared, the dimensional spiral will then expand in logarithmic fashion, with the dialectical matrix now growing from 4^2 to 5^2. In this wider turning of the spiral, a new and more dialectically intricate, *four*-dimensional order of the flesh will come into play beyond the Kleinian—a "meta-Kleinian" onto-topological action pattern laid out in 5×5 matrices. In this novel lifeworld, there should be an entirely new mode of functioning beyond thinking, feeling, sensing, and intuition, and each of the five functional families should consist of five sub-functions. We can conjecture as well that there will be an anatomical metamorphosis here that will give rise to a major new organ system surpassing the brain. Indeed, in this milieu, a whole new order of nature should unfold beyond the *minera, vegeta, anima,* and human *cogito.*

Let me underscore the task at hand with regard to the new dimension of Being. As I noted in closing the last chapter, it would be an exercise in futility to attempt to encompass the meta-cognitive dimension of the flesh via an exclusively cognitive analysis aspiring to even greater heights of mental abstraction. Rather than merely looking to extend the present modus operandi, we must begin working in earnest on the companion volumes to the written text. Proceeding in this way, the lower-dimensional self-significations will be made a reality, realized in the flesh. The fruits of this *coagulatio* will then be fed back into the *solutio*—assistance that will be much appreciated, given that the abstract analysis carried out in the written text is in fact incomplete, even in its own terms. What seems abundantly clear is that the several works of ontological self-signification can only reach genuine completion when carried out in conjunction with one another, since this ontobiography is indeed a collaborative effort. And through the harmony of topodimensional spheres—the coordination of self-nativities wherein dimensional Mothers move backward in synchrony through their own birth canals—the embryo of the meta-cognitive Mother will be created.

NOTES

PREFACE

1. *Webster's New Twentieth Century Dictionary,* 2nd ed., s.v. "abstract."

CHAPTER ONE

1. *Webster's New Twentieth Century Dictionary,* 2nd ed., s.v. "space."

2. *Webster's New Twentieth Century Dictionary,* 2nd ed., s.v. "place."

3. *Webster's New Twentieth Century Dictionary,* 2nd ed., s.v. "topology."

4. For a discussion of the broader cultural implications of modernism and postmodernity, see Rosen (2004).

5. I will explicitly address the role of time in due course.

6. Rosi Braidotti on Deleuze and Guattari, quoted in Grosz (1994, 162).

7. The concept and experience of paradox, of course, have been deeply explored in *non*-Western traditions. Though I do touch on non-Western strains of thought in subsequent chapters, in the present book I am largely concerned with posing a challenge for the West. Nevertheless, it will eventually be necessary to probe philosophy's *transcultural* interior horizon, the place where Western and non-Western worlds "moebiously meet."

CHAPTER TWO

1. It should now be clear that what makes this text "Kleinian" is not merely its signification of the Klein bottle in words, schematic drawings, or objective models in space. By "Kleinian text," I mean a text in which the Klein bottle serves as the medium for dimensionally extending these self-reflective words so that they can actually reenter their own prereflective ground. Here the higher-dimensional Klein bottle is embodied, not just as an object of signification, but as part of the signifying act itself.

PREAMBLE TO PART II

1. *Webster's New Twentieth Century Dictionary,* 2nd ed., s.v. "dissolve."

CHAPTER THREE

1. Actually, the paradox of the Klein bottle is that it is both closed and *open,* whereas the edges of the cylindrical ring and Moebius surface make them simply open structures. The torus, for its part, is simply closed.

2. Throughout this text, the term "proprioception" is used in the special, literalized sense defined in chapter 2.

3. The use of ratios here is to be understood in a radically non-quantitative sense, since it applies to the relationship between non-ontical terms. Quantitative ratios are ontical: they are mathematical objects embedded in a common field of measure, a continuum. For his/her part, the analyst is detached from the mathematical object-in-space. Quantitative ratios thus adhere to the classical formula of object-in-space-before-subject. In contrast, the dimensional ratios of table 3.2 express relationships between orders of Being, and between Being and nothingness.

4. We must keep in mind an important difference between ontical appearances and ontological reality. From the standpoint of the former, it seems as if higher-dimensional structures are created simply from scratch out of the fusion of lower-dimensional enantiomorphs. However, when grasped ontologically, lower-dimensional action does not follow the "fatherly" principle of creation *ex nihilo* but functions to catalyze the *self*-development of higher-dimensionality. Ontologically, then, rather than *n*-dimensional Being merely *arising* from the fusion of lower-dimensional enantiomorphs, it develops from action upon *itself* that is *facilitated* by selfless lower-dimensional action.

CHAPTER FOUR

1. *Webster's New Collegiate Dictionary,* 1975 ed., s.v. "instinctive."

2. *Webster's New Twentieth Century Dictionary,* 2nd ed., s.v. "instinctive."

3. The feminist philosopher Hélène Cixous speaks of the patriarchal structures of language and thought as covertly acting "for men as a surrogate umbilical cord, assuring them . . . that the old lady is always right behind them" (1980, 256).

4. In Merleau-Ponty's discussion of "nothingness," the term is not used in the absolute sense of meontological selflessness, but in the merely relative sense associated with the for-itself or subject.

CHAPTER FIVE

1. However, I must add that, in fact, the "radii" of *all* dimensional circulations are "zero," in the sense that stage transitions are not actually displacements in an extensive spatial continuum (though they appear as such in table 5.1), but rather—in the language of Heidegger—"*pre*spatial" gestures of "opening up" that first "provide space" (1962/1972, 14; emphasis added). In other words, the several dimensional circulations portrayed in the table cannot really have finite radii because, instead of taking place within a spatial field of reflection, they constitute the prereflective ontological actions from which the several fields of ontical reflection derive (see the next section for further discussion of dimensional action).

2. If dimensional process is indeed most essentially described not by a circle but by a spiral, we should expect that the story of Topogeny would not end with the realization of three-dimensional Kleinian Being. Instead there would be another logarithmic expansion that would mark the inauguration of a new round of maternal self-development, one characterized by the Topogeny of the 5 × 5 matrix of a four-dimensional ontological body. That is, a *new order of Being would grow organically.* This idea will be adumbrated further in due course.

3. Schwenk makes the interesting observation that jellyfish propel themselves in a vortical fashion in which they literally turn themselves inside out! Moreover, this fact is not presented as an isolated curiosity but as a variation on the archetypal vorticity that pervades all of nature. See Schwenk 1965, 57–58.

4. The quantized substage structure peculiar to the Moebius appears to gain support from additional ontical experimentation. The continuously curved Moebius strip may be flattened out. When this is done, we obtain the triangular form illustrated below:

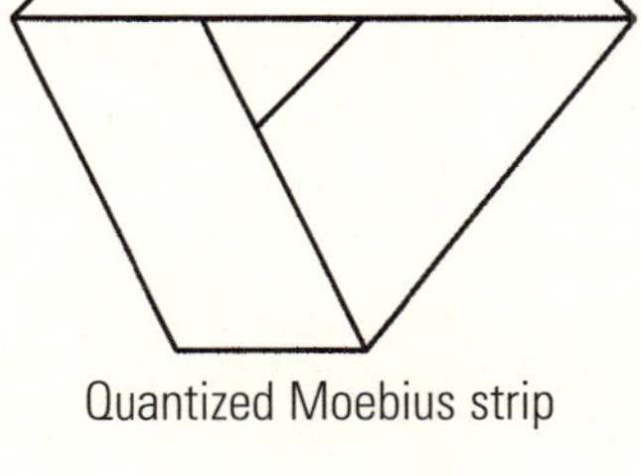

Quantized Moebius strip

Flattening the Moebius band brings out its "quantized infrastructure." Instead of each cycle consisting of a continuously curved rotation through 360°, now each is seen to be composed of three distinct sub-phases set off from one another by the creation of edges. Further preliminary research indicates that a quantized form of the Klein bottle would consist of four sub-phases per cycle, and that the quantized lemniscate would have two such sub-phases. All this accords with the account of the substages given in table 5.1.

5. See Persichetti's discussion of musical "mirror writing" (1961, 72–73) and Helmholtz's classical 1875 study of harmonic undertones.

CHAPTER SIX

1. The reader may have noticed a similar ambiguity within each perspective of figure 6.1(a). This speaks to the fact that there is actually a whole series of ambiguously reversible cubes, analogous to the topological bisection series described in chapter 3. For the purposes of this volume however, it will suffice to deal only with the member of the series displayed in figure 6.1(b).

2. We may again take note of the *esoteric* correlate of dimensional flesh. Whereas two-dimensional flesh is associated with the *astral body,* one-dimensional flesh would be *etheric* in nature. See Steiner's (1972) description of the "ether body," which he relates to the basic force of life, and to sleep.

3. *Webster's New Twentieth Century Dictionary,* 2nd ed., s.v. "animal."

4. *Webster's New Twentieth Century Dictionary,* 2nd ed., s.v. "vegetable."

5. We must keep in mind that while the lower-dimensional "holy object" surely would be *analogous* to the Kleinian container, it also would be different in most radical ways—not just quantitatively or qualitatively, but *ontologically* and *phylo-functionally.* Though a full understanding of the form that paradoxical containment would take in a noncognitive, nonhuman environment may be beyond the capacity of the human intellect, it appears we must make allowances for such containment nonetheless, since human cognition is not the only player in the onto-dimensional transactions being described. Therefore, whereas the glass vessel of alchemy "is an artificial human product and thus signifies the intellectual purposefulness and artificiality of the procedure" (Jung 1967, 197), it seems we need to entertain the possibility, say, of a vessel composed purely of *feeling* or *sensuality!*

CHAPTER SEVEN

1. Historically, there is reason to believe that the habit in question was reinforced around the time of the Renaissance, and was associated with the enhancement of depth perception, the increased use of perspective in art, advances in mapmaking, in mathematics and science, and so on.

2. In a popular book titled *Taking the Quantum Leap,* theoretical physicist Fred Wolf (1981) employed the Necker cube as his primary metaphor for the phase transitions of quantum mechanics.

3. See Gendlin and Lemke's (1983) critique of the conventional notion of linear time as it operates in contemporary physics.

4. See Tanabe's (1986) telling critique of the long tradition of "mystical intuition," as manifested in Eastern and Western philosophy and religion alike.

CHAPTER EIGHT

1. The term derives from mathematics, where it denotes a number that leaves unchanged the number on which it operates. For multiplication, this indispensable grounding element is the number 1. The iconic similarity between the mathematical "1" and the English "I" is noteworthy, especially when we take into account the distinctively *phallic* character of these two prime operators!

2. Contemporary brain research and phenomenological investigation alike indicate that cognitive activity cannot be treated in simple separation from emotions without risking an oversimplification of human behavior and experience (Ellis 1997; Suvin 1997).

3. For compelling accounts and demonstrations of how the journal format allows the writer to stand present in her text, see two works by my wife, Marlene Schiwy: *A Voice of Her Own* (1996) and *Simple Days* (2002).

4. This term is defined as the "species man regarded as an organismic whole in which the element or individual is a phylically integrated unit" (Burrow 1953, 531).

5. Robert Gibbons.

6. Recent innovations in stereoscopy are rooted in much older experimentation along these lines. However, an account of the nineteenth-century precursors of contemporary stereoscopic research is beyond the scope of this book.

7. Merleau-Ponty 1968, 145; emphasis added.

8. Note the similarity of this ritual to totemic initiation rites among native bands of the Pacific Northwest, as described in chapter 6.

9. The mathematical physicist David Bohm is a prime example of a thinker with such insight. See my dialogue with Bohm on this matter in Rosen 1994, chapter 13.

10. *Webster's New Twentieth Century Dictionary,* 2nd ed., s.v. "pungent."

11. For more information, see www.uni-ulm.de/uni/intgruppen/memosys/desn19.htm#SMELL.

Does scientific evidence exist to support the association between smell and the emotion of anger? First, there is a large body of research that links emotion to olfaction (e.g., Herz 1998). More specifically, Alaoui-Ismaili (1997) has demonstrated that acetic acid, a liquid characterized by its highly pungent odor, can directly evoke feelings of anger. It is obvious, however, that a more extensive study of the research in this field would be required to establish the relationship decisively. While the relationship *will* be further examined below, our deliberations will be carried out in the context of advancing our exploration of shamanism.

12. For readers not familiar with the psychological literature, the bracketed phrase is a variation on the term "regression in the service of the ego." This expression refers to the deliberate induction of an inchoate state of consciousness in the interest of the mature ego, as is done in the work of artists, creative writers, and the like.

13. Worwood (1998) uses this neologism in a different context.

14. The ability of plants to smell, to engage in chemoperception, depends on their capacity to operate over the short-range scale of chemical interactions at which smell takes place. Plants also interact with light, of course. They soak up photons in the process of photosynthesis. But this does not necessarily mean that they engage in *visual perception.* To be influenced by light is not necessarily to *experience* it. For the more distantiated, abstract kind of functioning entailed in visual experience, one must be able to operate over the long-range scale of electromagnetic interactions at which vision takes place. Plants cannot do this. See chapter 7 for a discussion of the correlation between the abstractness of sensory experience and the distance over which sensory interaction occurs. See also Gendlin 1991b, 137.

15. Beginning with Proust's famous account of how a simple aroma unleashed a cascade of memories from early childhood, there has been a wealth of anecdotal data suggesting that olfactory cues can have such an effect. The recent experimental research of Rachel Herz (1998) seems to lend credence to these reports. See also Michael Stoddard's *The Scented Ape* (1990).

16. In her allusion to "heavenly orders," von Franz is referring to archaic Chinese number archetypes constituting "nothing less than 'cosmic plans'" (1974, 22). "In the ancient Chinese view," says von Franz, "the entire time-space continuum of the universe was 'organized' according to numerical patterns of this kind" (24). The most primordial of these number arrays was the "sequence of older heaven" or *Ho-t'u,* which, according to von Franz (24–25) formed the basis of the trigram system of Taoism's *I Ching* (von Franz also relates the *Ho-t'u* to the cabalistic "Tree of Life"; cf. the Cosmic Tree of shamanism described by Eliade). The *Ho-t'u* sequence consists essentially of a double cycle of numbers, each number constituting a rhythmic action unto itself. There are four clockwise

number pulsations in the first cycle and four counterclockwise or retrograde pulsations in the second. The remarkable correspondence of this pattern to the one arising from our topodimensional analysis cannot be further explored within the scope of this book.

17. The mineral nature of this realm is evident from Eliade's account. The shamanic underworld is a hole lying deep within the earth, often entered through a cavern (1964, 51–52). In the Altaic tradition, the voyager hears "metallic sounds" during his descent, encounters "subterranean rivers," and finally arrives at a palace of "stone and black clay" where the Lord of the underworld dwells (201). In other shamanic traditions, magical stones or rock crystals frequently play a significant role in initiation rites involving descent, death, and resurrection (45–47, 50, 52).

18. Von Franz describes the total *coniunctio* realized in the *unus mundus* as a "fearsome mystery" (1974, 293).

19. In his "absolute critique" of Hegelian logic, Tanabe, following Kierkegaard, noted the need for a "paradoxical dialectic" (1986, 52), so that reason could be "resurrected" after its death.

20. Our phylo-functional analysis appears to require that we specify here a second, more primal region of the heart. Just as older regions of the brain are engaged in successive Proprioceptions, so too should there be older regions of the heart that are activated when the Proprioception of this organ is repeated more concretely. Investigation of this is another matter that is beyond the scope of the present book. Evidently, then, in addition to the need for companion volumes rendered in media that are denser than the written word, the written opus itself can stand to be extended!

BIBLIOGRAPHY

Abbott, Edwin. 1884/1983. *Flatland: A Romance of Many Dimensions.* Repr., New York: Barnes and Noble.

Abram, David. 1996. *The Spell of the Sensuous.* New York: Vintage.

Ackerman, Diane. 1990. *A Natural History of the Senses.* New York: Vintage.

Alaoui-Ismaili, Ouafae. 1997. "Odor Hedonics: Connection with Emotional Response Estimated by Autonomic Parameters." *Chemical Senses* 22:237–48.

Althusser, Louis. 1969. "Freud and Lacan." *New Left Review* 55 (May–June): 48–65.

Anfinsen, C. B. 1968. *Developmental Biology Supplement* 2. New York: Academic Press.

Applebaum, David. 1983. "On Turning a Zen Ear." *Philosophy East and West* 33:115–21.

Barfield, Owen. 1977. *The Rediscovery of Meaning and Other Essays.* Middletown, CT: Wesleyan University Press.

Barr, Stephen. 1964. *Experiments in Topology.* New York: Dover.

Berman, Morris. 1989. *Coming to Our Senses.* New York: Bantam.

Bigwood, Carol. 1993. *Earth Muse.* Philadelphia: Temple University Press.

Bohm, David. 1980. *Wholeness and the Implicate Order.* London: Routledge and Kegan Paul.

———. 1994. "The Bohm/Rosen Correspondence." In *Science, Paradox, and the Moebius Principle,* edited by Steven M. Rosen, 223–58. Albany: State University of New York Press.

Boller, Thomas. 1995. "Chemoperception of Microbial Signals in Plant Cells." *Annual Review of Plant Physiology and Plant Molecular Biology* 46:189–214.

Braine, Susan. 1995. *Drumbeat/Heartbeat: A Celebration of the Powwow.* Minneapolis: Lerner.

Braun, Allen R., Thomas J. Balkin, Nancy J. Wesensten, Richard E. Carson, Mary Varga, Paul Baldwin, Scott Selbie, Gregory Belenky, and Peter Herscovitch. 1997. "Regional Cerebral Blood Flow throughout the Sleep-Wake Cycle. An H2(15)O PET Study." *Brain* 120:1173–97.

Braun, Allen R., Thomas J. Balkin, Nancy J. Wesensten, Fuad Gwadry, Richard E. Carson, Mary Varga, Paul Baldwin, Gregory Belenky, and Peter Herscovitch. 1998. "Dissociated Pattern of Activity in Visual Cortices and Their Projections during Human Rapid Eye Movement Sleep." *Science* 279:91–95.

Bridges, Katherine M. B. 1931. *The Social and Emotional Development of the Pre-school Child*. London: Routledge and Kegan Paul.

Burrow, Trigant. 1932. *The Structure of Insanity*. London: Kegan Paul, Trench, Trubner and Co.

———. 1937. *The Biology of Human Conflict*. New York: Macmillan.

———. 1953. *Science and Man's Behavior*. New York: Philosophical Library.

Čapek, Milič. 1961. *Philosophical Impact of Contemporary Physics*. New York: Van Nostrand.

Cixous, Hélène. 1980. "The Laugh of Medusa." In *New French Feminisms*, edited by Elaine Marks and Isabelle de Courtivron, 245–64. Sussex: Harvester Press.

Classen, Constance. 1993. *Worlds of Sense*. London: Routledge.

Connor, Steven. 2002. "Topologies: Michel Serres and the Shapes of Thought." Steven Connor home page, Birkbeck College, University of London. http://www.bbk.ac.uk/eh/skc/skc.htm.

Crile, George W. 1913. "A Mechanistic View of Psychology." *Science* 38 (974): 283–95.

———. 1915. *The Origin and Nature of the Emotions*. Philadelphia: W.B. Saunders.

Darwin, Charles. 1872/1965. *The Expression of Emotions in Man and Animal*. Repr., Chicago: University of Chicago Press.

Deleuze, Gilles, and Felix Guattari. 1987. *A Thousand Plateaus: Capitalism and Schizophrenia*. Minneapolis: University of Minnesota Press.

de Quincey, Christian. 2002. *Radical Nature*. Montpelier, VT: Invisible Cities Press.

Descartes, René. 1628/1954. "Rules for the Direction of the Mind." In *Descartes: Philosophical Writings*, edited and translated by Elizabeth Anscombe and P. T. Geach, 153–80. London: Thomas Nelson and Sons.

Eliade, Mircea. 1960. *Myths, Dreams, and Mysteries*. New York: Harper and Row.

———. 1962. *The Forge and the Crucible*. New York: Harper and Row.

———. 1964. *Shamanism*. Princeton, NJ: Princeton University Press.

Ellis, Ralph. 1997. "Differences between Conscious and Non-Conscious Processing." Background paper for After-Postmodernism Conference, University of Chicago, November 14–16.

Flew, Antony. 1979. *A Dictionary of Philosophy*. New York: St. Martin's Press.

Franz, Marie-Louise von. 1974. *Number and Time*. Evanston, IL: Northwestern University Press.

———. 1975. "Psyche and Matter in Alchemy and Modern Science." *Quadrant* 8:33–49.

Freud, Sigmund. 1914/1957. "On Narcissism: An Introduction." In *Standard Edition of the Complete Works of Sigmund Freud,* vol. 14, edited by James Strachey, 117–40. London: Hogarth Press.

Fuller, B. A. G., and Sterling M. McMurrin. 1957. *A History of Philosophy.* New York: Henry Holt.

Galt, Alfreda. 1995. "Trigant Burrow and the Laboratory of the 'I'." *The Humanistic Psychologist* 23 (Spring): 19–39.

Gardner, Martin. 1979. *The Ambidextrous Universe: Mirror Asymmetry and Time-Reversed Worlds.* 2nd rev. ed. New York: Scribner.

Gebser, Jean. 1985. *The Ever-Present Origin.* Athens: Ohio University Press.

Gendlin, Eugene T. 1978. *Focusing.* New York: Bantam.

———. 1991a. "Crossing and Dipping: Some Terms for Approaching the Interface between Natural Understanding and Logical Formulation." In *Subjectivity and the Debate over Computational Cognitive Science,* edited by Mary Galbraith and William J. Rapaport, 37–59. Buffalo: Center for Cognitive Science, State University of New York at Buffalo.

———. 1991b. "Thinking Beyond Patterns: Body, Language, and Situations." In *The Presence of Feeling in Thought,* edited by Bernard denOuden and Marcia Moen, 27–189. New York: Peter Lang.

———. 1993. "Words Can Say How They Work." In *Proceedings, Heidegger Conference,* edited by Robert P. Crease, 29–35. Stony Brook: Department of Philosophy, State University of New York at Stony Brook.

———. 1996. "Making Concepts from Experience." Presented at International Focusing Conference, Gloucester, MA, May 2–6. [Unpublished transcript (33 pp.) available from the Focusing Institute, 34 East Lane, Spring Valley, NY 10977; or http://www.focusing.org.]

Gendlin, Eugene T., and Jay L. Lemke. 1983. "A Critique of Relativity and Localization." *International Journal of Mathematical Modelling* 4:61–72.

Graves, John C. 1971. *The Conceptual Foundations of Contemporary Relativity.* Cambridge, MA: MIT Press.

Grobstein, Clifford. 1973. "Hierarchical Order and Neogenesis." In *Hierarchy Theory,* edited by Howard Pattee, 29–48. New York: George Braziller.

Grof, Stanislav. 1985. "Modern Consciousness Research and Human Survival." *Re-Vision* 8 (1): 27–39.

Grosz, Elizabeth. 1994. *Volatile Bodies.* Bloomington: Indiana University Press.

Haase, Rudolf. 1989. "Harmonics and Sacred Tradition." In *Cosmic Music,* edited by Joscelyn Godwin, 91–110. Rochester, VT: Inner Traditions.

Harries-Jones, Peter. 1996. "Ecology and Economy after the Earth Summit: The New Concepts of Bioentropy and Recursive Logic." Presented at Sixth International Karl Polanyi Conference, Concordia University, Montreal, November 7–10.

Heidegger, Martin. 1927/1962. *Being and Time.* Translated by John Macquarrie and Edward Robinson. New York: Harper and Row.

———. 1929/1975. "What Is Metaphysics?" In *Existentialism from Dostoevsky to Sartre,* rev. ed., edited by Walter Kaufmann, 242–64. New York: New American Library.

———. 1946/1984. *Early Greek Thinking.* Translated by David F. Krell and Frank A. Capuzzi. New York: Harper and Row.

———. 1954/1968. *What Is Called Thinking?* Translated by J. Glenn Gray. New York: Harper and Row.

———. 1962/1972. "Time and Being." In *On Time and Being,* translated by Joan Stambaugh, 1–24. New York: Harper and Row.

———. 1962/1977. "Modern Science, Metaphysics, and Mathematics." In *Martin Heidegger: Basic Writings,* edited by David F. Krell, 247–82. New York: Harper and Row.

———. 1964/1977. "The End of Philosophy and the Task of Thinking." In *Martin Heidegger: Basic Writings,* edited by David F. Krell, 373–92. New York: Harper and Row.

Helmholtz, H. L. F. von. 1875. *On the Sensations of Tone as a Physiological Basis for the Theory of Music.* Translated by A. J. Ellis. London: Longmans, Green.

Herz, Rachel Sarah. 1998. "Are Odors the Best Cues to Memory? A Cross-Modal Comparison of Associative Memory Stimuli." *Annals of the New York Academy of Sciences* 855:670–74.

Hesse, Mary B. 1965. *Fields and Forces.* Totowa, NJ: Littlefield, Adams and Co.

Hillman, James. 1960. *Emotion.* Evanston, IL: Northwestern University Press.

Jantsch, Erich. 1980. *The Self-Organizing Universe.* New York: Pergamon Press.

Jaynes, Julian. 1976. *The Origin of Consciousness in the Breakdown of the Bicameral Mind.* Boston: Houghton Mifflin.

Jung, Carl Gustav. 1959. *Aion.* Vol. 9, part 2 of *Collected Works,* translated by R. F. C. Hull. Princeton, NJ: Princeton University Press.

———. 1967. *Alchemical Studies.* Vol. 13 of *Collected Works,* translated by R. F. C. Hull. Princeton, NJ: Princeton University Press.

———. 1968. *Psychology and Alchemy.* Vol. 12 of *Collected Works,* translated by R. F. C. Hull. Princeton, NJ: Princeton University Press.

———. 1970. *Mysterium Coniunctionis.* Vol. 14 of *Collected Works,* translated by R. F. C. Hull. Princeton, NJ: Princeton University Press.

———. 1971. *Psychological Types.* Vol. 6 of *Collected Works,* translated by R. F. C. Hull. Princeton, NJ: Princeton University Press.

Kaufmann, Walter, ed. 1975. *Existentialism from Dostoevsky to Sartre*. Rev. ed. New York: New American Library.

Kittelson, Mary Lynn. 1995. "The Acoustic Vessel." In *The Interactive Field in Analysis,* edited by Murray Stein, 89–105. Wilmette, IL: Chiron.

Kline, Morris. 1980. *Mathematics: The Loss of Certainty.* New York: Oxford University Press.

Kohl, James V., and Robert T. Francoeur. 1995. *The Scent of Eros.* New York: Continuum.

Lacan, Jacques. 1953. "Some Reflections on the Ego." *International Journal of Psychoanalysis* 34:11–17.

———. 1966/1970. "Of Structure as an Inmixing of an Otherness Prerequisite to Any Subject Whatever." In *The Languages of Criticism and the Sciences of Man: The Structuralist Controversy,* edited by Richard Macksey and Eugenio Donato, 186–200. Baltimore: Johns Hopkins University Press.

———. 1968. "The Function of Language in Psychoanalysis." In *The Language of the Self,* edited and translated by Anthony Wilden, 3–87. New York: Delta.

Lachelier, Jules. 1962. "Psychology and Metaphysics." In *The Search for Being,* edited and translated by Jean T. Wilde and William Kimmel, 153–69. New York: Noonday Press.

Lapedes, Daniel N., ed. 1978. *McGraw-Hill Dictionary of Physics and Mathematics.* New York: McGraw-Hill.

Lauer, Quentin. 1965. *Edmund Husserl: Phenomenology and the Crisis of Philosophy.* New York: Harper and Row.

Lavery, David. 1983. "The Eye of Longing." *Re-Vision* 6:22–33.

Leder, Drew. 1990. *The Absent Body.* Chicago: University of Chicago Press.

Levin, David Michael. 1985. *The Body's Recollection of Being.* London: Routledge and Kegan Paul.

Lind, Richard E. 2001. "Historical Origins of the Modern Mind/Body Split." *Journal of Mind and Behavior* 22 (1): 23–35.

MacLean, Paul D. 1968. "Alternative Neural Pathways to Violence." In *Alternatives to Violence,* edited by L. Ng, 22–34. Alexandria, VA: Time-Life.

———. 1990. *The Triune Brain in Evolution.* New York: Plenum.

Macquarrie, John. 1968. *Martin Heidegger.* Richmond, VA: John Knox.

Maquet, Pierre, Jean-Marie Péters, Joël Aerts, Guy Delfiore, Christian Degueldre, André Luxen, and Georges Franck. 1996. "Functional Neuroanatomy of Human Rapid-Eye-Movement Sleep and Dreaming." *Nature* 383:163–66.

Massumi, Brian. 1998. "Strange Horizon: Buildings, Biograms, and the Body Topologic." Indiana University Web site, http://www.indiana.edu/~thinkmat/strange.doc (accessed February 15, 2003; link currently dead). See Massumi,

Brian. 1999. "Strange Horizon: Buildings, Biograms, and the Body Topologic." In "Hypersurface Architecture II," edited by Stephen Perrella, special issue, *Architectural Design* 69, nos. 9–10:12–19.

Merleau-Ponty, Maurice. 1962. *Phenomenology of Perception,* translated by Colin Smith. London: Routledge and Kegan Paul.

———. 1964. "Eye and Mind." In *The Primacy of Perception,* edited by James M. Edie, 159–90. Evanston, IL: Northwestern University Press.

———. 1968. *The Visible and the Invisible,* translated by Alphonso Lingis. Evanston, IL: Northwestern University Press.

Metzner, Ralph. 1971. *Maps of Consciousness.* New York: Macmillan.

Millington, T. Alaric, and William Millington. 1966. *Dictionary of Mathematics.* New York: Harper and Row.

Montague, Ashley. 1974. *Körperkontakt.* Stuttgart: Klett.

Neumann, Erich. 1954. *The Origins and History of Consciousness.* Princeton, NJ: Princeton University Press.

Ong, Walter. 1977. *Interfaces of the Word.* Ithaca, NY: Cornell University Press.

Ouspensky, P. D. 1970. *Tertium Organum.* New York: Vintage.

Partridge, Eric. 1958. *A Short Etymological Dictionary of Modern English.* New York: Macmillan.

Pattee, Howard. 1973. "The Physical Basis and Origin of Hierarchical Control." In *Hierarchy Theory,* edited by Howard Pattee, 71–108. New York: George Braziller.

Perrella, Stephen. n.d. "Hypersurface Theory: Architecture >< Culture." http://www.hum.ku.dk/visuelkultur/efteraar2002/digvis/perellahypersurface.html.

Persichetti, Vincent. 1961. *Twentieth-Century Harmony.* New York: W.W. Norton.

Plato. 1965. *Timaeus and Critias.* Translated by Desmond Lee. New York: Penguin.

Ratner, Carl. 1989. "A Social Constructionist Critique of Naturalistic Theories of Emotion." *Journal of Mind and Behavior* 10:211–30.

Rosen, Steven M. 1985. *The Moebius Seed.* Walpole, NH: Stillpoint.

———. 1994. *Science, Paradox, and the Moebius Principle.* Albany: State University of New York Press.

———. 1995. "Pouring Old Wine into a New Bottle." In *The Interactive Field in Analysis,* edited by Murray Stein, 121–41. Wilmette, IL: Chiron.

———. 1997. "Wholeness as the Body of Paradox." *Journal of Mind and Behavior* 18 (4): 391–423.

———. 1999. "Evolution of Attentional Processes in the Human Organism." *Group Analysis* 32 (2): 243–53.

———. 2004. *Dimensions of Apeiron.* Amsterdam: Editions Rodopi.

Rucker, Rudolph. 1977. *Geometry, Relativity, and the Fourth Dimension.* New York: Dover.

Ryan, Paul. 1993. *Video Mind/Earth Mind.* New York: Peter Lang.

Sartre, Jean-Paul. 1943/1956. *Being and Nothingness: An Essay on Phenomenological Ontology.* Translated by Hazel E. Barnes. New York: Philosophical Library.

———. 1943/1975. "Self-Deception." In *Existentialism from Dostoevsky to Sartre,* rev. ed., edited by Walter Kaufmann, 299–328. New York: New American Library.

Schiwy, Marlene. 1996. *A Voice of Her Own.* New York: Simon and Schuster.

———. 2002. *Simple Days: A Journal on What Really Matters.* Notre Dame, IN: Sorin.

Schultz, Duane P., and Sydney Ellen Schultz. 1992. *A History of Modern Psychology.* New York: Harcourt Brace Jovanovich.

Schwenk, Theodor. 1965. *Sensitive Chaos.* London: Rudolf Steiner.

Serres, Michel. 1994. *Atlas.* Paris: Editions Julliard.

Serres, Michel, and Bruno Latour. 1995. *Conversations on Science, Culture, and Time.* Translated by Roxanne Lapidus. Ann Arbor: University of Michigan Press.

Sheets-Johnstone, Maxine. 1990. *The Roots of Thinking.* Philadelphia: Temple University Press.

Shipley, Joseph. 1984. *The Origins of English Words.* Baltimore: Johns Hopkins University Press.

Sloan, Lisa. 1999. "Shamanic Initiation: Map of the Soul." PhD diss., Pacifica Graduate Institute.

Sokal, Alan D., and Jean Bricmont. 1999. *Fashionable Nonsense: Postmodern Intellectuals' Abuse of Science.* New York: Picador USA.

Spiegelberg, Herbert. 1982. *The Phenomenological Movement.* The Hague: Martinus Nijhoff.

Spivak, Gayatri C. 1976. Preface to *Of Grammatology,* by Jacques Derrida, translated by Gayatri C. Spivak. Baltimore: Johns Hopkins University Press.

Steiner, Rudolf. 1972. *An Outline of Occult Science.* Spring Valley, NY: Anthroposophic Press.

Stoddard, Michael. 1990. *The Scented Ape.* Cambridge: Cambridge University Press.

Sulloway, Frank J. 1979. *Freud: Biologist of the Mind.* New York: Basic.

Suvin, Darko. 1997. "On Cognitive Emotions and Topological Imagination." Background paper for After-Postmodernism Conference, University of Chicago, November 14–16.

Tanabe, Hajime. 1986. *Philosophy as Metanoetics.* Translated by Takeuchi Yoshinori. Berkeley: University of California Press.

Varela, Francesco J., Humberto R. Maturana, and Ricardo Uribe. 1974. "Autopoiesis: The Organization of Living Systems, Its Characterization and a Model." *Biosystems* 5:187–96.

Vaughan, Frances. 1979. *Awakening Intuition.* New York: Doubleday/Anchor.

Washburn, Michael. 1988. *The Ego and the Dynamic Ground.* Albany: State University of New York Press.

Watson, John B. 1913. "Psychology as a Behaviorist Sees It." *Psychological Review* 20:158–67.

Wolf, Fred. 1981. *Taking the Quantum Leap.* New York: Harper and Row.

Worwood, Valerie A. 1998. *Scents and Scentuality.* Novato, CA: New World Library.

Wright, R. H. 1964. *The Science of Smell.* London: Allen and Unwin.

———. 1982. *The Sense of Smell.* Boca Raton, FL: CRC Press.

INDEX

Endnotes are indicated by the letter *n* following the page number(s).